Handbook of
Coffee Post-Harvest Technology

A Comprehensive Guide to the Processing, Drying, and Storage of Coffee

Flávio Meira Borém, Editor

Translated by Joel Shuler

John Outler, English Editor

Copyright © 2014 Flávio Meira Borém. All rights reserved.

Dedication

*Dedico
à memória
dos meus pais
e a minha esposa e
minhas filhas em quem
encontro sempre carinho,
apoio, e coragem.*

Agradeço aos autores e co-autores por terem, generosamente, compartilhado o seu conhecimento. À Universidade Federal de Lavras, escola de grandes mestres em cafeicultura. À Editora UFLA por ter cedido os direitos autorais da obra originalmente escrita em português. Agradeço ao Joel Shuler por ter se dedicado a realizar uma tarefa muito além de uma simples tradução, pois buscou expressar, em novo idioma, conhecimentos desenvolvidos por uma grande equipe do Brasil. Agradeço aos patrocinadores por terem depositado nessa obra o merecimento do seu apoio.

*I dedicate
this book to the
memory of my
parents and to my
wife and daughters
in whom I always find
support, affection, and
courage.*

I would like to thank the authors and co-authors for having generously shared their knowledge; the Federal University of Lavras, a school known for its coffee experts; Editora UFLA for ceding the rights of the original work, written in Portuguese; Joel Shuler for his dedication in bringing to fruition a work that is much more than a simple translation, but also an effort to bring this coffee knowledge developed by a great team in Brazil to another language and more diverse audience; and the sponsors, for recognizing that this work was deserving of their support.

Flávio Meira Borém
Editor

We are grateful to Editora UFLA and the
Federal University of Lavras for granting usage
rights to the original Portuguese text of
Pós-Colheita do Café.

© 2014 by Flávio Meira Borém.

All rights reserved.

This book, or parts thereof, may not be reproduced
in any form without permission. The scanning, up-
loading, and distribution of this book via the Internet
or via any other means without the permission of
the author are illegal and punishable by law. Please
purchase only authorized editions.

Published by Gin Press, Norcross, Georgia

ISBN 978-0-9915721-0-6

Book design by John Outler

First Edition

gin press

About the Authors

Adélcio Piagentini | Adélcio is an industrial chemical engineer and currently serves as the Technical and Industrial Director at Pinhalense S.A. Máquinas Agrícolas. He specializes in projects and equipment development for coffee drying, milling, sorting and wet-method processing as well as general seed and grain preparation. He holds various patents in these areas.

Antonio Teixeira de Matos | Antonio is an agricultural engineer with a master's in Agricultural Engineering and a PhD in Soils and Plant Nutrition from the Federal University of Viçosa (Universidade Federal de Viçosa). He is an associate professor at the Federal University of Viçosa. His areas of specialization are physical environmental quality and the agricultural reuse of waste products.

Eder Pedroza Isquierdo | Eder is an agricultural engineer with a master's degree and PhD in Food Science from the Federal University of Lavras (Universidade Federal de Lavras). He is a professor at the State University of Mato Grosso (Universidade Estadual do Mato Grosso). His areas of specialization are coffee post-harvest and coffee quality.

Ednilton Tavares de Andrade | Ednilton is an agricultural engineer. He received a bachelor's degree from the Federal University of Lavras (Universidade Federal de Lavras) and a master's degree and PhD in Agricultural Engineering from the Federal University of Viçosa (Universidade Federal de Viçosa). He is currently a professor at the Federal University of Lavras. His areas of specialization are the processing and storage of agricultural products, principally in the areas of quality, drying, warehousing, storage, energy, and workplace safety.

Edvaldo Aparecido Amaral da Silva | Edvaldo is an agronomist with a PhD in Plant Physiology from Wageningen University. He is currently a researcher at the Federal University of Lavras (Universidade Federal de Lavras). His areas of specialization are seed germination, dormancy, coffee anatomy, and the physiological and molecular biology of seed quality.

Evandro de Castro Melo | Evandro is an agricultural engineer with a PhD in Agronomy. He completed a post-doctorate fellowship at the University of Valladoid. He is an associate professor at the Federal University of Viçosa (Universidade Federal de Viçosa). His areas of specialization are drying, storage, and simulation.

Fabiana Carmanini Ribeiro | Fabiana is an agricultural engineer with a bachelor's degrees in Agronomy from the Federal University of Viçosa (Universidade Federal de Viçosa) and a masters's degree and PhD in Agricultural Engineering from the Federal University of Lavras (Universidade Federal de Lavras). She is currently an assistant professor at the University of Brasilia (Universidade de Brasília). Her areas of specialization are the processing, drying, storage and quality of agricultural products.

Fátima Chieppe Parizzi | Fátima is an agronomist and holds a PhD in Agricultural Engineering. She is currently a Federal Agricultural Inspector for the Ministry of Agriculture, Livestock and Food Supply. Her areas of specialization are grain quality and the standardization, classification, inspection, and processing of plant products.

Fátima Resende Luiz Fia | Fátima is an agricultural engineer with a bachelor's degree from the Federal University of Lavras (Universidade Federal de Lavras) and a master's degree and PhD in Agricultural Engineering from the Federal University of Viçosa (Universidade Federal de Viçosa). She is currently a professor in the Engineering Department at the Federal University of Lavras. Her areas of specialization are water quality, water supply treatment, and wastewater treatment.

Flávio Meira Borém | Flávio is an agronomist with a PhD in Plant Production and a post-doctorate in Science and Process Engineering of Agricultural Products from Wageningen University. He is currently an associate professor at the Federal University of Lavras (Universidade Federal de Lavras). His areas of specialization are coffee processing, drying, storage and quality.

Francisco Carlos Gomes | Francisco is an agricultural engineer with a PhD in Structural Engineering. He is currently an adjunct professor at the Federal University of Lavras (Universidade Federal de Lavras). His areas of specialization are the physical properties of stored products, warehouse structures, and wooden structures.

Geraldo Andrade Carvalho | Geraldo is an agronomist with a master's in Plant Science and a PhD in Entomology from the Luis de Queiroz School of Agriculture (Escola Superior de Agricultura "Luiz de Queiroz."). He completed a post-graduate fellowship at the Polytechnic University of Madrid and is currently an associate professor at the Federal University of Lavras (Universidade Federal de Lavras). His areas of specialization are selectivity, pesticides, pest control, biological control and plant extract insecticides.

Jadir Nogueira da Silva | Jadir holds a bachelor's degree in Mathematics, a PhD in Agricultural Engineering and completed a post-doctorate fellowship at the Liebniz Institute for Agricultural Engineering in Postsdam/Berlin. He is currently a tenured professor at the Federal University of Viçosa (Universidade Federal de Viçosa). His areas of specialization are biomass energy, agricultural product drying, gasification and biomass combustion, thermic comfort, and aviary heating.

Jayme de Toledo Piza e Almeida Neto | Jayme is a mechanical engineer with a PhD in Sciences. He is currently a professor in Mechanics and Mechanized Agriculture, a technical advisor at Pinhalense S.A. Máquinas Agrícolas and holds various patents for agricultural equipment.

Joel Shuler | Joel Shuler is the owner of Casa Brasil Coffees. He holds a BA in Philosophy and a BS in Economics, both from Georgia State University. He is a certified Q Grader by the Coffee Quality Institute and holds the Santos Port Association Classification License.

José Henrique da Silva Taveira | José is an agricultural engineer with a master's degree in Food Science and a PhD in Agricultural Engineering from the Federal University of Lavras (Universidade Federal de Lavras). His areas of specialization are agricultural product processing, post-harvest, and coffee quality.

Luisa Pereira Figueiredo | Luisa is a food science engineer with a PhD in Food Science from the Federal University of Lavras (Universidade Federal de Lavras). She is a professor at the Federal Institute of Inconfidentes (Instituto Federal de Inconfidentes – MG). Her areas of specialization are coffee chemical composition and coffee quality.

Paola Alfonsa Vieira Lo Monaco | Paola is an agricultural engineer with a PhD in Agricultural Engineering from the Federal University of Viçosa (Universidade Federal de Viçosa). She is a professor at the Federal Institute of Education, Science, and Techology of Espírito Santo (Instituto Federal de Educação, Ciência e Tecnologia do Espírito Santo). Her areas of specialization are water and environmental resources.

Paulo César Afonso Júnior | Paulo is an agricultural engineer with a PhD in Agricultural Engineering. He is a researcher with the Brazilian Enterprise for Agricultural Research (Empresa Brasileira de Pesquisa Agropecuária). His areas of specialization are coffee, preparation methods, hygroscopy, drying, storage, thermic and physical properties, physiological quality, and chemical composition.

Paulo Cesar Corrêa | Paulo is an agronomist with a master's in Agricultural Engineering and a PhD in Agronomy with a specialization in Agricultural Engineering from the Polytechnic University of Madrid. He is an associate professor at the Federal University of Viçosa (Universidade Federal de Viçosa) and the editor of the Brazilian Journal of Storage (Revista-Brasileira de Armazenamento). His areas of specialization are the engineering of agricultural product processing, physical properties of coffee and post-harvest quality.

Paulo Rebelles Reis | Paulo is an agronomist with a PhD in Entomology from the University of São Paulo (Universidade de São Paulo). He is a researcher both with the Minas Gerais Enterprise for Agricultural Research (Empresa de Pesquisa Agropecuária de Minas Gerais) and the National Council for Scientific and Technological Development (Conselho Nacional de Desenvolvimento Científico e Tecnológico), as well as a professor in the Entomology Department at the Federal University of Lavras (Universidade Federal de Lavras). His areas of specialization are agricultural acarology, agricultural entomology, ecological coffee pest management, biological pest control, chemical pest control, and selectivity.

Roberto Precci Lopes | Roberto is an agricultural engineer with a PhD in Agricultural Engineering. He is a professor at the Federal Rural University of Rio de Janeiro (Universidade Federal Rural de Rio de Janeiro). His areas of specialization are furnaces, heating air, agricultural energy use, charcoal, boilers, and drying.

Ronaldo Fia | Ronaldo is an agricultural and environmental engineer with a PhD in Agricultural Engineering with a focus on Water and Environmental Resources from the Federal University of Viçosa (Universidade Federal de Viçosa). He is currently a professor in the Engineering Department at the Federal University of Lavras (Universidade Federal de Lavras). His areas of specialization are the treatment of wastewater and solid agro-industrial and household waste.

Sára Maria Chalfoun | Sára is an agronomist with a PhD in Plant Science. She is a researcher with the Minas Gerais Enterprise for Agricultural Research (Empresade Pesquisa Agropecuária de Minas Gerais) and the National Institute of Coffee Science and Technology (Instituto Nacional de Ciência e Tecnologia) Her areas of specialization are coffee, good agricultural practices, processing, food safety, integrated control, and mycotoxins.

Silvio Luis Leite | Silvio is a master cupper and cupping judge. He has been working with coffee since 1990 and is considered to be one of the world's foremost coffee cuppers. He is one of the founders of the Cup of Excellence (COE) and currently serves as a COE head judge as well as a director at Proud Café. He leads specialty coffee lectures and professional cupping sessions worldwide.

Terezinha de Jesus Garcia Salva | Terezinha is a food science engineer with a PhD in Food Science from the State University of Campinas (Universidade Estadual de Campinas). She is a researcher at the Alcides Carvalho Coffee Research Center, part of the Agricultural Institute of Campinas (Instituto Agronômico de Campinas). Her areas of specialization are coffee processing and the chemical characterization of coffee.

Valdiney Cambuy Siqueira | Valdiney is an agronomist with a PhD in Agricultural Engineering from the Federal University of Lavras (Universidade Federal de Lavras). He is currently a professor at the Federal University of Mato Grosso do Sul (Universidade Federal do Mato Grosso do Sul). His areas of specialization are agricultural product processing, drying, storage, physical properties of grains, and simulation.

Vanúsia Maria Carneiro Nogueira | Vanúsia is a system engineer and business administrator. She is currently a PhD candidate in marketing, with a focus on competitive positioning in the coffee market. She has been a part of the Brazilian coffee industry since 2001 after working 15 years as a marketing consultant at PwC Consulting. She is currently the executive director of the Brazil Specialty Coffee Association (BSCA).

Vitor Hugo Teixeira | Vitor is an agricultural engineer with a PhD in Agricultural Energy. He is an adjunct professor at the Federal University of Lavras (Universidade Federal de Lavras). His areas of specialization are environmental studies and coffee processing infrastructure.

Table of Contents

Table of Contents

Foreword

Pós-Colheita do Café was originally published in Portuguese in 2008. A college-level academic text written by experts in their respective fields, this book provided a much needed, comprehensive resource on post-harvest coffee with up-to-date information on techniques and technologies. Through travel and conversations with growers across the globe, it became clear to editor Flávio Borém that no such comprehensive manual was available for the market outside of Brazil, and that there was a clear demand for a single resource, written by subject matter experts, that clearly defined and prescribed proper post-harvest coffee care for those intent on maximizing coffee quality.

Handbook of Coffee Post-Harvest Technology is intended, foremost, to be a pragmatic resource for coffee growers wishing to maximize the quality of their harvest. It provides straightforward, comprehensive, and accessible information on how to properly plan and conduct post-harvest processes. This new English version was translated and updated with a larger audience in mind. Roasters, baristas, and coffee aficionados will all find in this book a resource to better understand how coffee should be properly processed, dried, and stored on its way to the cup. To this end we have made every effort to maintain the academic integrity of the original text while making the book accessible to those whose passion for coffee may not be accompanied by a degree in coffee science.

Handbook of Coffee Post-Harvest Technology focuses of the anatomical characteristics of the coffee fruit, the chemical and biochemical phenomena that occur during the post-harvest, and the important role water plays in preserving coffee quality during storage. The reader can learn both the theoretical and practical aspects of coffee drying and milling, as well as energy-saving practices and post-harvest pest and fungus control. Naturally, because the book was first written in Brazil for the Brazilian market, most of the authors and examples are from Brazil. But this provenance makes even more sense given that Brazil is, far and away, the world's leading coffee producer by volume, with commensurate leadership in coffee post-harvest technology and academic research. Brazil has been a major player in the coffee industry since the early days of global coffee trade, and now, as it progresses towards first world status, so too must its coffee industry. Improving post-

harvest technologies and best practices are critical elements of this progress.

Coffee largely lives in the languages and traditions of those who produce it, and on top of making this book more accessible while maintaining its academic integrity, a major challenge has been decoupling the terms enveloped in these traditions and finding effective replacements in English, a language with a less robust coffee vocabulary. We hope that we have done the authors justice and provided you, the reader, with an accessible text that will help you better comprehend all that is involved in the coffee post-harvest.

Flávio Borém and Joel Shuler

Anatomy and Chemical Composition of the Coffee Fruit and Seed

Flávio Meira Borém

Terezinha de Jesus Garcia Salva

Edvaldo Aparecido Amaral da Silva

1.1 Introduction

To fully understand what happens to coffee *after* it is harvested—the physiological, physical, chemical, and biochemical phenomena—it is important to understand the anatomy and chemistry of the coffee fruit *before* it is picked. As the fruit develops on the tree, a great number of events and conditions will determine the characteristics of the mature coffee seed and, ultimately, cup quality. First, various parts of the seed begin to grow, then tissues form where reserve compounds will be deposited. Carbohydrates, lipids, proteins, minerals, and secondary metabolites are the principle reserve compounds that influence flavor in the cup.

After harvest, the coffee fruit can be dry processed, maintaining intact all of its anatomical components, or wet processed, where the skin and mucilage are removed, resulting in parchment coffee. Each process affects coffee chemistry in different ways, resulting in characteristic flavor notes in the cup. Each processing method also impacts the environment in different ways, most importantly through the quantity and content of the wastewater associated with it, and the disposal of this wastewater should be carefully considered.

> Parchment coffee is any wet process coffee where the endocarp is still intact.

1.2 The Fruit

The coffee fruit and seed grow and develop through a process of ordered cellular division and differentiation that begins in the flowering phase and continues until maturity. After pollination, the embryo is formed by the fusion of a male gamete with the egg cell; at the same time, another male gamete fuses with the polar nuclei to form the endosperm. After this double fertilization, a transitory perisperm tissue is formed by the rapid multiplication of the integument cells. From this process, the fruit increases in size and the two seeds acquire their characteristic form.

The first division of the endosperm occurs between 21 and 27 days after flowering, after which the endosperm develops rapidly, replacing the nucellar tissue, which is progressively consumed. Four months after pollination, the endosperm is completely formed, with only a few remaining layers of the nucellar tissue enveloping the seed.[1]

Figure 1.1 The fruit of *Coffea arabica* L., which is formed by the pericarp and seed.

The fruit is classified as a drupe, which can be simply characterized as an indehiscent, fleshy fruit with a pericarp that is clearly differentiated into an exocarp, mesocarp, and endocarp. In the species of the *Coffea* genus, the pericarp envelops a bilocular ovary that normally contains two seeds (Figure 1.1).[2] The fruit sometimes contain three or more seeds as a result of trilocular or plurilocular ovaries. In the mature fruit, the seeds are enveloped by a layer of transparent sclerenchyma cells called the perisperm, commonly known as the silverskin. The seed is primarily composed of the endosperm. The embryo is located near the convex surface of the seed.[3]

1.2.1 Exocarp

The exocarp, commonly referred to as the skin, is the outermost tissue of the fruit (Figure 1.2). In the species *C. arabica* L. and *C. canephora* Pierre, the exocarp consists of a single layer of compact, polygonal parenchyma cells with a varying number of chloroplasts and stomatal pores evenly distributed across its surface.

At maturation, the predominant colored pigment in the skin of red fruit is anthocyanin.[4] In the Colombia cultivar of Arabica coffee, the concentration of anthocyanin is 8.26 µ g g^{-1} in mature fruit, and 47.15 µ g g^{-1} in overripe fruit.[5] In the skin of yellow coffee fruit the pigment leucoanthocyanin replaces anthocyanin, allowing exposure of the yellow pigment luteolin.[6]

In this text, the scientific nomenclature *C. arabica* L. and *C. canephora* Pierre will often be referred to by their common names of Arabica and Canephora, respectively.

Arabica coffee includes such cultivars as Bourbon, Typica, Caturra, Catuai, Mundo Novo, and others. Canephora coffee includes the two main cultivars Robusta and Conilon.

In the process of coffee pulping, the exocarp is removed along with part of the mesocarp and the vascular bundle , resulting in what many technicians and researchers call "pulp" (Figure 1.3).

Pulp contains 6% to 8% mucilage.[7] In Arabica coffee, pulp represents 39% to 49%[8] of the weight of mature, fresh fruit. In *C. canephora* cv. Pierre, it represents 38% of this weight. However, it should be noted that these numbers may vary due to genetic, environmental, or processing factors.[9]

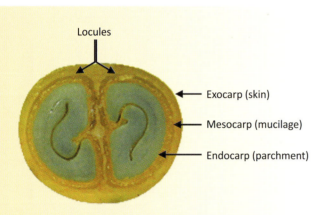

Figure 1.2 Diagram of the structure of the coffee fruit (*C. arabica* L.).

Locules

Exocarp (skin)

Mesocarp (mucilage)

Endocarp (parchment)

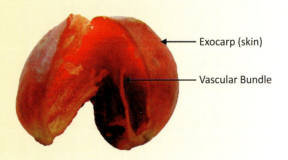

Figure 1.3 Exocarp, part of the mesocarp, and vascular bundle removed during the pulping process (*C. arabica* L.).

Exocarp (skin)

Vascular Bundle

The pulp of the coffee fruit is mainly composed of carbohydrates, crude protein (N x 6.25), crude fiber, and ash. The pulp of Canephora coffees contains more cellulose and tannins than the pulp of Arabica coffees. Measurements of acidity levels and quantities of solid materials of mature Arabica coffee fruit, Colombia cultivar, revealed the total titratable acidity of the pulp to be 11.4 mL of 1N NaOH 100 g^{-1}, the pH 4.5, and the content of soluble solids 17%.[10] The typical composition of Arabica and Canephora pulp is presented in Table 1.1.

Both the cellular structure and chemical composition of the exocarp constitute an important barrier to the transfer of water from the fruit to the air during the drying of coffee fruit. Therefore, removal of the exocarp results in faster removal of water from the fruit and greater drying efficiency. This contributes to lower energy consumption and reduces drying time on the drying patios and in the dryers, thus increasing a given infrastructure's post-harvest drying capacity. By contrast, the slower drying process that results when the exocarp remains intact increases the risk of fermentations that jeopardize coffee quality.

Table 1.1 Pulp composition of Arabica and Canephora Coffees.

Arabica coffee cv. Typica [I]		Canephora coffee cv. Robusta [II]	
Component	Content (db)	Component	Content (db)
Dry Material	93.07%	Non-nitrogen extract	57.90%
Carbohydrates	74.10%	Crude fiber	27.70%
Non-nitrogen extract	59.10%	Reducing sugars	12.40%
Crude fiber	15.10%	Crude protein	9.20%
Crude protein	8.25%	Pectic substances	6.50%
Ash	8.12%	Tannins	4.50%
Moisture Content	6.93%	Ash	3.30%
Tannins	3.70%	Non-reducing sugars	2.00%
Potassium	3.17%	Lipids	2.00%
Ether extract	2.50%	Chlorogenic Acids**	1.61%
Nitrogen	1.32%	Caffeine**	0.54%
Caffeine	0.75%		
Calcium	0.32%		
Phosphorus	0.05%		
Iron	250 ppm		
Sodium	160 ppm		
Chlorogenic Acids**	1.10%		

* Carbohydrates = Non-nitrogen extract + crude fiber | ** Clifford & Ramirez-Martinez (1991)
[I] Zuluaga (1999)[11]. | [II] Wilbaux (1961)[12].

On the other hand, there are certain advantages to maintaining an intact exocarp. Current technology for removing the exocarp and mesocarp produces water with elevated pollution potential as well as solid residues, although these residues can be transformed into energy or used as organic fertilizer. What's more, the exocarp acts as a barrier to the activity of microorganisms. When the skin of the fruit is damaged by machines or insects, fungi are more likely to develop, producing substances that contribute to coffees of inferior quality.

1.2.2 Mesocarp

The mesocarp, also called the mucilage, is made up of a tissue formed by parenchyma cells. The mucilage contains within it vascular bundles, which are composed of xylem internally and phloem externally, and which are distributed throughout the tissue. (Figure 1.4). This tissue is formed by about 20 layers of various sized cells, and in ripe coffee is found in concentrations that vary with growing altitude and plant cultivar.

In unripe coffee fruit, the mesocarp is composed of a hard tissue. During maturation, pectinolytic enzymes act upon pectic substances, leading to tissue disintegration by the breakdown of pectin chains into molecules of galacturonic acid. The final product of this enzyme action is a hydrogel that is insoluble, colloidal, hyaline, mucilaginous, and rich in sugars and pectins.[13]

As a general rule, the mucilage makes up 22%–31% of the mass of the dry fruit.[14] Both the overall quantity and moisture content of the mucilage diminish with an increase in growing altitude. Therefore, regions of higher altitude produce coffees with higher quantities of dry mucilage.[15] For example, the dry mucilage of Bourbon coffee cultivated at 250 m is 3.1% of the fresh weight of the fruit, while it is 5.5% of the same coffee cultivated at 1,700 m.

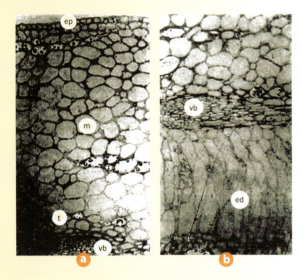

Figure 1.4 Characteristics of the mesocarp and endocarp of Arabica coffee fruit between 195 and 270 days after flowering. **a** Cells with higher tannin content (t), exocarp (ep), mesocarp (m), and vascular bundle (vb); **b** Endocarp (ed), vascular bundle (vb). (SALAZAR et al., 1994). [17]

The mucilage of the mature, fresh fruit of Arabica coffee comprises, on average, 88.34% (wb) water and 7.17% (db) total carbohydrates. The carbohydrates are made up of approximately 28.3% (db) pectin, 9% (db) cellulose, 15% (db) neutral non-cellulosic polysaccharides, and small quantities of arabinose, xylose, galactose, and rhamnose. Compounds including nitrogen, amino acids, caffeine, and enzymes make up 0.17% (wb) of the mucilage. The average total titratable acidity of mucilage is 17.6 mL 1N NaOH 100 g^{-1}, though levels can reach as high as 36.8 mL 1N NaOH 100 g^{-1}.[16]

Removing the mesocarp in coffee processing can increase a lot's homogeneity and mitigate risk. The ability of the mesocarp of unripe fruit to withstand force and pressure allows unripe fruit to be separated from ripe fruit during pulping when pulpers are regulated so that only ripe fruit are pulped. Pulping can mitigate risk since the high quantity of sugars in the mucilage of mature fruit increases the risk of fermentations that can compromise the quality of the coffee. The decision to keep intact or remove the mucilage, however, ultimately depends on the availability of sufficient infrastructure for processing and drying the coffee.

1.2.3 Endocarp

The endocarp, or parchment, is the innermost structure of the pericarp (Figure 1.5). It is made up of three to seven layers of irregularly shaped sclerenchyma cells that completely envelop the seed. With the development of the endosperm, the cells of the endocarp become sclereid, imparting the woody texture observed in mature fruit.[18] This hardening of the endocarp tissue limits the growth of the seeds due to mechanical restriction, and determines the potential final size of the bean, an important characteristic in the commercialization of coffee.

In a study calculating the averages of four cultivars of Arabica coffee produced for consumption, the endocarp weight of coffee dried to 11% (wb) moisture content was approximately 3.8% of the weight of the entire fresh fruit, while it was 16%–21% of the weight of parchment coffee. In Canephora coffee (Robusta cultivar), on average, these values were around 3.5% and 13.0%–14.5%, respectively. The parchment of Arabica coffee was essentially composed of a cellulosic material of 50% cellulose, 20% hemicellulose, and 20% lignin. It also contained about 1% ash and 1.5% crude protein. The parchment of Canephora coffee was richer in cellulose (60%) and ash (3.3%) than the parchment of Arabica coffee.[19]

1.2.4 The Seed

The coffee seed is made up of the silverskin, endosperm, and embryo. A normal seed is flat on one side, convex on the other (plano-convex), elliptical, and furrowed on the flat face.[20] When only one seed develops in the fruit, it forms a peaberry. When the fruit contains three locules with fertilized ovules, triangular beans are formed. When two seeds develop in one locule, one of the seeds, the core, becomes partially embedded in the other, the shell. During

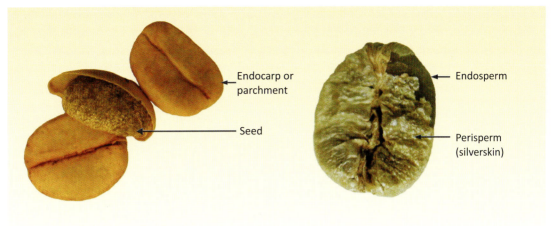

Figure 1.5 The seed and endocarp of Arabica coffee.

Figure 1.6 Arabica coffee seed with perisperm.

milling, these two beans can separate, resulting in the malformed and shell defects, respectively. Finally, withered beans result when the endosperm does not fully develop.

1.2.4.1 Silverskin

The silverskin is the outermost covering of the seed, formed by a layer of sclerenchyma cells (Figure 1.6). Contrary to what some authors have claimed[21], the silverskin is not formed by the integuments of the embryo sac, but rather from a layer of cells remaining from the nucellar tissue.[22] Therefore, the silverskin cannot be considered an integument of the coffee seed. Technically, it can be referred to as a perisperm or spermoderm. Its function is not well known. It contains chlorophylls a and b, and depending on the region of coffee production, it can acquire a dark or caramel coloration. Beans with this silverskin coloration are called "fox beans."

1.2.4.2 Endosperm

The endosperm is the principal reserve tissue, constituting the largest volume of the mature coffee seed. It is a single tissue formed by the fusion of a spermatic nucleus and two polar nuclei, resulting in a triploid (3n) tissue.

The endosperm is composed externally of small, oil-rich cells (Figure 1.7), and more internally of larger cells with slightly thinner walls. This distinction between the characteristics of these cell groups has lead some authors[23] to denominate the external part as "hard endosperm" and the internal part as "soft endosperm." Though not uniform, the endosperm is a single tissue.

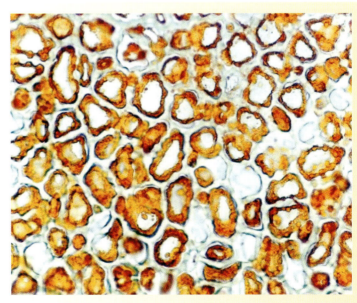

Figure 1.7 Light microscopy of the endosperm of an Arabica coffee seed, colored with Sudam IV dye. Note the oil distribution inside the cells.

The endosperm consists of polyhedral cells with thick walls (Figure 1.8) containing plasmodesmata, narrow threads of cytoplasm that establish connections between the cells and that can play an important role in the transport of water and other substances.[24]

The walls of the endosperm cells are polygonal (Figure 1.9). The more rectangular cells are located adjacent to the embryo and are more internally located in the endosperm, while the more convex, or rounded cells are located closer to the exterior. The endosperm of the coffee seed can be divided into the endosperm cap, or micropylar endosperm, and the lateral endosperm. The cells of the micropylar endosperm located in front of the radicle are smaller and have thinner walls than the cells of the lateral endosperm. Thicker cell walls indicate the accumulation of important reserve compounds that contribute to cup quality, and are a necessary source of energy during seed germination.

The chemical composition of the endosperm is of great interest as it represents the precursor elements to the flavor and aroma that will form in roasted coffee, directly influencing the quality of the beverage.

After coffee is harvested, the chemical composition of the endosperm is important since the exchange of water between fruit and atmosphere depends on the predominance of one component or another, given the stronger or weaker attraction of water to each component. Water is strongly attracted to carbohydrates and weakly attracted to lipids. Therefore, seeds rich in water-soluble carbohydrates can be stored with moisture content up to 14% (wb), while oily seeds should be stored with lower moisture levels, approximately 8%–9% (wb). Coffee is best stored at 11% (wb) moisture content since it is composed of both soluble and insoluble compounds.

The water soluble components in the coffee endosperm are caffeine; trigonelline; nicotinic acid (niacin); chlorogenic acids; mono-, di- and oligosaccharides; some proteins and minerals; as well as carboxylic acids. Known water-insoluble components, which make up 65%–74% of the endosperm weight, are cellulose, polysaccharides, lignin, and hemicellulose, as well as proteins, minerals, and lipids.[25]

Polysaccharides are located in the cell walls and constitute the largest percentage (50%) of the dry weight of raw coffee.[26] The main components of these polysaccharides are cellulose (15%), arabinogalactan proteins (25%–30%), mannan and galactomannan (50%), and pectin (5%)[27]. The concentration of starch in raw coffees is small, at around 0.5%[28]. The absolute concentrations of arabinogalactan and galactomannan polysaccharides are identical in *C. arabica* L. and *C. canephora* Pierre coffees[29]. The difference between the species seems to be that the ratio of arabinogalactan to galactomannan is higher in Canephora[30], and that the arabinogalactan in Canephora is more branched and has a longer side chain than in Arabica. The function of arabinogalactan is to aggregate the chemical components of the cell wall of the coffee bean, while mannan, besides being the principal reserve polysaccharide, is responsible for the thick and dense cell wall structure. The skeletal structure of this cell wall is made up of cellulose[31]. The seeds of Arabica have a particularly rigid endosperm due to the deposit of hemicellulose in the cell walls.

Sucrose is the most abundant of the low molecular weight carbohydrates in raw coffee. Its concentration in Arabica coffees produced for consumption varies from 5% to 12% (db), and in Canephoras from 4% to 6% (db). Sucrose concentration can reach up to 400 times the combined concentrations of glucose, fructose, and stachyose, other carbohydrates with low molecular weights found in relatively high concentrations in the seed[32]. Arabi-

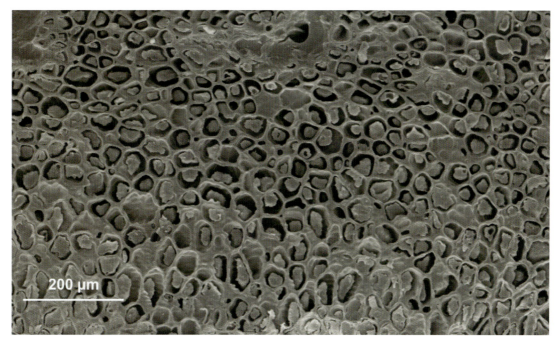

Figure 1.8 Scanning electron micrograph showing the endosperm of an Arabica coffee seed with 11% (wb) moisture content. Note that the cells have a polyhedral shape with a thick cell wall.

nose, galactose, isomaltose, rafinose, mannitol, and inositol phosphate are also found in Arabica and Canephora coffees[33].

After carbohydrates, lipids and proteins are the next most abundant compounds found in the raw coffee seed. The concentration of lipids extractable with a non-polar solvent is between 12% and 18% (db) in Arabica coffees, and between 9% and 13% (db) in Canephora coffees[34].

These lipids are found both in the cytoplasm of the reserve cells of the parenchyma, which compose most of the bean, and in the cytoplasm and the cell wall of epidermal cells, where

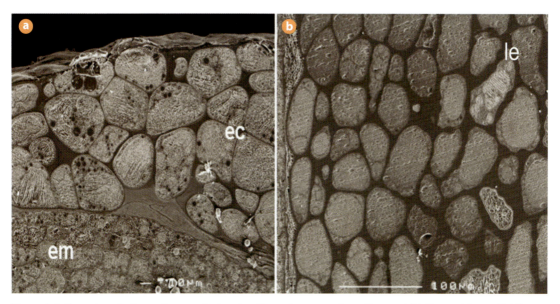

Figure 1.9 Scanning electron micrograph showing the endosperm of a coffee seed (*Coffea arabica* cv. Rubi). ⓐ Endosperm cap (ec), or micropylar endosperm, and embryo (em). ⓑ Lateral endosperm (le). Note that the cell walls of the endosperm cap cells are thinner than the cell walls of the lateral endosperm cells.

they form what is called *wax*[35]. This wax corresponds to about 0.25% (db) of the weight of raw coffee, and constitutes approximately 1.5%–2.5% of the lipid fraction[36]. The other components of the lipid fraction are generally present in the following percentages:[37]

- triglycerides 75.20%
- esters of diterpenic alcohol and fatty acids 18.50%
- diterpenic alcohol 0.40%
- sterol fatty acid esters 3.20%
- sterols 2.20%
- tocopherols 0.04–0.06%
- phosphatides 0.10–0.50%
- tryptamine derivatives 0.60–1.00%

The lipid fraction of coffee also contains free and esterified fatty acids in the form of triglycerides. In Arabica and Canephora coffees, the fatty acids present in the highest concentrations are:[38]

myristic (C14:0)	linoleic (C18:2)	gadoleic (C20:1)
palmitic (C16:0)	linolenic (C18:3)	behenic (C22:0)
stearic (C18:0)	arachidonic (C20:0)	lignoceric (C24:0)
oleic (C18:1)		

Raw Arabica coffees contain, on average, 9.2% (db) protein, while Canephoras contain 9.5% (db)[39]. This fraction contains both soluble and insoluble proteins[40] as well as free amino acids (FAA) and peptides[41]. Soluble proteins constitute half of the total protein in raw coffee[42], and include proteolytic enzymes and polyphenol oxidase (PPO)[43]. The proteins of the seed are bound to the cellular membrane (as with PPO), bound to the cell wall polysaccharides, or free in the cytoplasm[44].

The chlorogenic acids, normally caffeic, ferulic, and p-coumaric, are esters of quinic acid and cinnamic acid, and constitute a quantitatively important class of compounds in the coffee seed. The endosperm of raw coffee contains at least 18 chlorogenic acids divided into 5 classes: 3 caffeoylquinic acids, 3 p-coumaroylquinic acids, 3 feruloylquinic acids, 3 dicaffeoylquinic acids, and 6 caffeoylferuloylquinic acids[45]. The most abundant chlorogenic acids in raw coffee are the caffeoylquinic acids, which make up at least 80% (db) of the total chlorogenic acids found in the seed[46]. The most abundant caffeoylquinic acid in the coffee seed is 5-caffeoylquinic acid (5-CGA)[47], which composes 70%–85% of the total caffeoylquinic acid in the seed. Canephora and Arabica coffees differ in their concentration of total chlorogenic acids[48], as well as in the composition of these acids[49]. On average, chlorogenic acids in Arabica coffees are found in concentrations between 4.0% and 7.5%, while in Canephoras they are found in concentrations between 7.0% and 11.0%[50]. In the endosperm, the chlorogenic acids are found in the surface cells of the bean, associated with the wax, as well as in the parenchyma cells, where they are found in the cytoplasm near the cell wall[51].

Caffeine, one of the best-known compounds in coffee seeds, is present in *C. arabica* L. in concentrations between 0.53% and 1.45% (db)[52], with higher concentrations in *C. canephora* that vary between 1.94% and 3.04% (db) in the Conilon and Robusta cultivars, respectively[53]. Caffeine is found both free in the cytoplasm and attached to the cell wall, where it is most likely associated with chlorogenic acids[54].

Other compounds found in coffee seeds are trigonelline, nicotinic acid (niacin), other carboxylic acids, and minerals. Trigonelline concentration is higher in Arabica coffees than in Canephoras. In cultivars of *C. arabica* L., the concentration of trigonelline varies from 0.96% to 1.44% (db), and in cultivars of *C. canephora* from 0.81% to 1.09% (db). The level of nicotinic acid in raw coffee varies from 0.016% to 0.04%[55], with levels of 0.03% in Brazilian Arabica coffee and 0.02% in Canephora coffee from the Ivory Coast[56]. The carboxylic acid concentrations most commonly found in raw coffee seeds are as follows:

- acetic: traces–0.058% (db) in Arabica, traces–0.20% (db) in Canephora
- citric: 0.50%–1.58% (db) in Arabica, 0.33%–1.28% (db) in Canephora
- malic: 0.26%–0.67% (db) in Arabica, 0.18%–0.73% (db) in Canephora
- quinic: 0.33%–0.70% (db) in Arabica, 0.16%–0.86% (db) in Canephora
- succinic: 0%–0.74% (db) in Arabica, 0.013%–0.30% (db) in Canephora

In raw coffees, ascorbic acid can also be found in concentrations of 0.337% in Arabica and 0.308% (db) in Canephora[57].

Green coffee also contains phosphoric acid in concentrations between 0.107% and 0.147% (db) in Arabica and between 0.142% and 0.279% (db) in Canephora[58]. Low levels of glycolic and lactic acids can also be present in raw coffee[59].

Minerals compose approximately 4% of the dry mass of both raw Arabica and Canephora coffees[60], with potassium, magnesium, phosphorus, and calcium found in the highest proportions. Among these minerals, potassium is found in the highest concentration, averaging about 1.54% (db) in Arabicas and 1.71% (db) in Canephoras. Table 1.2 shows the mineral concentrations found in the two species.

Table 1.2 Average concentrations of minerals in raw Arabica and Canepehora coffees.

	Zn*	P*	Mn*	Fe*	Mg*	Ca*	Na*	K*	Cu*	Sr*	Ba*	Al**	Co**	Ni**
	Concentration × 10⁻⁴ (%)													
	Arabica													
Min	3.62	1,410	16.20	24.80	1,720	930	28.40	12,110	14.30	1.30	2.49	0.60	0.00	0.00
Max	61.27	1,700	50.00	55.10	2,060	1,370	118.00	18,820	76.90	11.60	7.85	32.20	1.20	44.20
Mean	10.40	1,520	32.30	33.90	1,867	1,081	52.00	15,426	18.63	5.00	4.80	13.00	0.30	4.90
	Canephora													
Min	5.38	1,720	14.50	28.80	1,600	950	18.20	15,450	15.20	3.30	1.65	....	....	....
Max	19.87	2,200	19.70	93.30	1,970	1,620	10.10	18,960	26.10	10.10	6.40	....	....	....
Mean	10.80	1,955	16.40	50.70	1,770	1,281	56.40	17,084	21.90	7.10	3.80	....	....	....

* % Dry Basis[61] | ** % Wet Basis[62]

Water in Coffee Fruit and Seeds

Flávio Meira Borém

Luisa Pereira Figueiredo

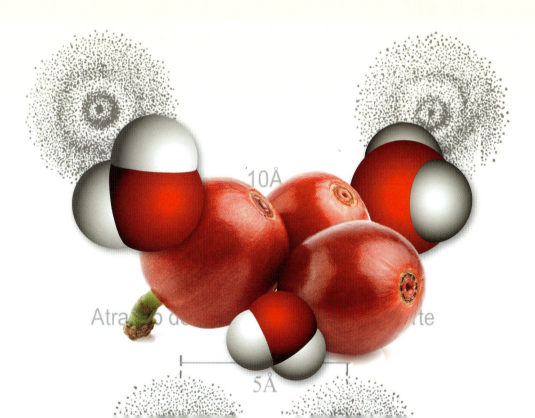

2.1 Introduction

Water is the most abundant component in the cells of living plants. While water is fundamental for the preservation of life, high moisture content levels in orthodox seeds are only required during germination and, in fact, can be detrimental to the long-term survival of the seed. As a result, unlike leaves, roots, fruits, and tubers, moisture content is low in mature orthodox seeds, making them one of the driest living tissues in plants. The moisture content of coffee depends on the state of maturation of the fruit (Table 2.1), the presence of various anatomic components, and the temperature and relative humidity of the air that is in contact with the coffee.

Table 2.1 Moisture content in coffee fruit.

State of maturation	Moisture content (% wb)
Unripe and semi-ripe	66–70
Ripe	50–65
Overripe	30–50
Dried*	11–30

Source: adapted from Wilbaux (1963).[1]

* The article by Wilbaux states "floaters and dried." However, in coffee phenology, "floater" is not a state of maturation. Floaters are defined not by their maturation state but by their density. Therefore, "floaters" has been removed here. Coffee is considered "half-dry" at 30% and below, and completely dry at 11%.

Water is the most abundant and best known solvent. As a solvent, it is the ideal medium for the movement of molecules within and between cells, decisively influencing the structure of proteins, membranes, and other cellular components[2]. Moisture content is the most critical parameter of coffee quality, as it governs the fermentation process as well as the growth of fungi during storage and transportation, which can result in unpleasant flavors and aromas as well as mycotoxins. As such, to achieve successful coffee processing and storage it is necessary to understand the properties of water, its interactions with the seeds, and its effect on the preservation of quality.

2.2 Structure and Properties of Water

The structure of water gives it unique properties. The water molecule is composed of an oxygen atom covalently bonded to two hydrogen atoms, which form a 105° angle (Figure 2.1). The oxygen atom, being more electronegative than a hydrogen atom, tends to attract electrons to the covalent part, resulting in a slightly negative charge near the oxygen and a slightly positive charge near the hydrogen. Since these slight charges are equal, water does not have a liquid charge. However, the partial separation of the charges as a result of the molecular structure makes the water molecule polar[3].

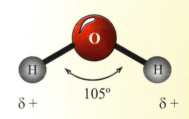

Figure 2.1 Diagram of a water molecule.

2.2.1 Hydrogen bonds

While covalent bonding brings chemical elements together, other connections allow for bonding between molecules. Through electrostatic attraction between the positive hydrogen charge and the electronegative side of the oxygen, the polarity of the water molecule allows for the creation of hydrogen bonds (Figure 2.2) between neighboring molecules[4]. Comprising two hydrogen atoms and one oxygen atom with two pairs of non-shared electrons, the water molecule has the perfect conditions, in terms of the number of hydrogen bonds, for each molecule to participate as a donor in two hydrogen bonds and as a recipient in two more bonds[5]. The result is a special tetrahedric arrangement (Figure 2.3) in the water molecule.

These connections give water uncommon physical properties. The hydrogen bonds allow for the grouping of water molecules in a liquid solution. In liquid state (Figure 2.4), transi-

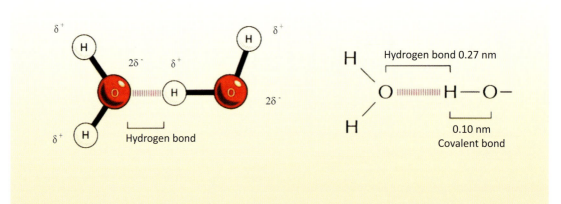

Figure 2.2 Schematic structure of water molecules showing formation of hydrogen bond[7].

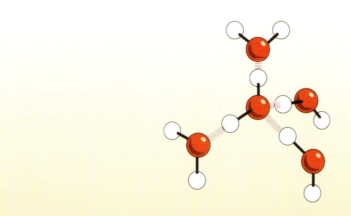

Figure 2.3 Geometric arrangement of water molecules with each molecule forming four hydrogen bonds[8].

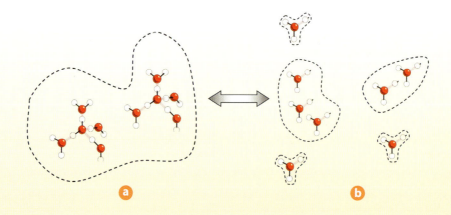

Figure 2.4 Schematic representation of water in liquid state, showing the transitory groups and the free monomeric molecules. **a** Transitory groups. **b** Due to constant thermal agitation of the water molecules, these groups have a short life span; they break apart and quickly form new, random configurations[9].

tory groups and monomeric species form quasi-crystalline structures that continually break apart and regroup[6].

2.2.2 Thermal, cohesive, and adhesive properties of water

The large number of hydrogen bonds between water molecules results in very important post-harvest thermal characteristics, which are particularly useful in the coffee drying process. Water has a high specific heat and a high latent heat of vaporization. (For a detailed discussion of specific heat, see section 3.4.2).

The amount of heat required to increase the temperature of a substance by 1 °C is referred to as that substance's heat capacity, which is proportional to its mass[10]. As the temperature of water is raised, the molecules vibrate faster, accelerating the breaking of hydrogen bonds between the molecules. Compared to other liquids, water requires a relatively higher amount of energy to increase its temperature. The latent heat of vaporization of water refers to the amount of energy necessary to change water molecules from a liquid state to a gaseous state at a constant temperature. During the coffee drying process, these characteristics are important as they will affect the total amount of energy required to remove water from the coffee.

Cohesion refers to the mutual attraction between molecules. Adhesion refers to the attraction of water molecules to solid surfaces, such as cell walls (Figure 2.5). These characteristics result in capillarity, also called capillary action. When coffee has very high moisture content, as is the case for unripe, semi-ripe, and ripe fruit immediately after harvest, water occupies the empty spaces inside the fruit and seeds and is retained by capillary action (Figure 2.6).

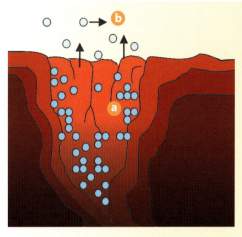

Figure 2.5 Representation of water movement from the porous space of the fruit or seed into the atmosphere. **a** Water molecules adsorbed on the surface; **b** water vapor molecules[11].

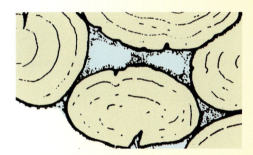

Figure 2.6 Schematic representation of water retained by capillarity[12].

2.2.3 Other types of bonds that occur in water

In addition to covalent bonds and hydrogen bonds, the water molecule is able to develop other bonds and intermolecular forces, such as the ion-dipole effect and van der Waals forces. Water molecules can be grouped into five classes or energy levels, depending on the number of hydrogen bonds that are formed[13]. The higher the energy level, the fewer the hydrogen bonds, and the greater the influence exerted by other types of weaker bonds (Figure 2.7).

Van der Waals forces (Figure 2.8) are created by the tendency of a positively charged nucleus of a molecule to attract the negatively charged electrons of a neighboring molecule. These forces are relatively weak and therefore only effective when the molecules are in close proximity.

The ion-dipole effect refers to the attraction of water molecules to cations or anions. The size of the ion is determined by the number of molecules bonded to it (Figure 2.9).

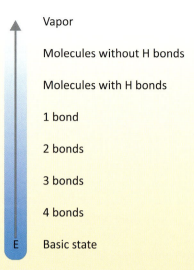

Vapor

Molecules without H bonds

Molecules with H bonds

1 bond

2 bonds

3 bonds

4 bonds

E Basic state

Figure 2.7 Schematic representation of energy levels of water molecules in the liquid state. In the vaporous state, pure water exists in the form of monomeric molecules, which are separated by a distance that makes it impossible for them to interact with neighboring molecules.

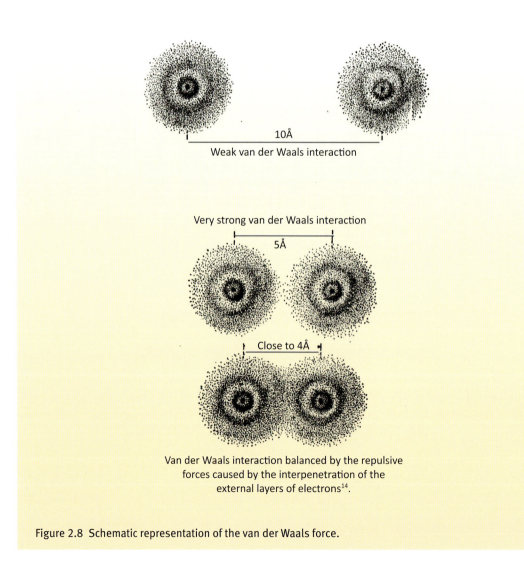

10Å

Weak van der Waals interaction

Very strong van der Waals interaction

5Å

Close to 4Å

Van der Waals interaction balanced by the repulsive
forces caused by the interpenetration of the
external layers of electrons[14].

Figure 2.8 Schematic representation of the van der Waals force.

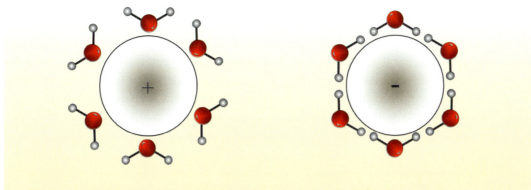

Figure 2.9 Distribution of water molecules in relation to ions: ion-dipole effect. Adapted by Fennema (1993).[15]

2.3 Interactions Between Water and Seed Components

The moisture content of the coffee fruit and seed is very high at the moment of fertilization and continuously diminishes throughout maturation. Until physiological maturity, the reduction in moisture content occurs due to the large accumulation of dry material in the fruit. In the stages that follow maturation, drying occurs as the fruit and the seed lose water to the environment.

Unlike most cereals and grains, when coffee is harvested the majority of the fruit still contain a large amount of water in the liquid phase. Once drying begins, water loses its mobility and only the strongly bonded water remains, with different thermal and physical characteristics from that of free water.

The water that is adsorbed, or strongly bonded, is present in tissues with high moisture content as well as in dry tissues. In fact, dry coffee does not contain water in the liquid phase but rather adsorbed water that is in balance with the water vapor in the environment.

2.3.1 Interactions at the molecular level

Water molecules can bond with the surface of a solid substrate in either the liquid phase or the gas phase. Water molecules interact[16] with other polar molecules through hydrogen bonds that have much higher bonding energy than the energy of the hydrogen bonds that join two water molecules. In seeds, the primary polar groups that are capable of forming hydrogen bonds with water are a, b, c, and d; to a lesser extent e and f:

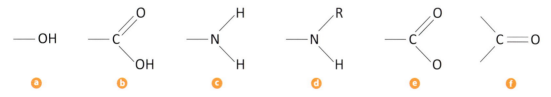

On the other hand, nonpolar groups, such as aliphatic chains or benzene rings, have no affinity with water. However, water molecules can organize themselves around nonpolar groups through hydrophobic interaction[17].

2.3.2 Interactions at the macromolecular level

The interactions between water molecules and polar groups also occur in macromolecules, such as proteins. In general, the affinity of a biological substance with water depends on its number of polar and nonpolar groups, its spatial orientation, and the water's accessibility to the molecule. Therefore, lipids and hydrocarbon chains with a high quantity of nonpolar groups have low affinity with water. Carbohydrates, which have many polar sites, and proteins to a lesser degree, are very hygroscopic[18].

Not all polar sites of a molecule are necessarily capable of bonding with water. In the case of yeast, for example, only the hydroxyls bond with water; in the case of proteins, only the polar sites of the primary chain and the hydrophilic amino acid chains can bond with water (Figure 2.10). With respect to lipids, only the polar extremity of the long, nonpolar aliphatic chain can bond with water, which explains its low hygroscopicity[19].

These interactions and the quantity of each of these components in the seeds will determine the total quantity of water that can be adsorbed in certain conditions of temperature and relative environmental humidity. These concepts are better understood when the difference between moisture content (% wb) and water activity (a_w) is established.

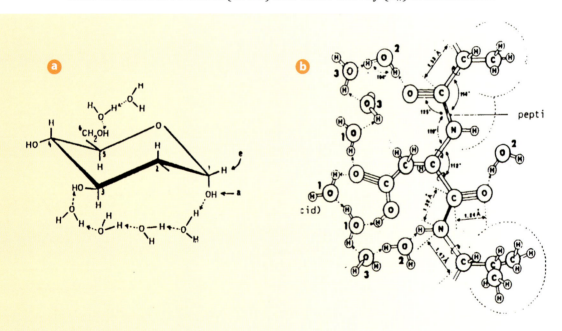

Figure 2.10 Water bonds with macromolecules. **a** Saturation of hydroxyls from a monosaccharide ring with water molecules. **b** Sorption in the primary chain and the protein amino acid chain[20].

2.4 Types of Water Found in the Coffee Fruit and Seed

Seed physiologists classify the water retained in coffee seeds into four types, based on the nature of the physicochemical bonds that exist between the components and the water molecules, and on the predominance of polar and nonpolar components.

The first type, constitutional water, is part of the composition of different biological molecules. Constitutional water is removed by reactions of oxidation of the molecule bonded to it, once there are only covalent bonds.

The second type is the water adsorbed into the molecular and macromolecular components of the coffee fruit and seeds. During the drying process, this type of water is partially removed, depending on the characteristics of the product and the drying conditions. Adsorbed water can be strongly bonded to certain molecular groups through hydrogen bonds, forming a monomolecular layer. It also exists in polymolecular layers bonded to a monomolecular layer through hydrogen bonds and very weak bonds, such as van der Waals forces. As the water molecules of the polymolecular layers distance themselves from the surface of adsorption, the number of hydrogen bonds is reduced, leaving van der Waals forces as the predominant bond.

The third type of water is found in fruit and seeds under osmotic tension. It is considered solvent water that retains different dissolved substances in the cells of the seed due to a large number of ion-dipole bonds. This type of water has biological functions that allow chemical reactions to occur and fungi to develop. As such it is a type of water that should be removed during the drying process.

The fourth type of water is composed of impregnation water or absorbed water. This type of water exists in the capillaries and in the empty spaces inside the fruit and seeds, and it is mechanically removed by capillary forces. It is considerably mobile and easily removed by the drying process. This type of water exists in unripe and mature fruit and is removed in the first hours and days of mechanical and sun drying, respectively. Since it is weakly bonded, it should be removed with sun drying, when possible, to avoid unnecessary expenses related to heated air. (Figure 2.11).

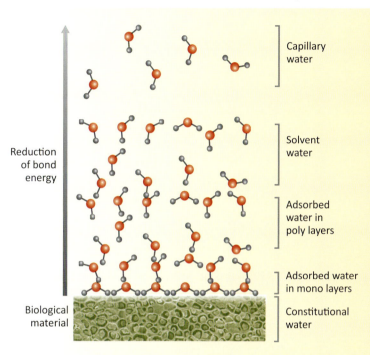

Figure 2.11 Schematic representation of different types of water.

2.5 Moisture Content and Water Activity

The water in seeds should be defined in both quantitative and qualitative terms. The moisture content of a substance indicates its degree of hydration. It can be expressed in dry basis (db) and wet basis (wb). Dry basis is preferred in mathematical simulations and in the study of the physical and thermal characteristics of seeds, as it expresses the relationship between water mass (m_{H_2O}) and the mass of the dry matter (m_{dm}).

$$MC\ (db) = m_{H_2O}\ m_{dm}^{-1} \tag{2.1}$$

in which: MC (db) = dry basis moisture content ($g_{H_2O}\ g\ m_{dm}^{-1}$)

However, moisture content is usually expressed as wet basis, a percentile ratio between moisture mass (m_{H_2O}) and the mass of the wet matter (water + dry matter):

$$MC\ (wb) = 100\ m_{H_2O}\ (m_{H_2O} + m_{dm})^{-1} \tag{2.2}$$

in which: MC (wb) = wet basis moisture content (%).

The concept of moisture content is not as simple as it appears, since the type of water to be used in the calculation must be specified, and zero must be defined on the scale of moisture content. Theoretically, only the quantity of adsorbed water, solvent water, and capillary water, as discussed previously, should be measured to determine moisture content. In this way the theoretical definition of moisture content is given as "the quantity of water that is lost by a substance in equilibrium, with gas pressure equal to zero in conditions where any reaction is avoided that could eventually facilitate moisture removal through oxidation reactions."[21] This is strictly a theoretical concept, as some components, particularly volatile ones, can be lost when moisture content is determined using routine methods that employ ovens with high temperatures.

The quantitative concept of moisture content is insufficient to understand the functional characteristics of seeds, particularly their biochemical viability and readiness to develop fungi. Therefore, to express the state of water in relation to seeds, the qualitative concept of water activity (a_w) is preferred.

Water activity is a thermodynamic parameter defined as the "chemical potential" of the water in seeds, referring to the state of energy of the molecule in the system. It expresses the potential availability of the water to participate in chemical and biochemical reactions and in the development of fungi.

There are various factors that affect water activity in a system:

- The type of bond between the water and the chemical components of the seeds. In this case, the higher the prevalence of hydrogen bonds, the lower the a_w, compared with systems where there are primarily hydrophobic bonds.
- The chemical groups that compose the seeds. Water activity will be higher in seeds rich in nonpolar compounds, compared with seeds rich in polar compounds, where moisture content is the same.
- Temperature alters the a_w value, since an increase in temperature reduces the stability of hydrogen bonds, increasing the values of water activity.

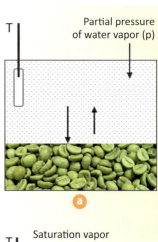

Water activity is therefore strongly related to a product's chemical composition and its temperature. For the same product under the same environmental conditions, the higher the moisture content, the higher the a_w. However, for different products higher moisture content does not always guarantee higher a_w. What's more, a product does not have to be rich in carbohydrates (known to be polar compounds) to more strongly retain water. Rather, it is necessary to identify the predominant type of carbohydrate. In green coffee, the predominant carbohydrates are cellulose and hemicellulose. Since these are insoluble in water they bond more loosely with water than carbohydrates such as sugars present in the mucilage of coffee, or yeast present in grains.

When a product that contains water is placed in an atmosphere that contains water vapor (Figure 2.12), energy and mass (water vapor) are exchanged between the seeds and the environment.

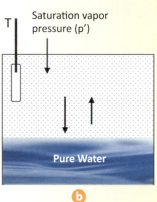

Figure 2.12 Equilibrium between the moisture content of seeds and the liquid and gaseous phases of air.

This physical property is known as hygroscopicity. The seeds will absorb or lose water to the atmosphere until hygroscopic equilibrium is established.

When this equilibrium is reached (Figure 2.12a), the seeds will have a specific quantity of water, called equilibrium moisture content (EMC). At constant temperature and pressure, the atmosphere has a maximum vapor pressure known as saturation vapor pressure (p') (Figure 2.12b). Since water vapor is a perfect gas, in a state of equilibrium a_w can be numerically defined as the ratio between partial water vapor pressure (p) and saturation vapor pressure (p') at the same temperature (T) and atmospheric pressure. This ratio is also known as the relative humidity of air (RH):

$$a_w = p / p' = RH/100 \tag{2.3}$$

in which: a_w = water activity (without dimensions)
                RH = relative humidity of the air (%)

Knowledge of the water activity of coffee fruit and seeds allows producers to better establish conditions that preserve coffee throughout processing, drying, and storage (Figure 2.13). When water activity is measured and controlled, it is possible to:

- predict which microorganism has the potential to develop, infect, and produce toxic metabolytes;
- maintain food product stability, minimizing non-enzymatic darkening reactions and lipid oxidation;
- prolong the activity of enzymes and vitamins in food products.

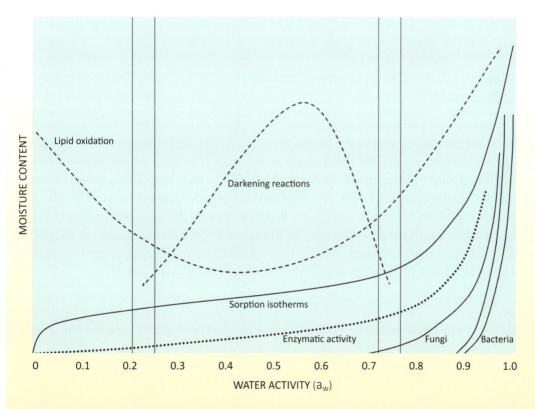

Figure 2.13 Diagram of the relationship between water activity (a_w) and the primary causes of deterioration in stored seeds. Adapted by Labuza (1970).[22]

2.6 Methods for Determining Moisture Content

There are various methods used to determine moisture content in agricultural products, such as toluene distillation, Brown-Duvel, and chemical and infrared methods, among others. However, the most commonly used methods for green coffee are the oven method and the indirect capacitance method.

The choice of method should be based on precision, accuracy, and repeatability. On the other hand, the choice of equipment should focus on simplicity, ease of use, operation time, and total cost of the analysis. The indirect method is usually chosen in the field and at cooperatives because of its simplicity, speed, and ease of use. The oven method is used in research projects or when there is a need to calibrate the equipment used in the indirect methods.

The oven method, internationally known for determining moisture content in seeds, is based on the drying of a sample of a known initial weight, then calculating moisture content as the difference between the initial and final weights. However, there are different methodologies available and the results depend on temperature, drying time, the state of the seed, and the atmospheric pressure under which moisture content is determined.

2.6.1 ISO Norms (International Standard)

2.6.1.1 ISO 1446

The ISO 1446 standard, edited in December 2001, outlines the basic reference method for determining moisture content in green coffee. The method is based on determining the loss of mass when the product, previously ground, reaches hygroscopic equilibrium under an anhydrous atmospheric condition established by phosphorous pentoxide (P_2O_5), a temperature of 48 °C ± 2 °C, and pressure of 2.0 kPa ± 0.7 kPa. The method states that, initially, moisture content should be determined according to the method specified in ISO 6673. The samples with moisture content above 11% (wb) should first be dried. In addition, grinding should be done carefully so as to prevent moisture loss caused by the increase in temperature in the grinder.

The ground samples should be weighed at least twice on an analytical scale using metal capsules that have been previously dried and weighed. The capsules containing the samples should be placed in a glass tube with phosphorous oxide. The tube should be connected to a vacuum pump with internal pressure reduced to 2.8 kPa, then placed in an oven to guarantee a constant temperature of 48 °C. Throughout the test, the phosphorous oxide should be renewed to maintain its activity level. After 80 to 100 hours, the sample should be weighed. Drying should continue until a constant weight is achieved, that is, a variation of less than 0.0005 g between two weighings with an interval of 48 hours. Under the described conditions, generally 150 to 200 hours are needed to achieve a constant weight.

2.6.1.2 ISO 1447

The ISO 1447 standard, edited in 1978, describes the routine method for determining moisture content in green coffee. The method is based on determining moisture content using an oven in two phases. A sample of approximately 5 g of green coffee should be dried in an oven at 130 °C ± 2 °C, with forced ventilation, for 6 h ± 15 min. The sample should then be removed from the oven, placed in a dehydrator until it reaches room temperature, then weighed. In the second drying phase, the sample should be placed again in the oven at 130 °C ± 2 °C, with forced ventilation, for 4 h ± 15 min. After the same cooling and weighing procedure, moisture content is calculated by adding the loss of mass from the first phase with half of the lost mass from the second drying phase.

2.6.1.3 ISO 6673

The ISO 6673 standard, first published in November 1983, specifies the method for determining the loss of mass at 105 °C for green coffee, which can be applied to both decaffeinated and caffeinated coffee seeds. According to the standard, this method can be used to determine moisture content, but it provides results close to 1% lower than those obtained with the methods outlined in ISO 1447 and ISO 1446.

The method described in ISO 6673 is based on moisture loss and a small quantity of volatiles. In this method, the sample of whole seeds is placed in the oven with forced air ventilation at a controlled temperature of 105 °C ± 1 °C for 16h ± 30 min. At least two samples of approximately 10 g of coffee seeds should be weighed and placed under the specified conditions. After the specified period, the samples should be removed from the oven, placed in a dehydrator to cool to room temperature, and weighed on a scale with a precision of 0.1 mg. Moisture content is expressed as a percentage and calculated using equation 2.4. The final value is the arithmetic average of the two results.

$$MC = (m_1 - m_2) \times 100 / (m_1 - m_0) \tag{2.4}$$

in which: m_0 = mass of container (g)
m_1 = mass of container and of sample before drying (g)
m_2 = mass of container and of sample after drying (g)
MC = moisture content (% wb)

Since there are three ISO standards for measuring the moisture content of green coffee, questions have been raised regarding the accuracy of the methods and which should be used. In 2005, a study was conducted to clarify the precision of ISO 1446, ISO 1447, and ISO 6673.[23] The authors used near-infrared spectrometry (NIR) and color measurement of the samples as references to support the conclusions about the precision of the methods. For comparison purposes, the authors considered real moisture content as that obtained using the method described in ISO 1446 and corrected by the residual moisture content obtained by NIR. The authors concluded that the results obtained through ISO 1447 and ISO 1446 methods were not similar. They observed that the ISO 1447 method provided higher values than the real moisture content, while the ISO 1446 method provided lower values than the real moisture content. They also concluded that the ISO 6673 method is the most appropriate for routine analyses since it does not require complete drying of the coffee seeds, control over the laboratory environment, or grinding of the samples, and it only uses one drying phase.

2.6.2 Brazilian Rule for Seed Analysis

The Rule for Seed Analysis (RSA), re-edited in 1992, is a publication of the Ministry of Agriculture, Livestock, and Supply[24]. It contains the rules and procedures for analyzing seed lots, including the standard for determining moisture content. According to the RSA, the oven method at 105 °C is the most common method used in Brazil and is recommended to determine moisture content for all seed species, not just coffee. Two other methods are outlined in the RSA and were adopted by the International Rules for Seed Analysis; however, they are not commonly used for coffee.

2.6.2.1 Oven method at 105 °C

Moisture content is determined by removing water from samples of whole seeds in a gravity convection oven at a temperature of 105 °C ± 3 °C over 24 hours. For this method, the rules do not recommend the use of ovens with forced air circulation.

Initially, the temperature of the oven is regulated using a calibrated thermometer with an accuracy of 0.5 °C. Next, glass or non-corrosive metal containers are dried for one hour in the oven at 130 °C, then cooled in a dehydrator, labelled and weighed. Two samples of approximately 50 g of whole seeds are weighed in the containers with their respective lids. Both samples are then placed in the oven for a period of 24 hours. When in the oven, the lid should be placed below the container so that the moisture can be released. The timer should not be started until the temperature returns to 105 °C. It is recommended that the oven have a temperature recuperation rate of approximately 15 minutes. After the drying period, the samples should be removed from the oven, quickly covered, and placed in the dehydrator to cool for about 10 to 15 minutes. The samples should then be weighed.

Moisture content should be calculated using equation 2.5, and the result, expressed as a wet-basis percentage (% wb), is the arithmetic average of the two repetitions. The difference between the results of the two repetitions should not exceed the established tolerance level of 0.5%, and the procedure should be repeated when the tolerance level is exceeded.

$$MC = ((m_i - m_f) / (m_i - t)) \times 100 \tag{2.5}$$

in which: MC = wet basis moisture content (% wb)
m_i = initial mass (mass of the container and its lid plus mass of the wet sample)
m_f = final mass (mass of the container and its lid plus mass of the dry sample)
t = tare (mass of container with its lid)

2.6.3 Dielectric method (capacitance)

In this method, moisture content is determined through seed characteristics that change with moisture content; therefore, it is considered an indirect method. One of these characteristics is the dielectric constant, defined as the relative ability of a material, in comparison with a vacuum, to store energy when subjected to an electric field. All of the components of a seed contribute to its dielectric constant. However, water molecules, due to their structure, exhibit a dielectric constant that is close to 20 times higher than that of most other seed components. Due to their structure, water molecules can be considered an electrostatic dipole[25].

When electrostatic dipoles are placed in an electric field, they tend to align themselves with the applied field in order to reduce the tension of the electric field within the material (Figure 2.14).

The higher the moisture content, the higher the dielectric constant of the seeds. The relationship between the dielectric constant and the moisture content of seeds is represented in equation 2.6.

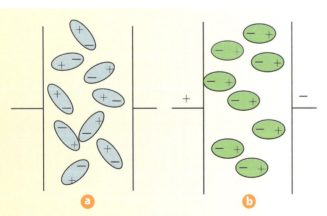

Figure 2.14 Arrangement of polar molecules, **a** without an electric field; **b** in the presence of an electric field[26].

MC = D × C (2.6)

in which: D = dielectric constant
            C = constant (depending on equipment, material, etc.)
            MC = moisture content

In equipment that employs this principle, samples of known weight are exposed to high voltage frequencies between 1 and 2 MHz. The effect of the contact between the seed mass and the high voltage in the circuit is measured as the dielectric constant.

This method can be very precise since the dielectric effect is a value independent of both surface conditions and moisture distribution within the seeds. Rather, moisture content is determined solely by the intrinsic properties of the seed mass. Thus the dielectric reading of a test cell is essentially a reading of the total amount of water present in the seeds.

Accurate temperature is essential in this type of equipment. It should be noted that a large number of technicians and producers make incorrect measurements because they do not consider the need for temperature accuracy. The key limiting factors for using the dielectric effect are the high cost of more precise, reliable equipment, and the difficulty in regulating and aligning different commercial models (Figure 2.15). In fact, studies have shown significant differences in moisture content measurements when comparing different commercial methods to ISO 1447[27] (ISO International Standard, 1978). However, the capacitance method did not exhibit significant differences between 12.45% and 14.18% (wb) moisture content. The results suggest that, before determining moisture content, calibration curves should be obtained to ensure accuracy.

Capacitance equipment should be tested every year with the standard oven method, or sent to factories for calibration.

Figure 2.15 Commercial models that determine moisture content using electrical capacitance.

Physical and Thermal Properties
of Coffee Fruit and Seeds

Paulo César Afonso Júnior

Flávio Meira Borém

Paulo César Corrêa

Valdiney Cambuy Siqueira

3.1 Introduction

Knowledge of the physical and thermal properties of agricultural products is essential for calculating the ideal size and capacity of the equipment and storage structures that will be used in post-harvest processing, and for the simulation of the various processes and phenomena to which the material will be subjected. Information about size, volume, porosity, and mass density, among other physical characteristics, is considered very important in studies involving heat transfer, mass, and air movement in the granular material of agricultural products. Along with moisture content, the parameters used to determine drying and storage conditions are mass density, porosity, and volume; these factors also help predict quality loss until the moment of sale[1]. What's more, it is necessary to determine the key thermal properties of a product—e.g., enthalpy of vaporization of moisture, specific heat, thermal conductivity and diffusivity—in order to predict its internal thermal changes as it is subjected to cooling, drying, and storage[2].

Information about size, volume, porosity, and mass density, among other physical characteristics, is considered very important in studies involving heat transfer, mass, and air movement in the granular material of agricultural products. Along with moisture content, mass density, porosity, and volume are parameters that can help to predict quality loss until the moment of sale.

3.2 Physical Properties

3.2.1 Mass density and porosity

Mass density can be defined as the ratio between the mass and volume of a given product. When this concept is applied to the mass and volume of a single coffee seed or fruit, it refers to the physical property *true density*. When applied to a specific quantity of a product, this characteristic is referred to as *bulk density*[3], that is:

$$\rho_t = \frac{m}{V} \tag{3.1}$$

$$\rho_b = \frac{m_p}{V_p} \tag{3.2}$$

in which: ρ = true density of the product (kg m^{-3})
ρ_b = bulk density of the product (kg m^{-3})
m = mass of a single coffee seed or fruit (kg)
m_p = mass of the product (kg)
V = volume of a single coffee seed or fruit (m^3)
V_p = volume occupied by product mass and the intergranular space (m^3)

This concept is applied to the measurement of silos, hoppers, dryers, storage bins, and transportation systems. In reality, it is also used to estimate moisture content of coffee during the drying process. This application should be used with caution, however, since mass density values vary among cultivars and also do not represent a linear relationship with moisture content reduction (Figure 3.1).

It is important to note that, unlike what happens with the majority of grains, the bulk density of both natural and parchment coffee increases at higher levels of moisture[4].

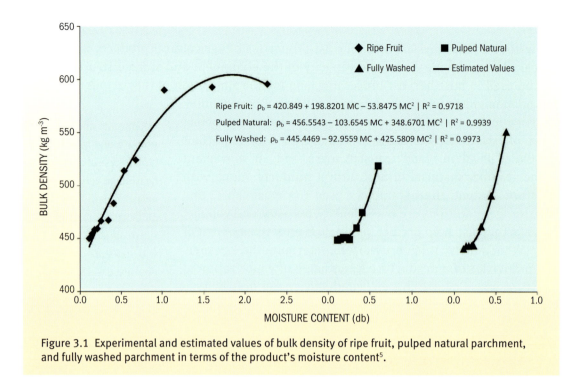

Figure 3.1 Experimental and estimated values of bulk density of ripe fruit, pulped natural parchment, and fully washed parchment in terms of the product's moisture content[5].

The porosity of a granular mass is defined as the relationship between the volume occupied by the air in the inter-granular spaces and the total volume of the mass. This property is usually represented in this way:

$$\varepsilon = \frac{V_p - V}{V_p} \qquad (3.3)$$

in which: ε = porosity of the product's mass (decimal)

Combining equations 3.1 and 3.2 results in the following equation:

$$\varepsilon = 1 - \left(\frac{\rho_b}{\rho}\right) \qquad (3.4)$$

The porosity of a grain mass is associated with the resistance of a layer of product to the movement of air, and is therefore widely used in projects involving drying and aeration equipment[6].

The fraction of empty space in a grain mass, its porosity, can be determined using the direct method or indirect methods. In the direct method, porosity is obtained by filling the empty spaces of the granular mass with a volume of a known liquid[7]. In the indirect method, porosity can be determined with a pycnometer or by using mathematical relationships, like the one shown in equation 3.4, that involve a product's true and bulk densities[8]. It should be noted that the pycnometer is the preferred means to determine porosity, since, unlike the volume completion method, it minimizes errors caused by superficial tension of the liquid used.

In general, porosity increases as the moisture content of a granular mass increases (Figure 3.2).

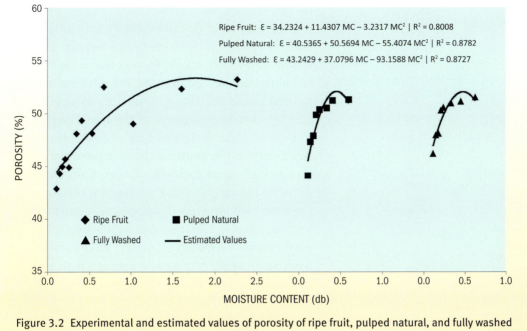

Figure 3.2 Experimental and estimated values of porosity of ripe fruit, pulped natural, and fully washed coffee in terms of the product's moisture content[9].

Table 3.1 represents the values of bulk gravity and porosity for ripe fruit, dry natural coffee, pulped natural coffee, fully washed coffee, and green coffee.

Table 3.1 Values of bulk density and porosity.

Coffee	Moisture content (%wb)	Bulk density (kg m^{-3})	Porosity (%)
Ripe fruit[10]	69.4	606.60	53.20
Dry coffee pods[10]	10.0	450.04	44.31
Fully washed parchment coffee[10]	38.3	550.92	51.51
Fully washed parchment coffee[10]	10.0	440.04	46.17
Pulped natural parchment coffee[10]	37.5	518.58	51.30
Pulped natural parchment coffee[10]	10.0	448.62	44.09
Green coffee[11]	11.0	722.05	44–45*

3.2.2 Shape, size, and volume

Information about shape, size, and volume, among other physical characteristics of agricultural products, is considered very important in studies involving heat transfer, mass, and air movement in granular products. It is also fundamental in the selection of screens used in the size grading of the product.

In general, seeds and fruits do not have a perfectly defined geometric shape; therefore, to resolve problems related to their geometry, it is necessary to use a known shape, which results in approximations and possible errors. For the majority of agricultural products, many of these solutions are obtained using geometric shapes such as spheroids or ellipsoids with three characteristic dimensions: length, width, and thickness (Figure 3.3).

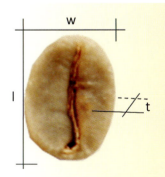

Figure 3.3 Characteristic dimensions of a coffee seed: length (l), width (w), thickness (t).

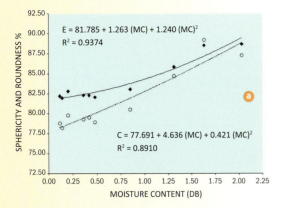

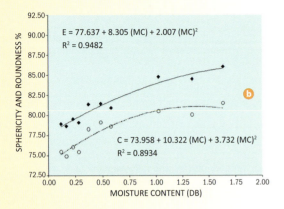

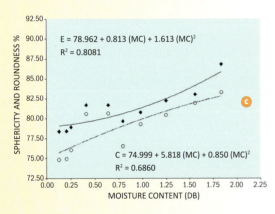

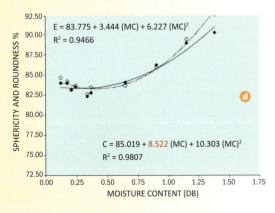

Figure 3.4 Calculated and estimated values of sphericity (s) and roundness (r) of coffee fruit as a function of moisture content: Red Catuai ⓐ, Mundo Novo ⓑ, Catimor ⓒ, Conilon ⓓ.

Knowledge of these characteristics is fundamental in the milling and classification stages. Peaberries can be separated from normal beans using screens with elongated perforations since normal beans are thinner than they are wide, and therefore the thicker, more rounded peaberries are retained while the normal beans fall through. For normal beans, screen size refers to the diameter of the round perforations in 64[ths] of an inch. In the case of peaberries, screen size refers to the width of the elongated perforation, also in 64[ths] of an inch. The screen size of a bean is given by the largest sized perforation through which the bean will not pass. For example, a normal coffee with screen size 17 is retained by a screen with round perforations 17/64" in diameter; screen size 13 peaberries are beans retained by a screen with elongated perforations that are 13/64" wide.

In addition to width, thickness, and length, the size and shape of grains can be characterized by their sphericity and roundness. Thus, a grain can be described by the degree to which it approximates the shape of a ball as well the approximation of its projected area to a circle.[12].

Variations in moisture content significantly affect the physical characteristics of coffee fruit in both Arabica and Canephora, causing notable changes in the shape (sphericity and roundness) and dimension of the fruit throughout the drying process (Figure 3.4). This reinforces the fact that changes in the dimensions of the coffee during its dehydration should not be neglected in studies related to energy transfer and mass[13].

Shrinkage of agricultural products during drying is not an exclusive function of moisture content, but also depends on drying conditions and the geometry of the product, as porous biological materials can contract differently in longitudinal, tangential and radial directions when dehydrated[14]. The volume of some agricultural products can be calculated using the geometry of an oblate spheroid (Figure 3.4), applying the following equation[15]:

$$V_p = \frac{\pi \times l \times w \times t}{6} \tag{3.5}$$

The length, width, and thickness values in equation 3.5 correspond to the major, medial, and minor axes, respectively (Figure 3.5).

Figures 3.6 through 3.8 represent, respectively, the estimated observed mean values of the surface areas of different cultivars, and the volume and unitary volumetric shrinkage of coffee fruit in different types of processing.

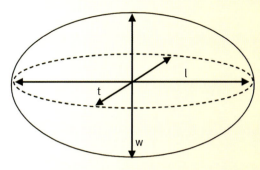

Figure 3.5 Schematic diagram of an oblate spheroid and its characteristic dimensions, length (l), width (w), and thickness (t).

Figure 3.7 shows the volume contraction of natural, pulped natural, and fully washed coffees during drying. Note that a reduction in moisture content of the coffee fruit from 2.27 (db), equal to 69.0% (wb), to 0.11 (db), equal to 9.9% (wb), contributes to a reduction in the volume of the product of approximately 39% when compared to its initial volume. For pulped natural and fully washed coffees the reduction in the moisture content of seeds from 0.60 (db), equal to 37.5% (wb), to 0.11 (db) was responsible for an approximate reduction of 12%–13% in volume in relation to initial volume.

3.2.3 Angle of rest

A mass of fruit, parchment, or seeds deposited on a horizontal plane will form a conical-shaped volume. The angle of rest is the maximum slope formed by an accumulated granular material in relation to the horizontal plane. Table 3.2 presents the values of the angle of rest for coffee fruit and seeds.

Note in Table 3.2 that the surfaces for gravity-related movement should have different angles for each type of coffee and product moisture content level. Therefore, to avoid manual labor upon receiving the coffee from the field, the walls of the hopper should have an incline greater than 40°. In general, when a producer has used equipment for natural coffee and then begins to work with pulped natural, the elevator discharge shafts become plugged because the angle of rest of the pulped natural parchment coffee is greater than that of natural coffee pods.

Table 3.2 Angle of rest for coffee fruit, parchment, and seeds (*Coffea arabica* L.).[16]

Coffee	Angle of Rest
Ripe fruit	40.3°
Pulped natural (wet)	40.7°
Fully washed	32.9°
Pulped natural (dry)	35.0°
Green coffee	27.8°
Natural dried coffee pod	31.7°

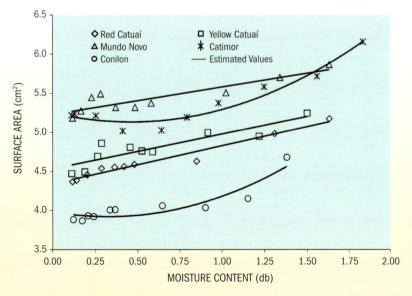

Figure 3.6 Estimated observed values of the surface area of coffee fruit for the cultivars analyzed, in terms of moisture content[17].

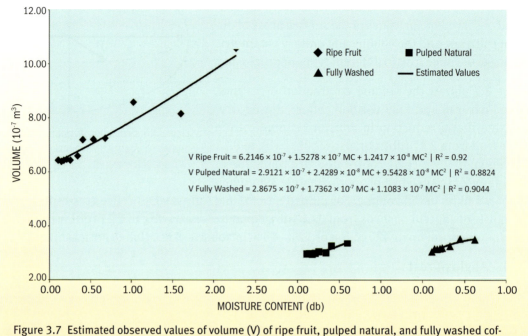

Figure 3.7 Estimated observed values of volume (V) of ripe fruit, pulped natural, and fully washed coffee, in terms of moisture content[17].

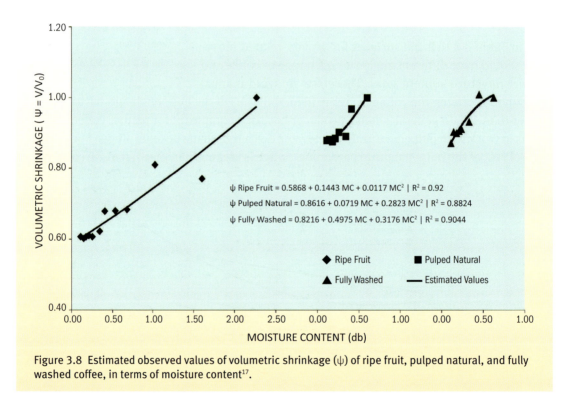

Figure 3.8 Estimated observed values of volumetric shrinkage (ψ) of ripe fruit, pulped natural, and fully washed coffee, in terms of moisture content[17].

Another situation that can occur in practice is the need to calculate the maximum distance for the location of a point of discharge so that the coffee moves freely by gravity out of rotary bin dischargers (Figure 3.9).

Figure 3.9 shows that for free movement by gravity, the incline of the tubing should be greater than the coffee's angle of rest (α1), thus minimizing the distance where the points of dis-

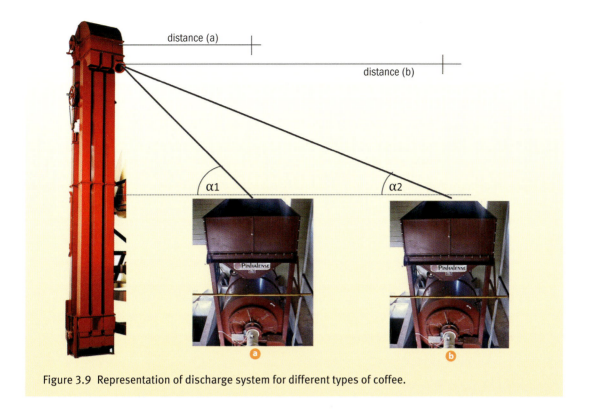

Figure 3.9 Representation of discharge system for different types of coffee.

charge are located. When the angle of rest α1 is higher than 35°, free discharge by gravity occurs for both dried pulped natural and natural coffees (Figure 3.9a). If the angle of rest α2 is equal to 32°, free discharge by gravity will only occur for dried coffee fruit and green coffee.

3.3 Hygroscopic Equilibrium

Like all hygroscopic material, coffee fruit and seeds release water to or absorb it from the environment to maintain constant moisture equilibrium with the surrounding air. Equilibrium moisture content (MCe), also called moisture content in hygroscopic equilibrium, is the level of moisture content at which the water vapor pressure in the product is equal to that of the air that surrounds it[18].

Establishing hygroscopic equilibrium curves is important for setting a product's dehydration limits, for estimating changes in moisture content under certain conditions of temperature and relative environmental humidity, and for defining ideal moisture content levels for agents that could cause the onset of product deterioration[19]. For coffee, it is important to note that adequate drying and storage conditions are essential to maintaining quality, given the elevated moisture content and presence of microorganisms at the time of harvest.

Equilibrium moisture content of an agricultural product depends both on environmental conditions and the way in which equilibrium is reached. For the same combination of temperature and relative humidity, there can be two isotherms, referred to as isotherms of adsorption and desorption. Coffee fruit and seeds that take on humidity to reach hygroscopic equilibrium fall into the category of adsorption isotherms; product that loses water to reach equilibrium is considered a desorption isotherm.

Countless authors have studied the hygroscopic equilibrium of various agricultural products using different methods to express equilibrium moisture content in terms of temperature and relative humidity of the air, and have established sorption isotherms for different products. However, to establish isotherms that represent this relationship of equilibrium, empirical mathematical equations are used as there is no theoretical model that can predict equilibrium moisture content with precision for a wide range of temperatures and relative air humidity.

Diverse models can be found in scientific literature to estimate equilibrium moisture content, such as BET[20], GAB[21], Harkins-Jura[22], Smith[23], Halsey[24], Aguerre[25], Roa[26], and Copace[27] Model. The models most often used to predict equilibrium moisture content for agricultural products are Henderson, Modified-Henderson, Chung-Pfost, and Oswin[28], due to their flexibility in accounting for experimental data. However, for each product the equation that provides the best fit must be found. Here is a brief look at important equilibrium moisture content models followed by the application of several of these models to coffee.

Henderson Model

Henderson is the model most used to portray the relationship of hygroscopicity between a product and the environment that surrounds it[29]. Using the Gibbs adsorption equation, the following equation was derived to explain the curves of equilibrium moisture content for biological products, including grains:

$$1 - RH = \exp(-a' \times MC_e{}^b) \tag{3.6}$$

in which: a', b = product-dependent parameter
             RH = Relative humidity
             MC_e = Equilibrium Moisture content

In equation 3.7, the effect of temperature was introduced through the thermodynamic Gibbs adsorption equation. Thus, the a' parameter was established as a function of temperature, converting the equation into its final version:

$$1 - RH = \exp(-a \times T_{abs} \times MC_e) \tag{3.7}$$

in which: T_{abs} = absolute temperature (K)

The Henderson model adequately describes the isotherms of equilibrium moisture content for relative humidity below 60%[30].

Modified Henderson Model

The Henderson equation was modified by Thompson & Shedd (1954), who fit the experimental data of equilibrium moisture content to other products, such as corn, resulting in the mathematical equation being expressed as equation 3.8[31].

$$1 - RH = \exp[-a \times (T + b) \times MC_e{}^c] \tag{3.8}$$

Various researchers confirm that the Modified Henderson Model allows for adequate estimates of values of equilibrium moisture content for various products for a wide range of temperature and relative air humidity[32]. In some studies this equation can also be evaluated to express the hygroscopic equilibrium of coffee.

Chung-Pfost Model

The Chung-Pfost Model[33], later modified[34], is based on the theory of the potential (free energy) to predict equilibrium moisture content. The modified equation is:

$$MC_e = a - b \times \ln[-(T + c) \times \ln(RH)] \tag{3.9}$$

With relative precision, this equation estimates the values of equilibrium moisture content in grains and cereals in the range of 20%–90% relative humidity[35].

Oswin Model

The Oswin Model[36], later modified[37], takes into consideration the dependence of the phenomenon of hygroscopicity on temperature, and is represented by the following equation:

$$MC_e = \frac{(a + bT)}{\left(\dfrac{1 - RH}{RH}\right)^{1/c}} \tag{3.10}$$

The Oswin model, modified by the effect of temperature, has exhibited satisfactory results in the prediction of adsorption isotherms for various agricultural products[38].

Harkins-Jura Model

The Harkins-Jura Model[39] is based on water potential and is presented in equation 3.11.

$$MC_e = \left[\frac{a}{b - \ln(RH)}\right]^{1/2} \tag{3.11}$$

Coffee-Specific Models

Various models have been adjusted to predict the equilibrium moisture content of green Arabica coffee. In 1982, Iglesias and Chirife developed the following equation for the desorption process of green Arabica coffee at a temperature of 25 to 35 °C and relative humidity of 45% to 90%[40]:

$$MC_e = a - \frac{b}{\ln(RH)} \tag{3.12}$$

in which the a and b constants assume, respectively, the values of 5.676 and 2.621 at a temperature of 25 °C, and 5.455 and 2.638 at 35 °C.

Further studies of the equilibrium isotherms of coffee fruit resulted in the following equation, which can be considered a modification of the Henderson equation:

$$MC_e = 1.1282 \left[\frac{-\ln(1 - RH)}{T + 40.2520}\right]^{0.5404} \tag{3.13}$$

However, equations for predicting the hygroscopic equilibrium of coffee depend on coffee processing, and coefficients must be adjusted in order to accurately predict the values of equilibrium moisture content for natural, pulped natural, and fully washed coffees. Figures 3.10 through 3.13 show equations as adjusted by Afonso Junior, and Tables 3.3 through 3.6 present the values of equilibrium moisture content (dry basis %) obtained using these formulas. For green coffee, the best fit is obtained using the Harkins-Jura equation in Figure 3.13 as modified by Afonso Junior.

Tables 3.3–3.6 Values of equilibrium moisture content (db) for different values of temperature and relative humidity (desorption).[41]

Table 3.3 Dried Coffee Pods

T °C	Relative Humidity (decimal)							
	.10	.20	.30	.40	.50	.60	.70	.75
25	.10	.10	.10	.10	.11	.12	.21	.42
30	.09	.09	.09	.09	.10	.11	.18	.36
35	.08	.09	.09	.09	.09	.10	.16	.32
40	.08	.08	.08	.08	.08	.09	.14	.27
45	.07	.07	.07	.07	.07	.08	.13	.23
50	.07	.07	.07	.07	.07	.08	.12	.20
55	.06	.06	.06	.06	.06	.07	.10	.18
60	.06	.06	.06	.06	.06	.06	.09	.16

Table 3.4 Pulped Natural Parchment Coffee

T °C	Relative Humidity (decimal)							
	.10	.20	.30	.40	.50	.60	.70	.75
25	.09	.09	.09	.09	.10	.12	.22	.40
30	.08	.08	.08	.08	.09	.11	.18	.34
35	.07	.07	.07	.07	.08	.09	.16	.28
40	.07	.07	.07	.07	.08	.08	.14	.24
45	.06	.06	.06	.06	.06	.07	.12	.21
50	.05	.05	.05	.05	.06	.07	.10	.18
55	.05	.05	.05	.05	.05	.06	.09	.15
60	.04	.04	.04	.04	.05	.05	.08	.13

Figures 3.10–3.13 Desorption isotherms of dried coffee pods, pulped natural parchment coffee, fully washed parchment coffee, and green coffee.[41]

Figure 3.10 Dried Coffee Pods

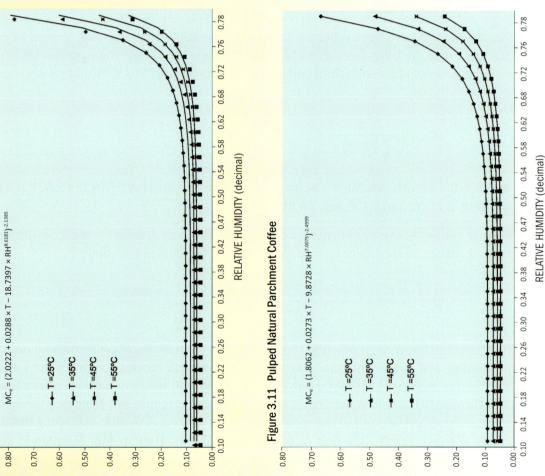

$$MC_e = (2.0222 + 0.0288 \times T - 18.7397 \times RH^{8.6181})^{-2.1385}$$

RELATIVE HUMIDITY (decimal)

EQUILIBRIUM MOISTURE CONTENT (db)

T = 25°C
T = 35°C
T = 45°C
T = 55°C

Figure 3.11 Pulped Natural Parchment Coffee

$$MC_e = (1.8062 + 0.0273 \times T - 9.8728 \times RH^{7.0075})^{-2.4999}$$

RELATIVE HUMIDITY (decimal)

EQUILIBRIUM MOISTURE CONTENT (db)

T = 25°C
T = 35°C
T = 45°C
T = 55°C

Figure 3.12 Fully Washed Parchment Coffee

$$MC_e = (2.9636 + 0.053 \times T - 10.7837 \times RH^{4.5136})^{-1.6503}$$

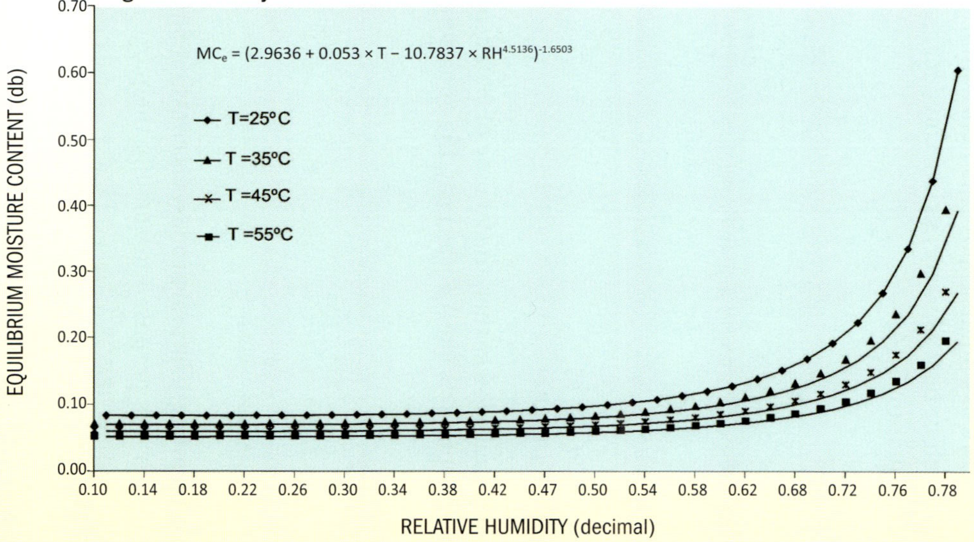

Table 3.5 Fully Washed Parchment Coffee

T °C	Relative Humidity (decimal)							
	.10	.20	.30	.40	.50	.60	.70	.75
25	.08	.08	.08	.09	.10	.13	.22	.38
30	.08	.08	.08	.08	.09	.11	.19	.31
35	.07	.07	.07	.07	.08	.10	.17	.26
40	.06	.06	.06	.07	.07	.09	.15	.22
45	.06	.06	.06	.06	.07	.08	.13	.19
50	.05	.05	.05	.06	.06	.08	.11	.17
55	.05	.05	.05	.05	.06	.07	.10	.14
60	.05	.05	.05	.05	.05	.06	.09	.13

Figure 3.13 Green Coffee

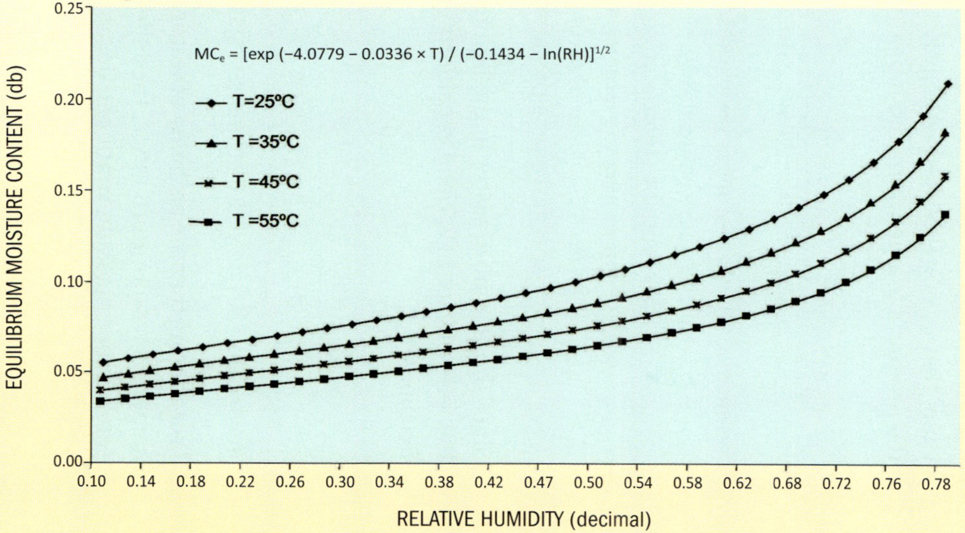

$$MC_e = [\exp(-4.0779 - 0.0336 \times T) / (-0.1434 - \ln(RH)]^{1/2}$$

Table 3.6 Green Coffee

T °C	Relative Humidity (decimal)							
	.10	.20	.30	.40	.50	.60	.70	.75
25	.06	.07	.08	.09	.10	.12	.16	.23
30	.05	.06	.07	.08	.10	.11	.15	.22
35	.05	.06	.07	.08	.09	.11	.14	.20
40	.04	.05	.06	.07	.08	.10	.13	.19
45	.04	.05	.06	.07	.08	.09	.12	.18
50	.04	.04	.05	.06	.07	.08	.11	.17
55	.03	.04	.05	.06	.07	.08	.10	.15
60	.03	.04	.04	.05	.06	.07	.09	.14

3.4 Thermal Properties

Understanding the thermal properties of coffee is fundamental in post-harvest processing for making correct measurements, optimizing equipment and required procedures, predicting heating and cooling rates, and simulating innumerable phenomena and processes.

It is necessary to determine the primary thermal properties of an agricultural product—enthalpy of vaporization, specific heat, thermal conductivity, and thermal diffusivity—in order to predict internal thermal changes as it is dried, cooled, and stored[42].

3.4.1 Enthalpy of vaporization

When seeking to conserve agricultural products using a partial dehydration method (such as drying coffee to 11% MC), an important property to be studied is enthalpy of vaporization, or the latent heat of vaporization of a product's water. This is defined as the amount of energy required to evaporate one unit of water mass contained in the product, under specific drying conditions[43].

The energy necessary to break the physicochemical bonds of adsorbed water found in agricultural products, in order to induce a liquid to vapor phase change, is significantly greater than the energy required to do so with free water (under the same conditions[44]). Clausius-Clapeyron developed studies based on thermodynamic theories with the objective of defining an equation that could represent the vapor pressure of a liquid-vapor system in equilibrium, that is, one that could quantify the values of latent heat of vaporization, considering the product's temperature and moisture content (Equation 3.14)[45].

$$\frac{\partial P_v}{\partial T_{abs}} = \frac{h}{(V_v - V_l) \times T_{abs}} \tag{3.14}$$

in which: P_v = water vapor pressure, for a specific temperature and a specific equilibrium
　　　　　moisture content (MC_e) (P_a)
　　　　T_{abs} = absolute temperature (K)
　　　　h = latent heat of water vaporization (kJ kg^{-1})
　　　　V_v= specific volume of saturated water vapor (m^3 kg^{-1})
　　　　V_l = specific volume of water in liquid state (m^3 kg^{-1})

Since the volume of water in the liquid state (V_l) is much lower than the volume of water in the form of vapor (V_v), and considering that water vapor acts like a perfect gas, the equation becomes[46]:

$$\frac{\partial P_v}{\partial T_{abs}} = \frac{h \times P_v}{R \times (T_{abs})^2} \tag{3.15}$$

in which: R = universal constant of gases (287 J kg^{-1} mol^{-1} K^{-1})

Considering that the enthalpy of water vaporization for agricultural products is constant for a specific temperature range, the following equation was proposed to quantify the partial water vapor pressure contained in capillary-porous materials[47]:

$$\ln(P_v) = -\frac{h_{lv}}{R} \times \frac{1}{T_{abs}} + C \tag{3.16}$$

in which: h_{lv} = latent heat of water vaporization of a product, at a temperature of equilibrium (kJ kg^{-1})

C = constant of integration

Applying the Clausius-Clapeyron equation for free water, which considers the value of vapor pressure as the saturated vapor pressure of free water, and combining this with the equations for a porous system, we obtain:

$$\ln(P_v) = \frac{h_{lv}}{h'_{lv}} \times \ln(P_{vs}) + C \qquad (3.17)$$

in which: h'_{lv} = latent heat of vaporization of free water, at a temperature of equilibrium (kJ kg^{-1})

P_{vs} = saturated vapor pressure of free water, at a temperature of equilibrium (Pa)

The curves of moisture content in hygroscopic equilibrium of biological products provide the necessary information to calculate latent heat of vaporization, since for a given moisture content in equilibrium at a specific temperature, there is a corresponding relative humidity in equilibrium[48]. By definition, relative humidity represents the ratio between existing water vapor pressure and saturated water vapor pressure:

$$RH = \frac{P_v}{P_{vs}} \therefore P_v = RH \times P_{vs} \qquad (3.18)$$

in which: RH = relative air humidity (decimal)

The calculated values of latent heat of water vaporization, with their respective relationships (h_{lv}/h'_{lv}), for three types of coffee processing, are represented in Table 3.7 in terms of the variables of moisture content and temperature.

Note that the energy required to evaporate the water of coffee fruit and seeds increases with a reduction of moisture content and temperature, regardless of the type of process used. In general, it has been verified that for higher levels of moisture content the value of latent heat of vaporization is lower than that observed in fruit and seeds with lower moisture content, where this value had the tendency to approximate the energy used to evaporate the free water.

3.4.2 Specific heat

The amount of heat necessary to increase the temperature of a body by 1 °C is termed the body's heat capacity, which is proportional to its mass. In turn, the heat capacity of a body per unit of mass is known as specific heat[50]. For example, the specific heat of one liter of water is the same as that for 1 milliliter of water, but the heat capacities of these two volumes of water are different. Various researchers have studied this thermal property in terms of the variation of moisture content in biological materials and observed that there is a relationship whereby an increase in moisture content provokes a rise in specific heat[51].

There is no common equation that can define the specific heat of ripe fruit for all coffee cultivars[52]. However, there are various methods to determine the specific heat of biological materials: method of mixtures, differential scanning calorimetry, calorimetric pump, and ice calorimetry[53]. These methods are based on the thermal equilibrium established between the material being studied and a second body of a known specific heat, generally water (4.186 kJ kg^{-1} °C^{-1}) or toluene (1.630 kJ kg^{-1} °C^{-1}). Another means of determining specific heat is to calculate the ratio between the thermal conductivity of a substance and the product of its thermal diffusivity and mass density, using Equation 3.19[54]:

$$C_p = \frac{k}{\rho_b \times \alpha}$$ (3.19)

in which: C_p = specific heat of the product (kJ kg^{-1} °C^{-1})
k = thermal conductivity of the product (W m^{-1} °C^{-1})
ρ_b = bulk density of the product (kg m^{-3})
α = thermal diffusivity of the product (m^2 s^{-1})

Considering the difficulty in determining specific heat using other methods, the routine method used to determine the specific heat of agricultural products such as grains is that of mixtures[55], which is based on establishing the thermal equilibrium of known masses of water and product at different temperatures and in calorimetry with known heat capacity.

Table 3.7 Latent heat of water vaporization of coffee fruit and seeds, in terms of moisture content in equilibrium and pre-established temperatures, for the different forms of coffee processing[49].

MCe	Latent Heat of Vaporization (kJ kg^{-1})					Angular Coef.
			Temperature (°C)			
(db)	25	35	45	55	65	(h_{lv}/h'_{lv})
Ripe Fruit						
0.12	2,858.5450	2,830.0970	2,801.6490	2,773.2010	2,744.7530	1.1707
0.30	2,547.4673	2,522.1152	2,496.7630	2,471.4108	2,446.0586	1.0433
0.60	2,520.8524	2,495.7651	2,470.6777	2,445.5904	2,420.5031	1.0324
0.90	2,512.7946	2,487.7875	2,462.7804	2,437.7732	2,412.7661	1.0291
1.20	2,508.8879	2,483.9196	2,458.9514	2,433.9831	2,409.0149	1.0275
1.50	2,506.2019	2,481.2604	2,456.3189	2,431.3774	2,406.4359	1.0264
1.80	2,504.4927	2,479.5682	2,454.6437	2,429.7192	2,404.7947	1.0257
2.10	2,503.2718	2,478.3595	2,453.4471	2,428.5348	2,403.6224	1.0252
Pulped Natural Parchment						
0.12	2,783.5840	2,755.8820	2,728.1800	2,700.4780	2,672.7660	1.1400
0.20	2,617.7890	2,591.7370	2,565.6850	2,539.6330	2,513.5810	1.0721
0.25	2,592.6400	2,566.8380	2,541.0360	2,515.2340	2,489.4330	1.0618
0.30	2,578.2330	2,552.5750	2,526.9170	2,501.2580	2,475.6000	1.0559
0.35	2,568.7100	2,543.1470	2,517.5830	2,492.0200	2,466.4560	1.0520
0.40	2,561.8740	2,536.3780	2,510.8820	2,485.3870	2,459.8910	1.0492
0.45	2,556.7460	2,531.3010	2,505.8570	2,480.4120	2,454.9680	1.0471
0.50	2,552.5950	2,527.1920	2,501.7890	2,476.3850	2,450.9820	1.0454
Fully Washed Parchment						
0.12	2,808.0010	2,780.0560	2,752.1110	2,724.1660	2,696.2210	1.1500
0.20	2,658.3220	2,631.8670	2,605.4120	2,578.9560	2,552.5010	1.0887
0.25	2,632.1960	2,606.0000	2,579.8050	2,553.6100	2,527.4140	1.0780
0.30	2,617.0570	2,591.0120	2,564.9670	2,538.9230	2,512.8780	1.0718
0.35	2,607.0460	2,581.1010	2,555.1560	2,529.2100	2,503.2650	1.0677
0.40	2,599.9650	2,574.0900	2,548.2150	2,522.3410	2,496.4660	1.0648
0.45	2,594.5930	2,568.7720	2,542.9510	2,517.1290	2,491.3080	1.0626
0.50	2,590.4420	2,564.6620	2,538.8820	2,513.1020	2,487.3220	1.0609
h'_{lv} (kJ kg^{-1})	2,441.74	2,417.44	2,393.14	2,368.84	2,344.54	

Figure 3.14 shows the experimental and estimated values of specific heat in terms of moisture content for three types of coffee processing[56].

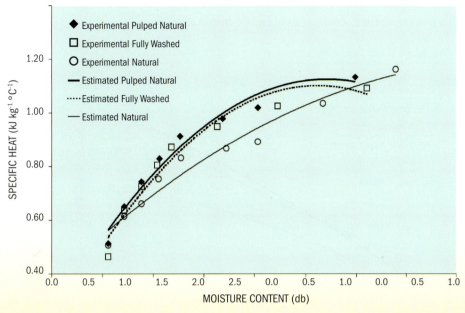

Figure 3.14 Experimental and estimated specific heat values of coffee, in terms of moisture content, for different types of processing.

3.4.3 Thermal conductivity and diffusivity

Many problems related to the drying and storage of agricultural products can be analyzed using the principles of heat transfer. This requires an understanding of the thermal conductivity and diffusivity properties of the product, in addition to its enthalpy of vaporization and specific heat.

Thermal conductivity defines the amount of heat that is transmitted per unit of time between two surfaces across one area unit, as a result of one temperature gradient, and can be described by equation 3.20[57].

$$q = -k\nabla T \qquad (3.20)$$

in which: q = heat flow (W m^{-2})
∇T = temperature gradient (°C m^{-1})

The negative sign in the equation indicates that the heat is transferred in the opposite direction to the temperature gradient. Heat transfer is a process where thermal energy flows from a region of high temperature to one of low temperature, within a medium (solid, liquid, or gas) or between different mediums in physical contact[58].

The conduction of thermal energy in agricultural products normally occurs during the warming or cooling process. However, this involves the accumulation or dissipation of heat, which results in variations in the material's temperature distribution over time.

The rate at which heat is diffused from inside to outside the material depends on the thermal diffusivity of the product[59]. Studies confirm that the magnitude of thermal diffusivity influences the drying kinetics of agricultural grains[60].

Heat transfer across fruits and seeds is a complex process. Due to the heterogeneity of these materials resulting from variations in cellular structure, chemical composition, and water and air content, variations are expected in their thermal conductivity and diffusivity. However, various researchers confirm that the values of these properties can also vary with the material's physical structure, mass density, and temperature[61].

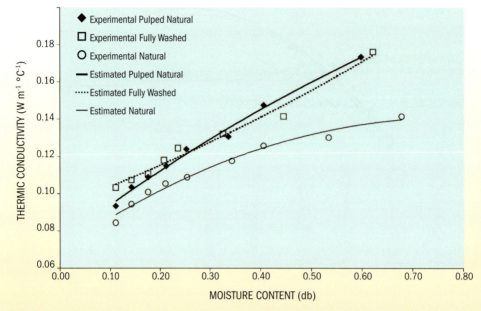

Figure 3.15 Experimental and estimated values of thermic conductivity of coffee in terms of moisture content for different processing methods.

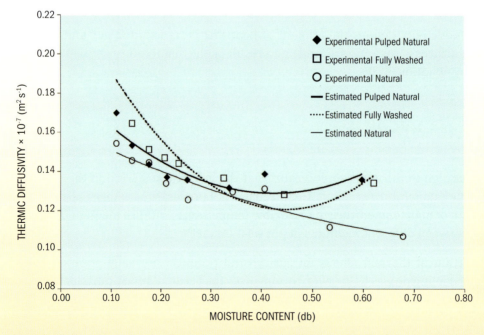

Figure 3.16 Experimental and estimated values of thermic diffusivity of coffee in terms of moisture content for different processing methods.

While there are many methods to evaluate the thermal properties of grains and seeds, the method that is most often used is the transient state method, as it requires less testing time and provides more precise results[62].

Figures 3.15 and 3.16 show the experimental and estimated values of thermal conductivity and diffusivity[63] in terms of moisture content for three coffee processing methods.

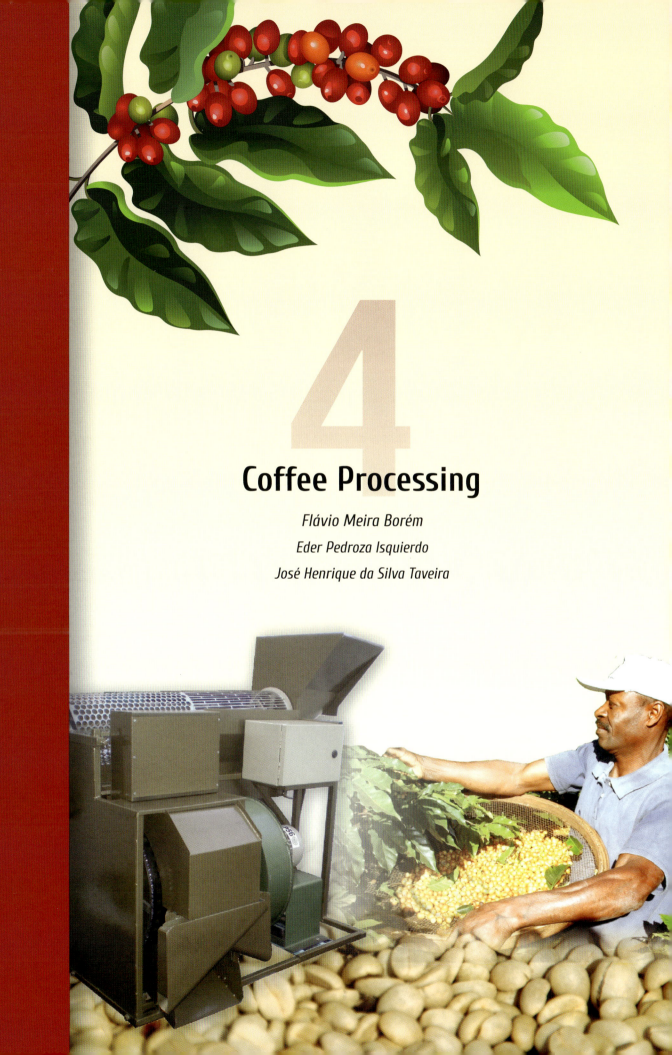

4

Coffee Processing

Flávio Meira Borém

Eder Pedroza Isquierdo

José Henrique da Silva Taveira

4.1 Introduction

The coffee crop can consist of ripe, unripe, semi-ripe, and overripe fruit, as well as dried fruit, leaves, twigs, dirt, sticks, and stones (Figure 4.1). The existence and proportion of each of these categories depend on the system and care taken in harvesting. Selective harvesting (Figure 4.2), recommended for producing coffees of superior quality, results only in the picking of mature fruit. Stripping, when done too early, produces coffees with a high percentage of unripe fruit, and when done too late results in a large quantity of dried fruit; in these two cases the resulting coffee will have a tendency to be of inferior quality.

Using a canvas (Figure 4.3) when strip picking coffee is preferred for two reasons. The canvas prevents contact between freshly picked fruit and the ground, and it prevents fruit that is already on the ground

Figure 4.1 Strip-harvested coffee containing coffee of various maturation states, as well as debris.

in a state of deterioration from being incorporated into the harvest. In addition to having a negative effect on quality, allowing coffee to come into contact with the ground should be avoided for hygienic and sanitary reasons.

Recently harvested fruit should not be stored in bags or silos for periods longer than eight hours as these conditions increase the risk of fermentation, and temperatures can reach more than 40 °C, potentially causing the sour bean defect[1]. In case of an emergency, the appropriate method for storing coffee for longer periods of time is to immerse the fresh fruit in water. It should be emphasized, however, that after harvesting, coffee should immediately be taken for processing to avoid storing wet fruit and to reduce the risk of quality loss caused by fungi and mycotoxins.

Figure 4.2 Selective coffee harvesting.

There are stark differences in the anatomy, chemical composition, and moisture content of coffee fruit in different stages of maturation. A high degree of heterogeneity, or diversity of maturity levels, of fruit coming from the field can both complicate coffee processing and compromise final coffee quality. The more homogeneous the harvested lot, the more efficient coffee processing becomes across all post-harvest procedures.

The choice of processing method directly affects the profitability of coffee production and depends on diverse factors such as regional climatic conditions; available capital, technology and equipment; consumer demand for specific quality characteristics; water usage rights; and the availability of technology for treating residual water. There are three fundamental consider-

Figure 4.3 Harvesting over a picking canvas or cloth.

ations in choosing a coffee processing method: the cost/benefit analysis of the production method, the need to adhere to environmental legislation, and the desired quality standard of the coffee.

Historically, the two methods employed to process coffee are dry processing and wet processing (Figure 4.4).

In dry processing, coffee fruit are processed whole, producing dry fruit pods known as natural coffee. In wet processing, parchment coffee is produced.

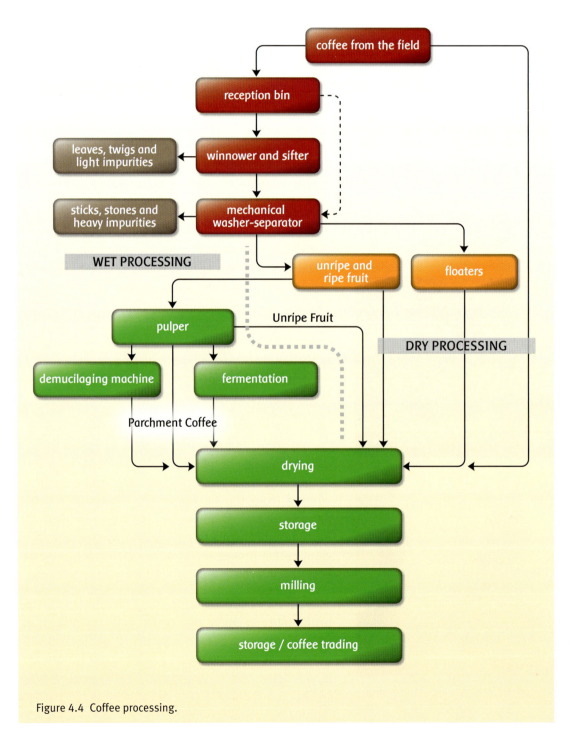

Figure 4.4 Coffee processing.

The dry process is the predominant process used for Arabica coffees in Brazil, Ethiopia, and Yemen, as well as for practically all Canephora coffees worldwide[2]. The wet process is the predominant method for Arabica coffees in Colombia, Costa Rica, Guatemala, Mexico, El Salvador, Kenya, and recently a small percentage of Canephora coffees[3].

Figure 4.5 Receiving the coffee.

4.2 Receiving

Coffee is received in hoppers (Figure 4.5), which should be constructed to facilitate unloading and moving the coffee crop. The hoppers, made of metal or stone, should be situated above the hydraulic separators, with walls inclined to 60° to allow gravity to move the coffee, thus eliminating the need for mechanical transport and extra labor.

Hoppers should also be made safe for the workers who operate them, since there is a higher risk of falling when facilities lack proper protection such as guard rails and stairs that are safe and properly constructed.

4.3 Winnowing and Coffee Separation

Manual or mechanical winnowing is performed to separate light impurities, such as leaves, sticks, and other debris, from the fruit. Manual winnowing is still employed by small producers, who conduct the winnowing in the fields using screens. Mechanical winnowing is done by mobile or stationary machines that move air across the winnowing surface through either suction or fan-blown air. Mechanical harvesters equipped with a ventilation system are able to harvest and winnow the crop simultaneously. To remove even more impurities before processing and after winnowing, a mechanical sifter may be used. This sifter is made of perforated vibrating screens and can be placed under the receiving bin to separate debris that is larger or smaller than the fruit.

After impurities are removed through winnowing, the coffee crop is ready for the hydraulic separator. It is very important to note that fruit that has fallen onto the ground both before and during harvesting, known as sweepings, should never be mixed with other lots since it is generally of inferior quality.

Hydraulic separation is one of the most important stages in coffee processing. It employs flotation to separate the more dense unripe and ripe fruit from the less dense fruit, known as floaters. It also removes material such as sticks and light impurities that were not removed in previous stages, as well as denser material such as soil and stones. Impurity removal is both essential

Figure 4.6 Hopper with adequate inclination for the coffee to slide by gravity alone.

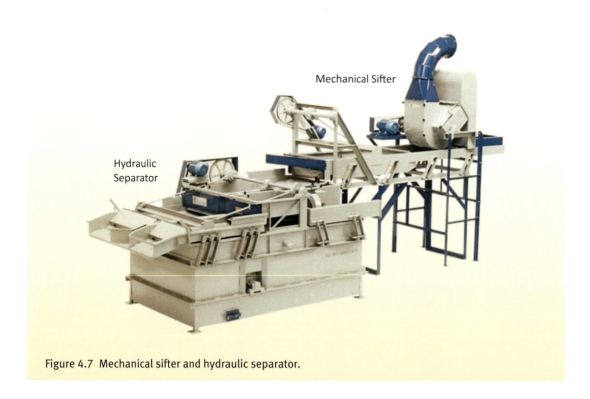

Figure 4.7 Mechanical sifter and hydraulic separator.

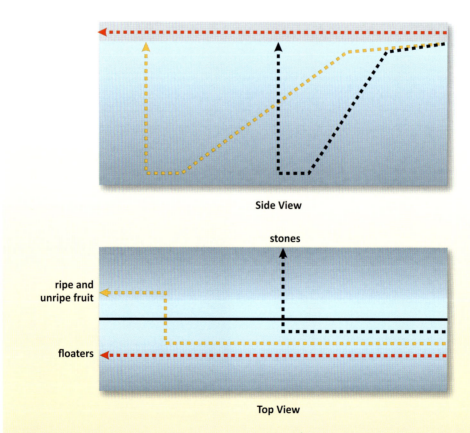

Side View

stones

ripe and
unripe fruit

floaters

Top View

Figure 4.8 Schematic representation of the separation of the most dense fruit (--), less dense fruit (--), and stones (--).

for hygiene, and pragmatic, as it increases the lifespan of equipment such as elevators, dryers, and milling machines used in later stages.

Hydraulic separators (Figure 4.7) consist of two water tanks connected at the bottom, a system to move the coffee and recirculate the water, two front exists, and one lateral exit.

Compared to older separators constructed of masonry, modern hydraulic separators consume much less water, between 0.3 and 1.0 liters of water per liter of coffee. After winnowing, the coffee is transported over a screen where the larger impurities are removed. The floaters—dry fruit; overripe fruit that is nearly dry ("raisins"); wrinkled underdeveloped beans; and both unripe and ripe fruit with only one developed seed or one or more bored seeds—float in the water and are transported to one of the front exits. The more dense fruit (unripe, semi-ripe, and ripe) sink and are swept away by an ascending water flow that forces them toward an opening linked to a secondary lateral water tank, allowing the fruit to return to the surface of the water and be transported to the second front exit (Figure 4.8). Stones that were not separated out during sifting (generally because their size is similar to coffee fruit) fall to the bottom of the hydraulic separator and are removed separately through the machine's lateral exit.

Sifting and hydraulic separation of coffee are particularly important when the coffee is strip harvested, as coffee harvested in this manner is more likely to be at different stages of maturation, with fruit that varies widely both in quality and moisture content (Table 4.1).

Separating is important, as drying time and uniformity depend both on the initial moisture content and size of the coffee. What's more, since the floaters are separated out, the quality of the ripe fruit is optimized. For this reason, a hydraulic separator should also be used for selectively harvested coffee; although correct harvesting ensures only ripe fruit, it is still necessary to separate out the fruit with underdeveloped or insect-damaged seeds.

Table 4.1 Moisture content in coffee fruit separated by density and size.

Coffee Type	Screen Diameter (mm)	Moisture Content (%wb)
Ripe and Unripe Fruit	11	65.4
	9	61.6
	7.8	57.8
Floaters	11	47.5
	9	45.8
	7.8	40.2
	<7.8	39.8

Source: Wilbaux (1963)[4].

In addition to separation by density, floaters can be further separated by size (Table 4.1), using a cylindrical sieve with circular perforations (Figure 4.9) that is placed just after the hydraulic separator. Both dry and smaller fruit pass through the perforations, while the larger and moister fruit are retained inside. At the beginning of the harvest this procedure is especially recommended since there are elevated levels of unripe and ripe fruit that float due to underdevelopment or insect damage. Toward the end of the harvest, small dried fruit pods and nearly-dry raisin fruit are separated out. Apart from its use to separate floaters by size, this cylindrical sieve can also be used with the non-floaters—the denser ripe and unripe fruit—to separate out small unripe fruit, which, due to their small size, pass through the screens of the pulper, reducing the quality of wet process coffees.

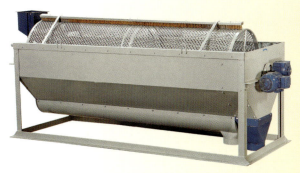

Figure 4.9 Cylindrical sieve used to separate floaters and denser ripe/unripe fruit by size (photo courtesy of Pinhalense S.A. Máquinas Agrícolas).

After sifting, hydraulic separation, and size separation, the coffee is then dried or pulped, depending on the processing method the producer has chosen.

4.3.1 Electronic Separation of Unripe Fruit

Growers now have the option of electronic separation to separate out unripe coffee fruit without using water or pulping equipment. An electronic separator (Figure 4.10) consists of a hopper, a screen to eliminate larger impurities, and a perforated, revolving cylinder in which the optical reader determines the color of the fruit. After the reading, the equipment will either eject or retain the fruit per the standard programmed into the machine.

Figure 4.10 Electronic separator of coffee fruit. (photo courtesy of Ventbras)

The capacity of models currently on the market is 1,500 liters of coffee fruit per hour. This equipment separates ripe and unripe fruit without using water, allowing for the production of high quality naturally processed coffees with low environmental impact.

4.4 The Dry Processing Method – Natural Coffee

The production of natural coffee, traditionally known as the dry method (Figure 4.11), is the oldest and simplest coffee processing method and entails drying the entire coffee fruit intact. It is largely used in tropical regions where the dry season coincides with the harvest period.

Traditional literature defines the dry method as the drying of *all* coffee fruit immediately following the harvest[5] with no lot separation based on maturation or coffee quality. While this is the most common way to perform the dry process, it is just one of the many processing options available and is generally the option chosen by producers with inadequate coffee processing infrastructure. In fact, all coffee, whether composed of ripe, unripe, overripe, dried coffee, or any combination thereof, is considered to be natural coffee if it was dried with its pericarp intact.

The final quality of a dry process coffee depends on myriad factors, including harvest method and care taken during processing and drying. When only ripe fruit are selectively harvested and then carefully dried to avoid any form of fermentation, it is possible to produce high quality dry process coffees.

Before dry processing begins, impurities that come from the field are removed and the fruit are separated by density in the hydraulic separator, resulting in lots with different moisture levels (as described earlier). The dry process is the processing method with the least effect on the natural state of the coffee, as the entire fruit remains intact during the drying stage. Furthermore, the environmental impact of this method is minimal as it produces only a small amount of residual solids and liquids and does not produce wastewater with elevated levels of organic material, as does the wet method.

Some physical changes observed during dry processing distinguish these coffees from wet process, or washed, coffees. The exocarp, originally red or yellow, becomes darker and forms

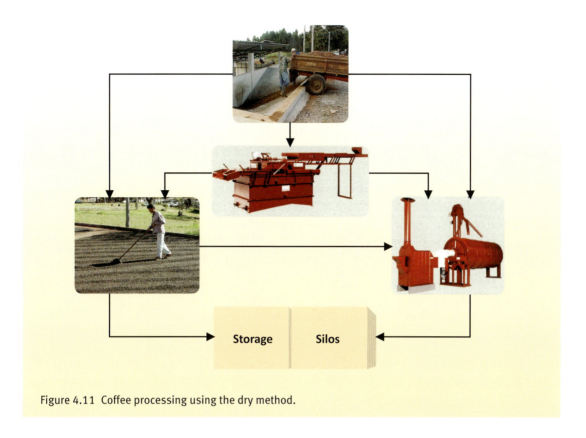

Figure 4.11 Coffee processing using the dry method.

a dry, hard outer layer of the coffee fruit, which once dried is called a coffee fruit pod. Fruit size is also greatly affected, with a ripe fruit contracting to about 40% of its initial volume. (The actual volumetric contraction of an entire coffee lot will depend on the proportion of ripe fruit, unripe fruit, and floaters.)

According to some authors, when a coffee fruit dries intact, part of the mucilage may be transferred to the seed, causing the silver skin both to darken and to adhere more to the endosperm as compared to wet process coffees. This, however, has not been completely verified. Another distinguishing physical change is bean coloration: dry process coffees tend to be more yellowish-brown, while wet process coffees are bluish-green[6]. In general, dry process coffees exhibit sensory characteristics that distinguish them from wet process coffees, such as lower acidity and more body[7], essential elements in espresso coffee.

However, the final quality of a dry process coffee depends on myriad factors, including harvest method and care taken during processing and drying. Picking fruit by strip harvesting and drying without care or separation will result in coffees that are, at best, astringent (also called *hard cup* coffees). On the other hand, when only ripe fruit are selectively harvested and then carefully dried to avoid any form of fermentation, it is possible to produce high quality dry process coffees.

The lower quality often seen in dry process coffees can be explained mainly by two factors. First, a lack of care during harvest can result in the presence of unripe, bored, or fermented fruit. Second, fermentation can occur due to the elevated levels of sugar in the mucilage as well as slower drying times caused by the presence of the fruit skin. However, these apparently conclusive explanations are not sufficient to explain the difference in flavor profile between coffees produced by the two methods, even when only ripe fruit are processed with similar care and under controlled conditions.

4.5 The Wet Processing Method

The first known use of the wet processing method was in 1730, in what is now Indonesia[8]. Today, this method can be carried out in three distinct ways:

1. *Fully washed* coffees are wet process coffees in which the fruit skin is removed mechanically and the remaining mucilage is removed by means of biological fermentation. The resulting clean parchment coffee is then dried. This is the traditional and most common of the three wet process methods throughout the world.

2. *Pulped natural* coffees are wet process coffees in which the fruit skin and part of the mucilage are removed mechanically. The remaining mucilage, however, is not removed and is dried intact with the parchment coffee. This method is commonly used in Brazil where such coffees are called *cereja descascado*, or *cd* coffees. Recently, other countries have adopted this method, referring to these coffees as *honey coffees*.

3. *Semi-washed* coffees are wet process coffees in which the skin and all of the mucilage are removed mechanically. Semi-washed coffees are also called demucilaged or mechanically demucilaged coffees.

The wet processing method was developed in equatorial regions with continual precipitation during the harvest period, a condition not appropriate for dry processing. In these regions, the dry process would almost always result in coffee of inferior quality. The quality of dry processed coffees was evaluated in Colombia[9] and it was determined that this method of processing produced coffee with a vinegar-like odor that was not fit for consumption. Given the climatic conditions in Colombia, 20–25 days were needed to dry the coffee to 11%–12% moisture content levels, indicating inadequate conditions for dry processing.

> The wet processing method was developed in equatorial regions with continual precipitation during the harvest period, a condition not appropriate for dry processing.

The wet processing method will generally yield good quality coffee if only ripe fruit are harvested, if the skin and mucilage are properly removed, if biological fermentation is controlled, and the if coffee is carefully dried. In Brazil, wet process coffees make up a small part of total production. However, they are becoming more common every year, not only in regions with climatic limitations for the dry process, but also as a means to produce higher quality coffees, even in regions with climates appropriate for the dry method. However, using the wet process does not guarantee coffee quality. In a study of wet process coffees on 32 properties in the Sul de Minas region of Brazil[10], 75% of the samples were classified as hard beverage, corresponding to an SCAA score of 70 to 75. Although none of the samples exhibited the Rio taint, the study showed that pulping alone does not guarantee the production of specialty coffees corresponding to an SCAA score of 80 or higher.

Independent of the type of wet processing method adopted, the pulping operation is common to all of

Figure 4.12 Coffee Pulper Machine

Figure 4.13 Elongated perforated screens

them. Pulping—the removal of the exocarp—is a procedure carried out by coffee fruit pulping machines, like that picture in Figure 4.12.

The pulping operation is based on the fact that ripe fruit have lower resistance to pressure than unripe fruit. All of the models of coffee pulpers on the market perform two basic operations: separation of unripe from ripe fruit and pulping of the fruit. The order of these operations varies by model.

When separation of unripe fruit is the first pulping step (Figure 4.12), this separation occurs when both ripe and unripe fruit are forced through a cylindrical, perforated screen with lateral exits. The mesocarp of an unripe fruit is still rigid and resistant to pressure, causing it to move through the screen and into the lateral exits intact and unpulped. A ripe fruit, with its mucilagenous mesocarp, separates into two seeds, which, along with the pulp, pass through elongated, perforated screens (Figure 4.13). This mixture of seeds and pulp is transported to the pulper, a mammillated cylinder that separates the pulp from the seeds. The pressure to which the fruit are subjected as well as the timing of the process can be regulated by weights located near the lateral exits, and by the rate of water flow (Figure 4.14).

Other pulping machine models reverse this process, first pulping the coffee, then passing the pulped coffee through the unripe coffee separator. This sequence is more common on equipment in Central America and Colombia.

Traditional equipment used in pulping requires a large quantity of water, close to 6,000 liters per hour, and produces highly contaminated wastewater. As a result, some companies have begun to offer equipment that uses less water to minimize the impact of this process.

Figure 4.14 Weights for adjusting pulping pressure.

Figure 4.15 Separation of unripe fruit during pulping.

The pulping process produces optimal results when coffee is harvested selectively, but it is even more important in the processing of strip-harvested coffee as a means to separate out unripe fruit (Figure 4.15).

Pulping quality and efficacy depend on the proportion of unripe and ripe fruit that enter the pulper. It is recommended to not exceed 30% unripe fruit as this can reduce pulping capacity, damage equipment, and decrease product quality. What's more, a high proportion of unripe fruit increases the risk of unripe fruit being cut in half and passing through the screen, thus compromising the quality of the resulting parchment coffee.

It is also recommended to frequently monitor four settings on the pulper: fruit influx, water volume, screen perforation size (Figure 4.16), and the positioning of weights on the retention arm.

Both fruit influx and water volume should be regulated to optimize equipment efficiency without damaging the product or using excess water. The screen should be selected according to the average size of the coffee crop, and the retention pressure for unripe fruit should be adjusted to prevent their passage through the screen. Pulping pressure should be calibrated to allow some ripe fruit to exit along with the unripe fruit; this will result in a better quality of both unripe coffee and parchment coffee lots.

Other equipment can be used for pulping coffee fruit, such as disc pulpers, but today this equipment is not commonly used in most coffee producing nations.

4.5.1 Fully Washed Coffee (Fermentation)

In the processing of fully washed coffees, mucilage removal occurs immediately after the pulping of the coffee fruit through controlled fermentation. The pulped coffee is transported to fermentation tanks (Figure 4.17) where it remains for periods ranging from 12 to 48 hours, depending on the altitude and temperature of the processing locale.

Figure 4.16 Pulper screen.

To optimize quality after pulping and before the fermentation stage, lighter weight seeds should be separated from the heavier seeds[11]. Fermentation of the lighter seeds occurs more quickly and can cause chemical changes that will negatively affect the aroma and flavor of the heavier seeds, which, in general, are of higher quality and contain a larger percentage of ripe coffee. While separation by density is an effective process, it is a slow operation that requires significant manual labor and a large quantity of water.

The purpose of controlled fermentation is to break down the mucilage through hydrolysis, which facilitates mucilage removal in the final washing stage. After fermentation, the coffee must be washed, manually or mechanically, to finalize removal of the mucilage. Proper fermentation is important because, depending on drying conditions, any mucilage that remains and adheres to the parchment while drying can increase the risk of undesirable fermentation during the drying process, which decreases product quality.

Figure 4.17 Fermentation tank.

The hydrolysis of pectin in the fermentation process is caused by the biochemical action of the pectinase still present in the seed. This reaction can be accelerated by different microorganisms, such as saccharomyces, which also contain pectinolytic properties. The speed of hydrolysis depends on temperature, and it is necessary to adjust fermentation time based on different ambient conditions. Some microorganisms can cause the development of undesirable flavors, especially in prolonged fermentations. For this reason, it is important to avoid the development of fungi, which can lead to the development of acidogenic species. Controlling pH is important to avoid the formation of acids, such as propanoic acid[12].

Fermentation can be done either "dry," or using water immersion. In dry fermentation, the coffee goes directly from the pulper to the tanks without any additional water. In this case, the fermentation processes are faster, attaining lower pH levels in shorter times. The more traditional method involves immersing the seeds in water, and requires longer fermentation times. While dry fermentation is recommended for colder regions, water immersion is the process of choice for hotter regions. In some cases, mixed fermentation can be used, where dry fermentation is employed in the first hours to quickly acidify the environment and prevent the development of fungi, followed by fermentation with water. This option is also known as double fermentation.

Washed coffee generally loses around 1% of its dry material during fermentation and some soluble substances are leached[13]. The partial loss of some compounds can improve coffee quality by reducing astringency and bitterness in the cup[14].

4.5.2 Semi-washed (Mechanical Demucilaging)

Like fermentation, mechanical demucilaging occurs shortly after pulping and serves to remove the mucilage that adheres to the seeds. In mechanical demucilaging, a mucilage remover is used to generate friction both between the individual seeds and between the seeds and a metal cylinder (Figure 4.18).

Water is added in small quantities to both lubricate and clean off the mucilage. The parchment coffee exits through the upper part of the machine while water exits through the lower part. The use of water in this equipment varies depending on the configuration of the water regula-

Figure 4.18 Mucilage Remover (courtesy Pinhalense S.A. Máquinas Agrícolas)

Figure 4.19 Pulped Natural Coffee.

tor. Despite producing lower quantities of wastewater compared to pulping, mechanical mucilaging produces a higher concentration of organic material, resulting in liquid residue that is extremely harmful to the environment.

A big advantage of the semi-washed process is that it removes all or part of the mucilage without the use of fermentation tanks, thus diminishing the overall amount of wastewater. Furthermore, when the coffee is placed on the patio for drying, unlike pulped natural coffees the semi-washed beans do not clump together, thus facilitating the work of raking, rotating, and drying.

4.5.3 Pulped Natural

In the production of pulped natural coffees, the mucilage that sticks to the parchment after pulping is maintained (Figure 4.19) and the coffee is sent directly to dry. Recently, these coffees have been referred to as "honey coffee" due to the dried honey or caramel look of the remaining mucilage on the parchment.

The mucilage that remains on the seeds impedes the raking and rotating of the drying coffee, and due to elevated moisture and sugar levels also increases the risk of fermentation. Because of this, the pulped natural process is not recommended for regions that are humid and hot. Furthermore, given the presence of the mucilage, producers must take greater care in the first days of drying to ensure that the mucilage is quickly dried. This can be done by spreading the parchment coffee on the patio in microrows two to three centimeters high, and rotating the coffee as much as possible throughout the day. In smaller scale production, raised beds can be used.

4.6 Processing Unripe Fruit

Recently, pulped natural coffees have been referred to as "honey coffee" due to the dried honey or caramel look of the remaining mucilage on the parchment.

In large-scale production, the presence of the immature defect in commercial lots is one of the principal challenges to producing coffees of higher quality. The immature defect comes from harvesting unripe fruit and is characterized by a green spermoderm that adheres to the endosperm. When unripe fruit are pulped, the spermoderm detaches from the endosperm upon drying, thus eliminating the way these defects are identified. But because the predominant harvesting method is stripping and the predominant processing method is the dry method, the immature defect remains a challenge. However, studies have shown that even with washed and semi-washed coffees, the presence of 2.5% unripe fruit at harvest has led to the disqualification of 30% of the cups tested due to undesirable flavors[15].

For large-scale coffee production, as is common in Brazil, it is practically impossible to complete a harvest with less than 2.5% unripe fruit. Doing so would require waiting longer to conduct the harvest, which would, in turn, result in a larger percentage of floaters and sweepings, lowering coffee quality. In certain situations where crop maturation levels vary greatly, mechanical or manual harvesting can be done in parcels, reserving the less-ripe sections of the farm for later harvesting. This results in increased product quality. In Brazil, except in instances of selective harvesting, the harvest will likely include variable quantities of unripe fruit. Separating these out is essential to achieving higher quality in the remaining ripe fruit.

A reduction in the percentage of black, immature, and sour defects in pulped unripe coffee increases the value of the product and thus increases the overall economic viability of processing coffees using the wet method.

The most viable technique for removing unripe fruit is the pulping process. After pulping, two lots are formed. The first lot consists of mainly ripe fruit and therefore has higher quality potential. The second lot consists primarily of unripe fruit and has very low potential for producing a quality product. If the value of the second lot is so low that separating the coffee into two lots actually renders a net loss, then the pulping process is not economically viable.

However, the quality of unripe coffee can be improved depending on the way it is processed and the care taken during drying[16]. Unripe fruit can be dried immediately, using the dry processing method, or pulped, using the wet method. Though not common, pulping is an alternative that can improve the final quality of unripe coffee (Figure 4.20).

During the pulping of a crop containing both ripe and unripe fruit, the weight that controls pulping pressure should be removed, permitting the free flow of unripe fruit through the lateral exit of the pulper. This will allow some ripe fruit to go through the lateral exit along with the unripe fruit. However, it will minimize the volume of unripe fruit in the portion of pulped ripe fruit, optimizing the quality of the lot. The lot of unripe fruit should then be pulped immediately. When this is not possible, the unripe fruit should be temporarily stored in silos and pulped within no more than 20 hours. To pulp unripe coffee fruit, the weights should be adjusted to mid-level pressure[17]. After processing and drying, pulped unripe coffee contains, on average, only 2.8% black, immature, and sour defects, and a cup quality of hard cup or better. This quality is significantly better than unripe coffee processed using the dry method, is similar to coffee processed using the traditional natural process method (where unripe and ripe fruit are not separated), and is inferior to pulped natural coffees (which normally contain only ripe fruit)[18].

Figure 4.20 Pulping unripe coffee.

The better quality of pulped unripe coffee over natural unripe coffee can be explained by the reduction of fermentation risk and the facilitation of uniform drying. Although pulped unripe coffee is normally classified as having a cup quality of hard cup (corresponding to SCAA 70–75), it does not exhibit any signs of the phenolic and ferment defects.

It is also interesting to note that pulped unripe coffee contains fewer immature defects than the same lot dried using the natural process. As noted at the beginning of this section, the pulping process causes the

Figure 4.21 Pulped unripe coffee.

green spermoderm to detach during drying. Since the presence of an attached green sper-moderm is the means of classifying the immature defect, unripe coffee that undergoes pulp-ing is less likely to be classified as such.

These results are promising, since a reduction in the percentage of black, immature, and sour defects in pulped unripe coffee increases the value of the product and thus increases the overall economic viability of processing coffees using the wet method.

4.7 The Cost of Processing Coffee

As described above, there are several distinct ways to carry out coffee processing, and the choice of processing method, which is key to the profitability of coffee production, will de-pend on diverse factors. Other variables that will affect the cost/benefit ratio of each meth-od include the percentage of ripe and unripe fruit, as well as premiums and discounts at the time of trade. Given all of the variables, it is common for a producer to struggle in choosing between wet and dry processing. The answer is not simple and requires computational tools to facilitate decision-making. To make a decision, the producer must be able to calculate the cost of each processing method, taking into consideration both the fixed and variable costs of each method. The cost of processing includes the following:

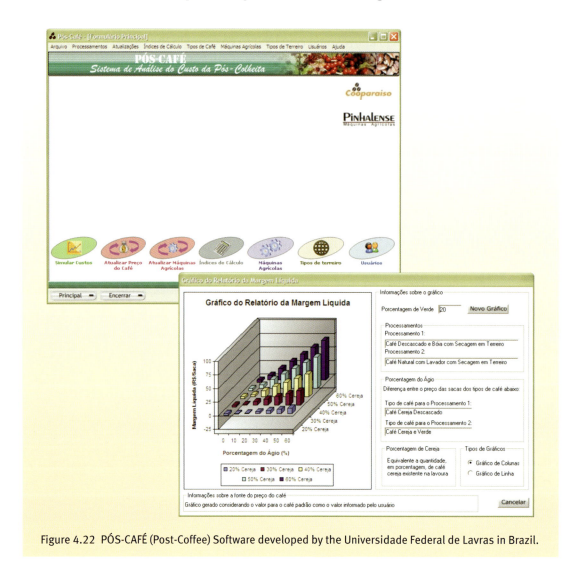

Figure 4.22 PÓS-CAFÉ (Post-Coffee) Software developed by the Universidade Federal de Lavras in Brazil.

a. *Fixed Costs* are all of the "overhead" costs that do not vary with the quantity of coffee processed. Depreciation is also included among the fixed costs and is defined as the costs associated with replacing assets that can no longer be used due to wear and tear or obsolescence. In other words, depreciation is the financial reserve accumulated by the producer to replace the asset at a later date.

b. *Variable Costs* are all costs that vary with the quantity produced during a production cycle. Labor and electrical energy costs are examples of variable costs.

c. *Total Cost* is the defining parameter for determining processing method. It is the sum of variable costs and fixed costs.

d. *Unit Cost* is obtained by dividing the total cost of drying by the quantity of coffee processed.

The Federal University of Lavras (UFLA), in Brazil, in partnership with a business software consulting group, developed a software program called *PÓS-CAFÉ* (Post-Coffee). The program covers all aspects of post-harvest coffee, including wet and dry processing and drying systems. It consists of three modules: data input, cost calculation, and reports.

Figure 4.22 shows the main screen of the software program. On the bottom are links for cost forecasting, coffee price updating, agricultural equipment updating, calculation indexes, asset/equipment inventories, and user information. Using the coffee cost forecaster, the user can choose two types of processing methods. The software then opens a data entry form where the user enters total production, harvest time, percentage of ripe and unripe fruit, labor and energy costs, type of drying surface, coffee market price, and premiums and discounts. In the next window, the user can choose the model of hydraulic separator, pulper, and mucilage remover.

Next, the software presents spreadsheets specific to each processing type to be analyzed where the user enters drying time, layer thickness when drying, and the labor and per diem costs associated with running all of the equipment involved in processing. The software then calculates and presents a cost-benefit analysis for each processing method analyzed and provides comparative net margins between the methods. The user can also view the net margin graphic representation (Figure 4.22), which shows the break-even point. This facilitates the user's choice of processing type given the characteristics of their crop (percentage of ripe and unripe).

When price is a limiting factor, the software can also help determine the optimum time to sell coffee on the market given different proportions of ripe and unripe fruit.

4.8 Physiological and Biochemical Alterations During Coffee Processing and Their Impact on Quality

Coffee quality is determined principally by the flavor and aroma developed during roasting, and is based on diverse chemical compounds that are the precursors, found in green coffee, to these aromas and flavors. While the formation and presence of these precursors depend on genetic, environmental, and technological factors during the growing phase, the transformation of these precursors into aromatic compounds depends on the conditions and control of the roast. Furthermore, between growing and roasting, there are diverse post-harvest stages such as method of processing, drying, storage, milling, and transport that also influence coffee quality.

The sensorial attributes of coffee are aroma, acidity, bitterness, body, flavor, and overall beverage/cup impression. The intensity and balance of these characteristics define the sensorial quality of a coffee.

Depending upon the processing method used, a coffee will exhibit distinct characteristics in final beverage/cup quality. Generally, dry process coffees have more body and wet process coffees exhibit accentuated aroma and light, pleasurable acidity. There is a considerable amount of literature that compares the quality of dry process coffees and wet process coffees[19].

There are innumerable physiological and biochemical events that occur in coffee during the post-harvest period that can result in characteristics that distinguish one coffee from another. Despite this fact, the traditional explanation for differences between dry process and wet process coffees is the undesirable fermentation and the lack of care in harvesting and drying that are common to dry process coffees. In the dry process, as previously stated, harvesting is traditionally done by stripping, and even if the fruit pass through a hydraulic separator, the lots of coffee will contain fruit at different stages of maturation. By comparison, wet process coffees are usually harvested through selective picking, resulting in lots that are more uniform and composed only of ripe fruit. However, these distinctions do not explain the quality differences between dry and wet process coffees when, for both methods, only ripe fruit are processed with the same care and under the same controlled conditions.

> Depending upon the processing method used, a coffee will exhibit distinct characteristics in final beverage/cup quality. Generally, dry process coffees have more body and wet process coffees exhibit accentuated aroma and light, pleasurable acidity

In fact, during the processing and drying of coffee, physical, chemical, biochemical, and physiological changes occur, resulting in different coffee qualities[20].

Processing methods differ in the way they alter both the chemical composition and sensorial characteristics of green coffee. In a study comparing wet and dry process Canephora, it was found that wet process Canephora contained[21]

- lower levels of free carbohydrates (fructose and glucose), organic acids (quinic and oxalic acid), minerals (K^+, Ca^{2+}, Mg^{2+}, Cu^{2+}), and trigonelline;
- higher levels of chlorogenic acids and polysaccharides in the cell walls (arabinogalactan and mannan), and lipids;
- less astringency, bitterness, and rubber and wood flavors;
- a flavor that is lighter, fruitier, and more acidic.

Variations in chemical composition can be attributed to degradation mechanisms such as

- hydrolysis of polysaccharides that form free glucose and fructose;
- breakdown of chlorogenic acids, which results in quinic acid;
- degradation of phospholipids, forming phosphoric acid.

The authors of this study concluded that since dry process coffees have a lower evaporation rate than wet process coffees, and this entails a longer drying time, dry processing results in more intense degradation mechanisms that cause greater chemical changes.

The two processing methods also differently alter the chemical composition of Arabica coffee. Compared to wet process coffees, dry process Arabica seeds accumulate more free amino acids (γ-aminobutyric acid, also called GABA) and have different levels of fructose and glucose[22]. The higher GABA accumulation in dry process Arabica coffee, relative to the wet process, shows that, apart from degradation mechanisms, different metabolic reactions can occur in coffee fruit during post-harvest, and their intensity levels vary by processing method[23]. In plants, GABA is formed by the decarboxylation of glutamic acid, and in coffee its accumulation represents a reaction to stress during drying. In dry processing, the time available for the decarboxylation process to occur is greater, since the drying rate is slower.

Possible alterations in the coffee seed related to physiological events, such as germination, are not normally considered in scientific studies due to the short drying time and rapid dehydration during processing. Studies examining whether germination occurs with the different processing methods have reached similar conclusions regarding wet process coffees, but no consensus has been reached regarding dry process coffees.

There is evidence that various biochemical processes occur in coffee seeds and that germination does, in fact, begin and continue throughout the processing period[24]. This can be seen by monitoring the metabolism related to germination, both through the expression of enzymes specific to germination—isocitrate lyase—as well as through the reactivation of cellular division determined by the accumulation of β-tubulin. The time necessary for the germination process, as well as its extent, depend on the processing method. Apparently, removing the skin and mucilage provides the necessary stimulus to activate the seed germination of wet process coffees. Surprisingly, metabolism related to germination occurs even in dry processed seeds, probably due to the activation of endogenous factors. In this case, germination time is different from that observed in wet process coffees. Other studies have shown that while germination does occur in wet processing, in dry processing it does not, due to the presence of inhibitors in the mucilage[25].

Therefore, the removal of the mucilage in pulped coffee permits the unfolding of diverse reactions related to germination. The mobilization of reserves and other physiological reactions results in different metabolic profiles compared to dry process coffees. Even though this mobilization of reserves and the synthesis of low molecular weight compounds have not been completely clarified and accepted by the scientific community, it is accepted that the formation of aroma precursors present in green coffee depends on the type of processing used[26].

New techniques have been used to discover the highest possible number of metabolites in a particular extract to obtain an impartial, conclusive evaluation concerning the metabolism of a biological system. Recent research, using a metabolomics platform, has been conducted to describe the metabolic profiles of green coffee processed by dry and wet methods, then sun-dried with air temperature between 40 °C and 60 °C. The analysis was done using high-precision liquid chromatography–mass spectrometry (LC-QTOF-MS). More than 1,700 mass signals were obtained, representing hundreds of metabolites. Noticeable changes were observed in the metabolic composition of green coffee during drying. An analysis of key components revealed distinct groups depending on the processing method, indicating a distinct variation in the metabolic composition of dry process coffees compared to wet process coffees[27].

In light of these results, perhaps new concepts will arise. If the metabolic profile of coffee changes during processing and drying, could it be possible in post-harvest to actually *improve* the quality of coffee over that attained in the field?

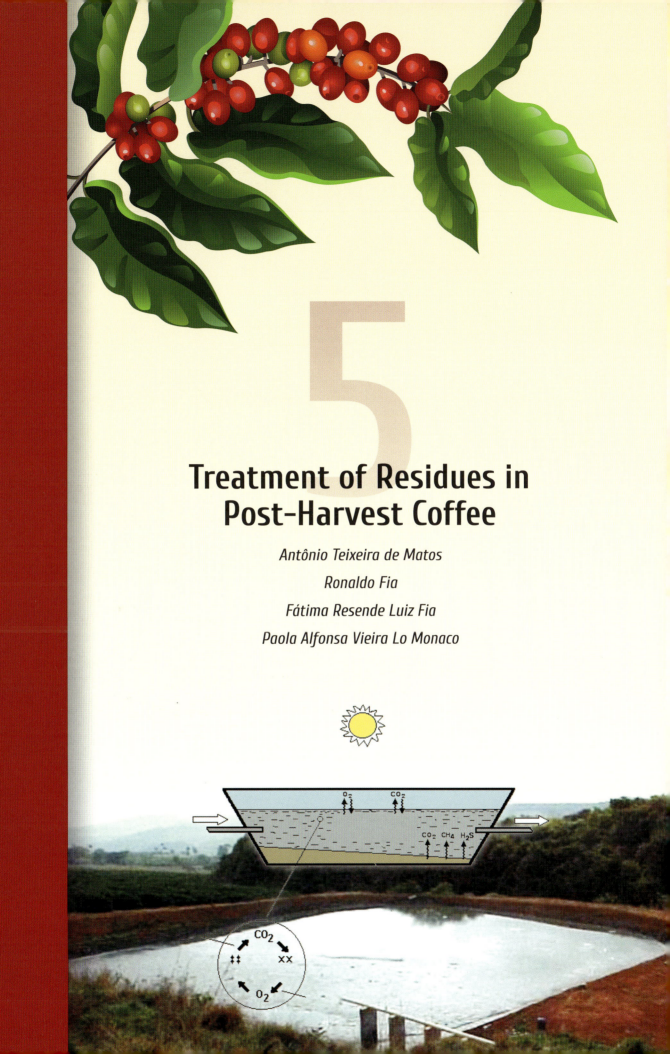

Treatment of Residues in Post-Harvest Coffee

Antônio Teixeira de Matos

Ronaldo Fia

Fátima Resende Luiz Fia

Paola Alfonsa Vieira Lo Monaco

5.1 Introduction

Hydraulic separation and pulping of coffee fruit generate large volumes of residual solids and liquids that are rich in organic and inorganic materials. If improperly disposed of, these materials can cause severe environmental problems, degrading or destroying flora and fauna, and compromising water and soil quality[1]. In coffee producing regions, these residues have become a major environmental problem, generating significant demand for simple treatment systems with low cost installation and operation[2].

The principal effect of organic pollution on a body of water is decreased availability of dissolved oxygen. Aerobic bacteria use the dissolved oxygen in a medium to carry out metabolic processes. This, is turn, makes possible the degradation of organic material within the medium. A decrease in the concentration of dissolved oxygen in water can be fatal to fish and other aquatic animals, in addition to producing unpleasant odors.

Because of this potential harm, many legislative bodies in producing countries are increasingly regulating wastewater contamination. An example of this is the Environmental Legislation of the State of Minas Gerais, Brazil (Normative Deliberation COPAM no. 10/86), which established that to release wastewater into bodies of water, either the biochemical oxygen demand (BOD, a measure of the average quantity of organic material present) must be 60 mg L^{-1}, or the efficiency of the wastewater treatment system for the removal of BOD must be greater than 85%, as long as the standards established for the receiving watercourse are not surpassed.

With respect to solid residues, special attention must be given to the accumulation of hulls where coffee milling takes place. An evaluation of soils where coffee hulls were stored found elevated concentrations of nitrogen in the form of ammonium (free ammonia), as well as potassium, in locations where the hulls had been stored for periods of up to three years. The surface and subsurface soil contamination caused by leaching of the hulls rendered the area inadequate, at least temporarily, for agricultural use, and raised the risk of groundwater contamination[3].

5.2 Types of Residues

5.2.1 Solid Residues

Pulp is the first residue generated in the processing of coffee fruit, representing about 39% of its fresh mass and 29% of its dried mass. The quantity of pulp in the ripe fruit depends on maturation level, climatic conditions during fruit development, and the type of coffee cultivated. The parchment represents about 12% of the coffee fruit in terms of dry material. Due to the large volume of solid residues generated in the milling of coffee fruit (Figure 5.1), alternative methods of hull disposal and re-use have been examined.

Figure 5.1 Solid residues (hulls) generated in the processing of coffee fruit.

According to the data in Table 5.1, coffee pulp is rich in potassium and other nutrients, and as such can be used as an agricultural organic fertilizer, either pure or after composting. This is an interesting alternative to inorganic fertilizers, from both environmental and economic perspectives.

Composition	Content		
	(1)	(2)	(3)
C-total (g kg⁻¹)	529.5		
N- total (g kg⁻¹)	14.7	13.2	18.9
P- total (g kg⁻¹)	1.7	0.5	2.1
K (g kg⁻¹)	36.6	31.7	47.0
Ca (g kg⁻¹)	8.1	3.2	3.0
Mg (g kg⁻¹)	1.2	--	2.9
S (g kg⁻¹)	1.4	--	--
Mn (mg kg⁻¹)	125.0	--	--
Zn (mg kg⁻¹)	30.0	--	4.4
Cu (mg kg⁻¹)	25.0	--	18.7

Source: 1: Matos et al. (1998)[4]; 2: Vasco (1999)[5]; 3: Brandão et al. (1999)[6]

Storage or treatment of wastewater from coffee fruit processing creates a thick layer of scum that must be removed. This results in a solid waste that must also be properly disposed of. Scum forms on the wastewater after a 24-hour resting period and comprises 32,268 mg L⁻¹ total solids, 19,730 mg L⁻¹ suspended solids, 7,184 mg L⁻¹ total fixed solids, and 25,084 mg L⁻¹ total volatile solids[7]. The mass densities of scum in various forms, after drying in the open air for 25 days, are 0.965 g cm⁻³ in its fresh "in natura" paste form, 0.084 g cm⁻³ in its dried solid form, and 0.503 g cm⁻³ when the solid scum has been triturated. The mass density of scum is greater when in paste form (as when present in coffee wastewater), since it has higher moisture content. As the scum dries, its mass density decreases so that the residual material, which is primarily organic, has a lower mass density. Grinding the dried scum increases its mass density, as the lower granulometry of the ground material enables the scum to organize itself more easily than when in its post-drying state.

Table 5.2 shows a chemical analysis of waste water scum resulting from coffee processing. As the water evaporates, the concentration of elements increases due to the decrease of moisture content in the material.

Table 5.2 Chemical analysis of wastewater scum resulting from the hydraulic separation and pulping of Conilon coffee fruit.

Samples	pH	C-organic	N-total	P	K	Ca	Mg	Cu	Fe	Mn	Zn
						mg L⁻¹					
1	5.37	10,973	1,040.0	102.80	799.0	673.8	21.80	1.60	70.80	n.d.	n.d.
2	5.19	14,850	-	194.19	-	1,108.3	35.30	2.70	97.70	n.d.	1.200
						g kg⁻¹					
3	5.11	618	32.5	4.43	26.6	23.8	0.88	0.05	2.84	0.082	0.058

Sample 1: 24 hours after collection in liquid form | Sample 2: 2–7 days after collection in liquid form
Sample 3: 3–25 days after collection in solid form | n.d.: not detected | Source: Lima et al. (2004)[8]

The samples demonstrate that coffee wastewater scum can be used to fertilize agricultural crops, including coffee, given that each kilo of scum contains 32.5 g of nitrogen, 4.43 g of phosphorous, 26.6 g of potassium, and 23.8 g of calcium, as well as micronutrients. The carbon/nitrogen (C/N) ratio of dried scum is 19.0, indicating that it can be composted without adding more nitrogen. However, since the C/N ratio is less than 30, it can be useful to add other residues with higher C/N ratios to produce a higher quantity of organic compounds.

5.2.2 Liquid Residues (Wastewater)

In coffee fruit processing, water is used in hydraulic separation, pulping, and demucilaging. In hydraulic separation, water consumption can be reduced by using the highest water circulation potential available in the hydraulic separators. In the pulping and demucilaging stages, when there is generally no water recirculation, consumption tends to be higher. Water discarded in these three processes constitutes wastewater.

Using more recent equipment, the amount of wastewater generated during hydraulic separation is about 0.1 to 0.2 L per liter of processed fruit, depending on the size of the hydraulic separation tank and the number of times the water is changed during the processing day. In pulping and demucilaging, about 3 to 5 L of water are used for each liter of fruit. In cases where the water is recirculated during pulping and demucilaging, this quantity of water drops to about 1 L for each liter of processed fruit.

In traditional wet processing, 3 tons of byproducts are generated and 4 m^3 of water are used for every ton of green coffee processed[9]. The wastewater from coffee pulping is composed mainly of carbohydrates and sugars (fructose, glucose, and galactose) along with proteins, polyphenols (chlorogenic acids, caffeic acid, tannins, and caffeine) and small quantities of natural dyes such as anthocyanins[10].

Tables 5.3 and 5.4 illustrate the physical, chemical, and biochemical properties of the wastewater generated in the processing of two species of the *Coffea* genus: *Coffea arabica* L., and *Coffea canephora* Pierre, Conilon cultivar. The elevated levels of biochemical oxygen demand (BOD) and chemical oxygen demand (COD) shown in Table 5.4 indicate that the pulping wastewater contains high levels of organic material, which, as explained previously, can cause problems if released into bodies of water before proper treatment. Wastewater contains elevated levels of total solids, composed chiefly of total volatile solids (TVS), which can mostly be removed through biological treatment.

Table 5.3 Results of physical and chemical analyses of wastewater samples from hydraulic separation of coffee fruit.

Fruit	Recirculation	Water/ Fruit Ratio	pH	EC dS m⁻¹	SP mL L⁻¹	TS	TSS	TDS	TFS	TVS	COD	BOD	NT	PT	KT	NaT
Conilon	No	-		0.259	17	1,069	380	689	390	679	1,520	411	77	5	41	26
Arabica	No	-	4.9	-	130	18,134	6,200	11,934	3,546	14,588	-	-	-	-	-	-
Arabica	Yes/with dilution*	0.15:1	5.5	0.344	50	3,255	867	2,388	984	2,271	5,604	514	55	12	49	16
Arabica	Yes/with dilution*	0.15:1	5.5	0.599	80	5,038	2,430	2,608	898	4,140	6,582	1,887	75	15	77	23

pH: Potential Hydrogen; EC: Electrical Conductivity; SP: Settleable Solids; TS: Total Solids; TSS: Total Suspended Solids; TDS: Total Dissolved Solids; TFS: Total Fixed Solids; TVS: Total Volatile Solids, COD: Chemical Oxygen Demand; BOD: Biochemical Oxygen Demand; N_T: Total Nitrogen; P_T: Total Phosphorus; K_T: Total Potassium; Na_T: Total Sodium

* Dilution was carried out by adding "clean" water whenever the condition of the water was not adequate for hydraulic separation of the fruit. | Source: Matos (2003)[11] and Rigueira (2005)[12].

Table 5.4 Results of physical and chemical analyses of wastewater from pulping and demucilaging of coffee fruit.

Fruit	Recirculation	Water/ Fruit Ratio	pH	EC dS m⁻¹	SP mL L⁻¹	TS	TSS	TDS	TFS	TVS	COD	BOD	NT	PT	KT	NaT
Canephora	No	3:1	4.75	0.585	0	4,889	850	4,039	126	4,763	5,148	2,525	106	9	115	45
Canephora	1	3:1	4.10	0.718	180	5,504	1,888	3,616	706	4,798	10,667	3,184	125	11	154	58
Canephora	2	1.8:1	4.10	0.992	330	6,403	2,336	4,067	848	5,555	11,000	3,374	160	14	205	77
Arabica		--	3.50-5.20	0.550-0.950	0-45	2,100-3,700	--	--	370-530	1,800-3,200	3,430-8,000	1,840-5,000	120-250	4-10	315-460	2-6
Arabica	1	--	--	--	--	14,000-18,200	--	--	--		18,600-29,500	10,500-14,340	400	16	1,140	17
Arabica	Yes/with dilution*	3:1	5.40	1.090	850	16,507	2,647	--	1,406	15,101	18,680	6,384	168	23	157	46
Arabica	Yes/with dilution*	1.8:1	5.30	0.800	900	14,827	2,780	--	1,210	13,617	18,066	5,006	163	22	157	58

* Dilution was carried out by adding "clean" water whenever the condition of the water was not adequate for pulping. | Source: Matos (2003) and Rigueira (2005).

The data demonstrates that when "clean" water is not added during the process, the chemical and physical properties of the wastewater are significantly altered by water recirculation, which is done to reduce water consumption. In light of the impact of recirculating water on the quality of green coffee, this practice is viable only if the water undergoes preliminary treatment, followed by primary treatment, before being pumped into the system for recirculation.

As an agricultural fertilizer, wastewater from the processing of both species of the genera *Coffea* contains relatively high concentrations of nitrogen and potassium (Table 5.4). Released without proper treatment, this wastewater can encourage the development of vegetation that is harmful to the aquatic ecosystem.

5.3 Treatment and Agricultural Applications for Solid Residues

Coffee fruit pulp has myriad uses—agricultural applications being the most recommended—because of its practicality and fertilization power. The pulp can be used in its pure form or after composting.

5.3.1 Agricultural applications for pure residues

Harvest byproducts such as hulls, straw, silver skin, pits, bagasse piths, cobs, etc. can be returned to the field to be distributed as a covering or, better yet, spread at the bottom of a furrow or hollow[13]. The hull in its pure form can be used as organic fertilizer for a wide array of crops, including coffee, since it can both condition and fertilize the soil (Figure 5.2). Using the hull of the coffee fruit is a good option to correct soils with potassium deficiency. As seen in Table 5.1, the coffee hull is rich in nutrients; one kilo can contain about 47 g of potassium (K), 2 g of phosphorus (P), and 19 g of total nitrogen (N). To take advantage of its fertilizing potential, particularly because of its potassium content, this residue should be returned to the coffee field.

Figure 5.2 Coffee hull piles deposited near coffee plants as fertilizer.

When used as mulch, hulls decrease runoff and the impact of rainfall on the soil, thus helping to prevent erosion. They also decrease thermal fluctuations and moisture loss through evaporation. It should be noted that the organic material produced by composting has the capacity to retain larger quantities of water in the soil than uncomposted coffee hulls, as the composted material has greater surface area and higher surface charges, giving it higher water adsorption capacity.

Special caution should be given to phytotoxins that are produced when coffee hulls decompose naturally in the soil (allelopathic effect), as these toxins can inhibit the growth of some plants.

The annual amount of dry hulls that should be applied to agricultural crops, taking into consideration their total available potassium, can be determined using equation 5.1[14]:

$$D_s = \frac{K_{rec}}{(C_k \times PR \times T_m)} \qquad\qquad (5.1)$$

In which: D_s = Dose of coffee hulls to be applied (t ha^{-1})

K_{rec} = Recommended dose for the crop (kg ha^{-1} year^{-1})

C_k = Concentration of K in the residue (g kg^{-1})

PR = Proportion of K recuperation by the crop (kg kg^{-1})

T_m = Annual rate of mineralization of coffee hulls in the soil (kg kg^{-1} year^{-1})

The proportion of potassium recuperation is variable depending on the crop; nevertheless, it should be between 60% and 90% (PR between 0.6 and 0.9). The mineralization of the coffee hull after one year in the soil should be higher than 80%, thus the T_m value should be between 0.8 and 1.0.

5.3.21 Composting

Composting is the process by which controlled biological aerobic decomposition of organic waste is obtained, transforming it into a partially humidified substance. For quick and successful composting, it is necessary to combine materials that have low carbon/nitrogen (C/N) ratios with materials that have high C/N ratios.

Coffee hulls, like animal manure, have a low C/N ratio and should be combined with agricultural waste products with high C/N ratios (dead leaves, grass clippings, straw, cobs, sawdust and wood chips, fruit fiber, hulls, etc.) to obtain a final product that is a good soil conditioner and fertilizer.

In Brazil, an organic compost was produced by using coffee hulls to filter swine production wastewater. After mesophilic decomposition, the compost was determined to be a very useful agricultural fertilizer. Besides the necessary C/N ratios and total nitrogen concentration, it met Brazilian legislative requirements, allowing it to be sold as a compost fertilizer[15].

Another compost was produced using a mixture of coffee hulls and parchment as a filtering substance for the treatment of 2,000 L of coffee processing wastewater[16]. The temperature variation in the material, measured throughout the composting period, is illustrated in Figure 5.3. The physical and chemical properties of the organic compound produced are presented in Table 5.5. The table also shows the minimum Brazilian legislative requirements for an organic compost to be used as an agricultural fertilizer[17]. The organic compost produced did not meet Brazilian legislative requirements for commercial sale as an organic fertilizer due to its C/N ratio. This indicates that the mesophilic decomposition period should have been longer.

Given the chemical analysis presented in Table 5.5, there is low risk of heavy metal contamination by the organic compost when it is used as a fertilizer for agricultural crops.

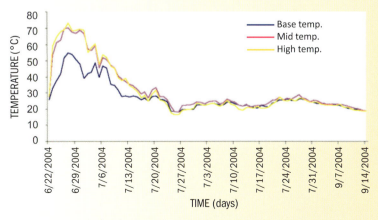

Figure 5.3 Variation in temperature of the compost material, taken in different locations of the mixture, throughout the composting period.

5.4 Treatment of Recycled Water in the System

The purpose of recycling water is to reduce overall water consumption in coffee fruit processing. However, the increased amount of suspended and liquid material in the recycled water can become a vehicle for fungi and contaminants, which compromise final product quality. To optimize product quality and conserve water, technology that can quickly remove suspended solids from recycled water should be implemented.

The process of coagulation/flocculation is one option for wastewater treatment. It is used for water in which the mucilage does not settle, necessitating the introduction of coagulants to bring about the flocculation/sedimentation of suspended solids. Coagulation is the process of neutralizing the electrostatic charge of particles, which allows them to agglomerate into larger particles that more readily enter the process of sedimentation. Various salts can be used for this process, among them aluminum sulfate, chlorinated ferrous sulfate, ferric chloride, and the less-known moringa (*Moringa oleifera*) seed extract, a vegetal species of the legume family.

Table 5.5 Physical and chemical characteristics of the produced organic compost and the minimum requirements for a matured organic compost to be used in agriculture, according to Brazilian legislation governing organic fertilizers.

Attribute		Organic Compost	Minimum Requirement
Density	g cm^{-3}	0.49	-
Moisture Level (drying 65 °C)		19.90	< 40%*
Organic Matter		56.40	>40**
C Total		32.70	22
Ca Total		0.86	>5*
Mg Total	dag kg^{-1}	0.10	>0.50*
K Total		0.55	>0.70*
P Total		0.37	>0.14*
N Total Kjeldahl		1.68	1.0
C/N Ratio		19.50	<18**
pH	-	6.68	>6.0**
Cu Total		29.60	<300*
Zn Total		78.30	<1000*
Mn Total		80.00	-
Fe Total	mg kg^{-1}	4,376.00	-
Cd Total		n.d.*	<5*
Ni Total		2.60	<50*
Pb Total		1.10	<500*
Cr Total		14.90	<150*

Source: * Gonçalves (1997)[17]; ** Brasil (1993)[18]

Aluminum salts are the most common chemical coagulants used in water treatment, and work well in pH levels between 5.5 and 8.0. Ferric salts are also widely used as coagulation agents in water treatment. They react to neutralize the electrostatic charge of colloids by forming ferric hydroxide. Due to the low solubility of ferric hydroxide, it can work over a wide range of pH levels. When coagulated, flocs form faster because of the higher molecular weight of this chemical compared to aluminum; the result is denser flocs and significantly reduced sedimentation time[19].

The extract produced by triturating moringa seeds is also an option as a coagulation agent in water purification. Many studies have shown its efficiency in the treatment of diverse types of wastewater, including the wastewater from the hydraulic separation and pulping of coffee fruit[20], from dairy production[21], textile production[22], and public water supplies[23]. Compared to chemical coagulants, moringa seed extract was a promising alternative in the physicochemical treatment of these wastewaters. When used as an auxiliary to primary treatment, it provided an increase in the efficiency of the decanters used in the removal of suspended solids.

In a study conducted to obtain the best combinations of coagulant and pH concentrations to maximize the removal of suspended solids in coffee processing wastewater[24], aluminum sulfate, chlorinated ferrous sulfate, and ferric chloride were added to the recycled water on five occasions during the coffee pulping process. These coagulants, in concentrations of 3 g L^{-1}, with a suspension pH of 7.0 to 8.0 for the first two coagulants and 4.5 to 5.0 for the last, provided the greatest removal of suspended solids from the wastewater. However, Moringa seed extract, which was also evaluated as coagulant, removed the most suspended solids from the wastewater in a pH range of 4.0 to 5.0 and a dose size of 10 mL L^{-1}, the equivalent of about 0.15 g of seeds per liter of wastewater.

Another study added the coagulants aluminum sulfate, ferrous sulfate, and moringa seed extract to remove suspended solids from wastewater. When maintained in sedimentation columns of 1 m in depth, the results illustrated in Figure 5.4 were obtained[25]. With the addition of aluminum sulfate, the time necessary to remove the total suspended solids, with a range of 60% to 90% efficiency and at a depth of 0.8 m, was more than 120 minutes. Aluminum sulfate and ferrous sulfate demonstrated efficiency in the clarification of the layer at 0–0.2 m and 0–0.6 m, respectively, in a time interval of 60 to 120 minutes; however, they were not effective in removing the suspended solids from lower layers. In a shorter period of time, the moringa seed extract was able to clarify more of the entire column of wastewater, providing the best result in terms of wastewater clarification among the available coagulants.

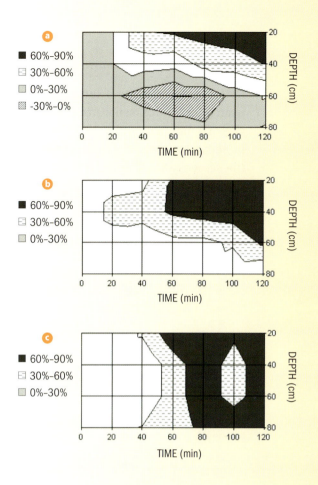

Figure 5.4 Isoefficiency curves in the removal of suspended solids from coffee processing wastewater in the sedimentation column, measured in terms of time and depth. **a** Aluminum sulfate, **b** ferrous sulfate, **c** moringa seed extract.

5.5 Wastewater Treatment and Agricultural Applications

Due to the serious environmental damage that wastewater can cause, it must be treated before being released into soil or watercourses. Wastewater treatment can be divided into preliminary, primary, and secondary levels of treatment. In preliminary treatment, coarser solids are removed by mesh screens. In primary treatment, solids subject to sedimentation are removed. Anaerobic degradation of the organic material in suspension may also occur in this phase. During secondary treatment, in cases where there has been previous treatment of wastewater in anaerobic systems, treatment of the organic material in suspension continues. Otherwise, biological, aerobic, or facultative degradation of the suspended organic material begins. In this stage, the removal of suspended organic matter in the wastewater occurs through the work of microorganisms that develop in liquid mediums or in soil-plant systems.

It is important to note that the water from the hydraulic separator, which is usually discarded once or twice per day, should be captured and sent for treatment separate from the wastewater generated by pulping the coffee fruit. This is due to the different properties of these waters, as seen in Table 5.4, which necessitate different treatment techniques and systems. After each of these types of water has undergone its respective primary treatment, the waters can be mixed together to be used in field fertigation or to undergo further biological treatment.

5.5.1 Preliminary Treatment

Both screens that are inserted into the water channel for treating the wastewater coming from the hydraulic separator and from the coffee pulper should have a maximum perforation size of 5 mm. This will remove leaves and twigs that might have passed through the hydraulic separator, as well as coffee fruit pulp coming from the pulper. The correct dimensions of the screens should be calculated based on the flow of the wastewater to be treated. Given normal operating conditions, it is recommended that the screen be inserted into a concrete channel 0.2 m wide by 0.4 m high. The screen should be installed at a 45° angle to facilitate periodic cleaning. If the use of manual labor to clean the screen is not desired, mechanical systems for pulp removal should be installed, the most common being "screw conveyors."

Screening allows for the removal of approximately 140 L of coffee pulp for every 1,000 L of gross effluent. In this stage, only small quantities of suspended organic material are removed, which means that the removal of BOD and COD is insignificant.

5.5.2 Primary Treatment

In this stage, various options are available, depending on the final destination of the water. If the effluent will be released into a body of water, a sedimentation tank should be constructed, as well as an anaerobic pond and a facultative pond, which will be discussed later. If the effluent is to be used in fertigation, or to be treated using a soil-plant system, it should be filtered using preliminary treatment, and if desired can be treated in an anaerobic pond or a facultative pond to avoid possible clogging of the irrigation system.

Sedimentation or decantation tanks serve to retain the solid material as the water flows through the hydraulic structure. The wastewater flows slowly and the suspended solids with a higher absolute density than the liquid being treated gradually settle and form sediment on the bottom of the tank.

The sedimentation tank for wastewater coming from the pulper should have adequate dimensions for the removal of finer organic material. Its dimensions, therefore, will be larger than those of the tank that receives the water from the hydraulic separator. These tanks can be rectangular or cylindrical and should be made of concrete. The width and length should be determined in relation to the flow of water that will be treated, and this can be estimated based on the total volume of coffee fruit processed each day.

A drainage system should be installed in the bottom of the sedimentation tank to remove the sludge that accumulates. This system can be composed of a tube and a valve that opens to remove the sediment. After drying in the sun, the sludge that is removed from the sedimentation tank can be returned to the fields along with other removed material, as long as these materials are composed primarily of vegetative matter and soil.

The effluent from the pulping wastewater sedimentation tanks should then be sent to an anaerobic pond or treated with organic filters if the effluent will be used in drip fertigation systems. Since the wastewater coming from the pulper is rich in suspended and dissolved solids, the use of conventional sand filters, which are used to treat water for human consumption and for drip application, is not recommended because of the rapid colmatation, or clogging, that can occur in the surface layers, reducing the flow rate of the wastewater. The use of agricultural or industrial byproducts, such as those that result from processing coffee using the wet method, is an interesting alternative to sand in the treatment of wastewater that is rich in organic material[26]. In fact, many solid residues generated in the production or milling of agricultural products can be reused on the farm, minimizing problems caused by disposing of them in the environment. When wastewater is treated through filtering, the material used to form the filtering layer can consist of agricultural waste products such as sugarcane bagasse pith, ground corn cobs, sawdust, or even coffee parchment. The filtering material should be removed from time to time since the pores in the upper layers of the filter will gradually become obstructed, diminishing the speed at which the liquid is filtered.

For greater efficiency in removing suspended solids while avoiding a rapid decrease in the filtering speed, filtering materials, such as sugarcane bagasse pith and sawdust, should be compressed[27]. The volume of sawdust should be reduced by 5% to 10% and sugarcane bagasse pith by 10% to 15%. Reductions of 10% to 15% in the volume of a filter made solely of ground coffee parchment were sufficient to obtain satisfactory removal of suspended solids from wastewater, while the volume reduction for filters made from non-ground coffee parchment should be more than 25%[28].

It is recommended that one of the lateral sides of the filtration tank be removable to facilitate the addition and removal of filtering material. The diameter of the filtering material granules should be 2.5 to 5.0 mm[29]. The bottom and top of the filtration tank should have layers, 20 cm thick, of material with larger granulation (between 5 and 10 mm) to facilitate water drainage, and in the lower layer a tube, 100 mm in length with perforations 5 mm in diameter, should be inserted to capture the filtered water (Figure 5.5).

The application rate of wastewater from coffee fruit pulping depends on the quantity of water used in the process and whether or not the water is recycled. For this reason, trials are necessary to determine how much water can be applied to the surface of the filter. Where trials are not feasible, non-recycled wastewater from the pulping process can be applied at a rate of 1.5 to 3.0 $m^3\, m^{-2}\, h^{-1}$.

This type of filter should be used for between 60 and 100 minutes. After this, the filter should either be given a period of rest (in cases were alternate filtration structures are available) so that its filtration capacity can be re-established, or the filter material should be replaced (in cases where there is only one filtration structure).

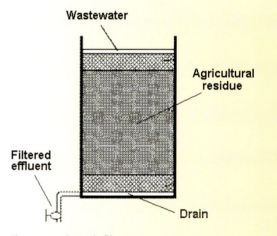

Figure 5.5 Organic filter.

A study compared sawdust and sugarcane bagasse pith filters for the treatment of wastewater coming from hydraulic separation and pulping[30]. The sawdust filter removed 60% to 70% of the total solids. When the filter was made from sugarcane bagasse pith, the removal rate was lower, reaching 40%. Both filters removed 100% of the solid sediment, 75% to 85% of the total nitrogen, and 50% of the total phosphorous. Furthermore, the sawdust filter removed more than 60% of the BOD from the wastewater. Another study of these two filter types obtained removal of suspended solids in swine wastewater in the range of 90% to 99% with sawdust filters, and 80% to 96% with sugarcane bagasse pith filters[31].

Two studies, one using sawdust and sugarcane bagasse pith and the other using coffee parchment, examined the effectiveness of these filter materials in the primary treatment of hydraulic separation and coffee pulping wastewater. None of the filter materials was effective in the removal of sodium and potassium; in fact, the concentration of potassium was higher after filtration than in the original effluent. Similar results were obtained when using sawdust as a filtering material in the primary treatment of coffee processing wastewater[32]. This low removal efficiency can be attributed to the fact that potassium is not a constituent of the organic material in the wastewater. Nonetheless, these filters are highly efficient because they remove suspended solids, permitting the localized application (dripping or micro aspersion) of effluent in the fertigation of agricultural crops with low risk of clogged emitters when the system is adequately operated[33].

After the filtering material is removed from the filter, to prevent it becoming another residue capable of causing environmental damage it should either be used as an organic surface fertilizer (without mixing it into the soil) or it should be composted.

5.5.3 Primary/Secondary Treatment

Some wastewater treatment systems are part primary and part secondary. In other words, they both physically remove pollutants and contribute to the anaerobic degradation of suspended organic material. The most important type of primary/secondary system is the anaerobic reactor, also call anaerobic filter and anaerobic pond.

5.5.3.1 Anaerobic Reactors

An anaerobic reactor is a treatment system in which the wastewater is filtered into a porous medium where colonies of microorganisms develop around a solid support. The mucilage formed by these colonies is called biofilm.

The layer of solid support material in an anaerobic filter retains the biological solids inside the reactor, either through adherence of the solids to the surface of the material, as with biofilm formation, or by entrapment of the solids in the interstitial spaces of the layer in an accumulated form, such as flakes or granules. The immobilization of the microorganisms comes from the solid or suspended support and is influenced by cell-to-cell interaction through the presence of polymers in the surface and the composition of the medium[34], as well as the characteristics of the support material, such as roughness, porosity, and pore size[35]. This layer also acts as a separation mechanism for gases and solids, which helps promote uniform flow throughout the reactor and better contact between the residue constituents (substrate) and the biomass contained in the reactor[36].

Supports that are capable of immobilizing the active biomass allow an increase in the average time the biomass can remain inside the reactor. Supports also facilitate methanogenic populations and increase the capacity of the reactor to resist load shocks (alterations in the quantity of organic material flowing in the system that change the characteristics of the substrate and the presence of compounds that help inhibit microorganism activity)[37]. Because of this, more stable systems can be created and controlled, and the systems can achieve high degradation efficiency of the organic suspended material, even when operated with reduced hydraulic retention time.

The following solid supports have been evaluated and used in different reactors for the anaerobic treatment of effluents: bamboo rings[38], coconut shells[39], blast furnace cinders[40], various types and sizes of rock[41], porous ceramic[42], polyurethane foam[43], nylon fibers[44], charcoal pieces, PVC[45], and glass spheres[46].

Studies were conducted to evaluate the operation and efficiency of fixed-bed anaerobic reactors (Figure 5.6) filled with different support mediums (blast furnace cinders, polyurethane foam, and crushed stone) for treating coffee processing wastewater. The study concluded that the reactor with the highest porosity (foam) performed better than the other reactors, both in the "start up" and in the equilibrium of the process. The medium's porosity had a significant impact on the performance of the reactors. The foam (porosity of 0.950 m³ m⁻³)

performed better than the crushed stone (porosity of 0.484 m³ m⁻³) and the blast furnace cinders (porosity of 0.533 m³ m⁻³). The foam was able to foster more microbial biomass adhesion to the support medium, presenting an average concentration of total volatile solids of 1,301 mg g⁻¹ of foam[47].

Figure 5.6 Fluidized-bed anaerobic reactors made of different solids support materials.

The study also verified that the reactor containing blast furnace cinders as the support material generated effluent with statistically lower concentrations of phenolic compounds than those obtained in the other reactors. Given their toxic nature, these compounds are resistant to biodegradation in the environment[48]. That said, some toxic and recalcitrant compounds can be completely mineralized or have their toxicity decreased in biological treatment systems that are adequately planned and properly operated, thus avoiding the formation of secondary pollutants[49]. The largest removal of phenolic compounds coming from coffee processing wastewater was in the reactor filled with steel blast furnace cinders. This could be due to the aluminum oxide (Al_2O_3) contained in this material, which presents a potential for adsorption of phenolic compounds[50].

Effluent from the hydraulic separator that has passed through the sediment tank can be mixed with effluent from a coffee pulper that has passed either through a sediment tank or an organic filter for treatment in anaerobic ponds, or these effluents can be treated separately. These materials are treated in ponds through biological methods that depend on the work of microorganisms that develop in a liquid medium.

Anaerobic ponds (Figure 5.7) can be used for the primary/secondary treatment of wastewater that comes from both the hydraulic separator and the pulper. They should be constructed on soil with low permeability, in isolated areas where phreatic waters (groundwater aquifers) are deep. The bottom can be compacted with a layer of argillaceous material, butyl rubber, or PVC plates to avoid the infiltration of the wastewater into the soil and subsequent risk of groundwater contamination.

Anaerobic ponds should be situated where no residential areas are located downwind since improperly constructed and/or managed ponds can release unpleasant odors that may travel up to 1.5 km. The area of a pond will depend on the flow and the BOD concentration of the wastewater, which should be detained in the pond for 3 to 6 days[54]; however, wastewater that is rich in suspended organic material should remain longer. The depth of an anaerobic pond should be between 3 and 5 m, a requirement for the development of

Figure 5.7 Anaerobic pond.

anaerobic conditions. Construction of an anaerobic pond should follow the same building criteria as the facultative pond or the secondary treatment pond.

It is important to emphasize that the degree of purification achieved with these ponds is insufficient for wastewater to be released into natural bodies of water. Therefore, effluent from anaerobic ponds must undergo treatment in facultative ponds or be placed in soil for treatment using overland or fertigation methods.

5.5.4 Secondary Treatment

The system of biological treatment in stabilizing ponds generally consists of an anaerobic pond followed by one or more facultative ponds (Figure 5.8). In mountainous regions, locating sufficient and appropriate areas for constructing these ponds can present quite a challenge.

The facultative pond receives effluent from the anaerobic pond. During the flow of wastewater into the facultative pond, which should take several days, part of the suspended organic material (particulate BOD) tends to settle, forming a layer of sludge on the bottom. This sludge undergoes a decomposition process by anaerobic microorganisms and is converted

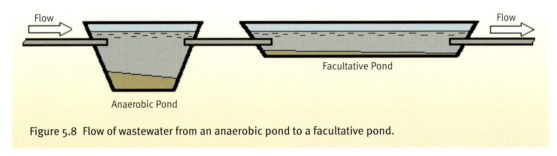

Figure 5.8 Flow of wastewater from an anaerobic pond to a facultative pond.

into carbonized gas, water, methane, and other compounds, leaving only an inert fraction of material that is resistant to biodegradation. The very small suspended matter (fine-particulate BOD) and the dissolved organic material (soluble BOD) are decomposed by facultative bacteria, that is, bacteria that have the capacity to survive with or without oxygen. This process further purifies the wastewater. The oxygen necessary for the respiration of aerobic bacteria is principally provided by photosynthesis, which is carried out by the algae that develop in the liquid medium.

Since sufficient isolation is necessary for the algae to fully develop, these ponds (Figure 5.9) need a large amount of surface area to function efficiently. The layer of sludge that forms on the bottom of the ponds should be removed every 5 to 6 years so that the accumulated material does not interfere with the efficiency of the purification process. After drying in the sun, the sludge that is removed from both the anaerobic and facultative ponds can be applied to crops as organic fertilizer.

To determine the correct dimensions for facultative ponds, it is necessary to know the rate of aerobic degradation of the organic material contained in the wastewater. Experiments conducted by the Department of Agricultural and Environmental Engineering at UFV, in Brazil,[51] indicate that the rate of degradation of the organic material in coffee processing wastewater is relatively low, probably due to the toxic effects of some of the chemical constituents in this type of wastewater on microorganisms. As a result, these treatment ponds need larger surface areas and longer treatment periods (usually more than 20 days) than ponds used for the treatment of the same organic load from swine wastewater or domestic wastewater. The value of the coefficient of decay for wastewater, at a temperature of 20 °C (K_{20}), from the pulping process is 0.128 d^{-1} [52]. The efficiency of residue removal in facultative ponds can be greater than 90% for suspended solids and 85% for BOD.

In situations where the available area is too limited for the construction of facultative ponds, aerated facultative ponds are an alternative. Aerated facultative ponds can be smaller in dimension, and require less detention time to pu-

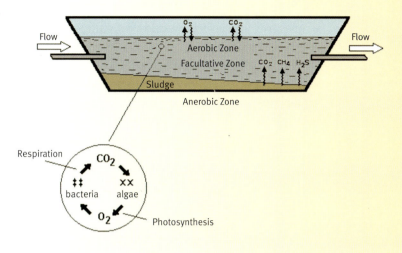

Figure 5.9 Degradation process of organic material in a facultative pond.

rify the wastewater. In this system, air is injected into the liquid medium to be treated, using compressors and gas injectors, or using surface aeration. Injection, or mechanical, aeration increases the concentration of oxygen and accelerates the process of stabilizing the suspended material.

The adoption of pond aeration should be based on thorough economic analysis since, depending on the investment required, this wastewater treatment process can be economically disadvantageous compared to traditional methods.

5.5.5 Secondary/tertiary treatment

5.5.5.1 Soil-based wastewater disposal methods

Another alternative for the treatment and final disposal of effluent from sedimentation tanks, organic filters, or anaerobic and facultative ponds is soil-based disposal. This technique has several advantages, including the use of the nutrients present in the effluent to fertilize crops, low cost of installation and operation, and low energy consumption. It is estimated that the cost of this treatment method is 30% to 50% of the cost of conventional systems.

Soil-based methods of wastewater disposal can be divided into infiltration/percolation (rapid infiltration process), overland flow (runoff), and fertigation (slow rate process), as well as release and treatment in wetland systems.

5.5.5.1.1 Infiltration/Percolation

With the infiltration-percolation method, also called rapid infiltration, the objective is to use the soil as a filter for wastewater. This method entails the percolation of wastewater through a porous filtering medium before it makes its way into phreatic water. Wastewater is placed in flat basins (trays) or infiltration ditches (Figure 5.10), where the soil is highly permeable, allowing for infiltration and percolation. Systems that use this method can be designed so that the percolated liquid 1) reaches groundwater supplies, 2) is recuperated by surface drainage, or 3) is pumped into aquifers.

Advantages of the infiltration/percolation method:

- Requires less area than other wastewater disposal methods.
- Uses very little energy.
- Requires less declivity in the contour of the land.
- Has lower start up and operating costs (i.e. construction, operation, and maintenance).
- Does not generate sludge.
- Facilitates the application of wastewater year-round (in cases where winter is not intense) and the recharging of groundwater aquifers.

Figure 5.10 Disposal of wastewater into the soil using the infiltration/percolation method.

Disadvantages:

- Can produce bad odors.
- Can create insect and pest problems.
- Depends on soil characteristics.
- Can contaminate the groundwater aquifer with nitrates or other constituents with greater soil mobility. In the case of wastewater from coffee processing, these include potassium, sodium, and phenols, which can cause physical and chemical changes in the soil.

For these reasons, it is recommended that the disposal of coffee processing wastewater produced by infiltration/percolation be carried out using precise technical criteria and that the soil and groundwater quality be monitored in the area where wastewater disposal takes place.

5.5.5.1.2 Overland flow

In the overland flow method (Figure 5.11) wastewater is applied to the soil on slopes planted with undergrowth, generally grasses. The wastewater is applied in rates that are higher than the capacity of the water to infiltrate into the soil. Some of the wastewater that runs along the ground is evapotranspirated and a small part infiltrates into the soil, leaving the rest to be collected in troughs. The soil-plant system, together with the microorganisms that develop in this medium, constitute a natural filter, allowing for the degradation of part of the organic material and the chemical and physical retention of a large part of the dissolved inorganic constituents.

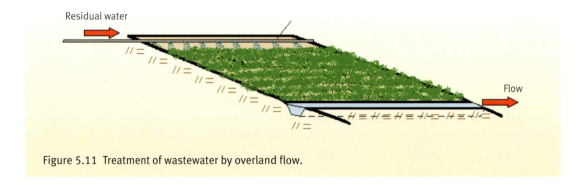

Residual water

Flow

Figure 5.11 Treatment of wastewater by overland flow.

The significant advantage of the overland flow method is that it allows for the treatment of large volumes of wastewater in small areas, while taking advantage of the fertilizing capacity of the wastewater. The efficiency of this method in removing pollutants is approximately 70%–85% for BOD and COD, 60%–80% for phosphorous, and 50%–90% for total nitrogen.

A. Area selection

Treatment slopes should be located in areas of low soil permeability. In mountainous coffee regions, this corresponds to the lower third of the hillside or plateau. The slopes should be 30–60 m in length, with varying width depending on the amount of wastewater to be treated. For proper wastewater runoff, the surface of the soil should have 2% to 5% declivity.

Studies were performed to test the effect of declivity on COD and BOD removal. A consistent influent rate of coffee processing wastewater was applied to runoff slopes with declivities between 5% and 15% that were cultivated with black oats[53] and ryegrass[54]. The removal rates were generally higher for treatment slopes where the declivity was 5%. Only on slopes cultivated with ryegrass were the results similar for all declivities for removing BOD from the wastewater.

The overland flow treatment system has the same advantages as the infiltration-percolation system, in addition to offering lower risk for groundwater contamination and producing pasture grass for use as animal feed or as a "green" fertilizer. This system's major disadvantage is its dependence on natural characteristics of the soil, such as permeability and declivity.

B. Selection of plant species to be cultivated

The selection of plants to be cultivated on the treatment slopes is fundamental to the success of this treatment method. Ideal species are perennials that are tolerant to conditions of low oxygen and elevated salinity at the root level, possess an elevated capacity for nutrient extraction, are resistant to pests and diseases, and allow for successive cuttings.

In Brazil the production of coffee processing wastewater coincides with the period of lowest annual temperatures, and therefore winter perennial species are recommended. Research conducted by the Department of Agricultural and Environmental Engineering of UFV (Figure 5.12) indicates that black oats and ryegrass are suitable winter species.

The runoff slope should be continuously cultivated to facilitate constant removal of soil nutrients to create the conditions for ideal chemical equilibrium. After the harvest, when wastewater production stops, plants with higher summer growth rates can be replaced with winter plants. Grasses of the *Cynodon* genus (such as tifton 85 and coastcross) can be interplanted amongst the winter species.

C. Application rate

When coffee processing wastewater is disposed of on planted slopes, it is recommended to apply an organic load of no more than 250 kg ha⁻¹ d⁻¹ of BOD, as larger applications can stunt the growth

Figure 5.12 Experimental slopes cultivated with black oats for overland flow treatment. The plant cover can be used in its fresh form or transformed into hay or silage for use as animal feed or green fertilizer.

of the plants[55]. Longer periods of rest are recommended for the area to allow for oxygenation to take place, which is critical for the oxidation of the organic material applied. Periodic trimming should be planned for the plants on the slopes, generally monthly, even during periods when the plants are no longer receiving wastewater. This promotes the removal of excess salts brought about by the application of wastewater to the area.

Wastewater can be applied to the soil using irrigation sprinklers of low to high pressure or using gated pipes. It can be applied intermittently or continuously (4 to 24 h d^{-1}), with a frequency of 5–7 days per week.

D. Dimensions of the treatment slopes

Method from the University of California

$$\frac{C_S}{C_0} = a.e^{(-k \times z)} \tag{5.2}$$

in which: C_s = Concentration of BOD at point z (mg L^{-1})
C_0 = Initial concentration of BOD (mg L^{-1})
a = constant
k = constant
z = distance traversed on slope (m)
q = rate of application (m^3 h^{-1} m^{-1})

Per Table 5.6, this equation has been adjusted for coffee processing wastewater at different declivities for different grasses cultivated[57].

Table 5.6 Equations of decay for BOD and COD in wastewater from coffee fruit processing, treated on slopes of different declivities and cultivated with black oats and ryegrass.

Variable	Declivity	Black Oats	Ryegrass
BOD	5%	$C/C_0 = 1.0388.e^{-0.0954.z}$	$C/C_0 = 0.9431.e^{-0.0363.z}$
	15%	$C/C_0 = 1.0369.e^{-0.0825.z}$	$C/C_0 = 0.9229.e^{-0.0383.z}$
COD	5%	$C/C_0 = 1.0351.e^{-0.0726.z}$	$C/C_0 = 0.9662.e^{-0.0099.z}$
	15%	$C/C_0 = 0.9422.e^{-0.0651.z}$	$C/C_0 = 0.96.e^{-0.0089.z}$

Source: Matos et al. (2001)[57]

Observations:

- Under these conditions, since the daily application time is very short, it is not necessary to give a rest period to the wastewater treatment area.
- When the soil is very dry it is possible to have no drainage of effluent into the collection trough positioned at the bottom of the slope due to retention of water in the soil and evapotranspiration of the plants.
- Regularly cutting the plants every 30–45 days is fundamental to the system effectively removing the nutrients in the wastewater.

5.5.5.1.3 Fertigation

Fertigation is a technique of wastewater deposition via irrigation systems, whose application rate is based on the nutrients present in the wastewater, providing supplementation for chemical fertilization in cultivated agricultural areas. Nutrients such as nitrogen, potassium, and, principally, phosphorus are fundamental for crop cultivation in nutrient-poor soils.

Fertigation with wastewater (Figure 5.13), if properly conducted, can decrease environmental pollution as well as improve the chemical, physical, and biochemical properties of the soil, resulting in increased productivity and enhanced quality of harvested products.

Figure 5.13 Fertigation with wastewater.

A. Selection of plant species to be cultivated

It is recommended that wastewater fertigation be applied to plant species that grow year-round. Some forage grasses with abundant and deep root systems can be very useful from an environmental perspective, as they can remove a large quantity of macro- and micronutrients from the soil, decreasing the risk of contamination of rivers, lakes, and groundwater. Wastewater from coffee fruit pulping can also be applied to staple crops such as coffee, corn, sorghum, wheat, fruits, and vegetables, as well as reforested areas.

B. Application rate

The application rate of wastewater used for fertigation should be sufficient to supply adequate nutrients and should never be superceded, at the risk of provoking salination or other forms of contamination of the soil, surface water, or groundwater. To avoid this problem, wastewater should never be applied in quantities equal to the water required by the plants for their moisture needs.

Potassium, the macronutrient found in the highest concentration in coffee processing wastewater, should be used as a reference for calculating the amount of wastewater that can be safely applied to coffee fields without compromising either the environment or coffee productivity. Special attention should be given to the disposal of these wastewaters into the soil, since high concentrations of potassium in relation to calcium and magnesium can cause clay dispersion. This dispersion, in turn, causes loss of soil structure and therefore porosity, thereby decreasing permeability[58]. Besides this, the disequilibrium of nutrients can impair crop development.

Increasing doses of wastewater from coffee fruit pulping (66.4, 99.6, 132.8, 166.0, and 199.2 grams per plant of potassium) were applied to a coffee field of four-year-old *Coffea arabica* L., cv. Catuaí, with 0.8 m between plants and 2.2 m between rows[59]. Significant increases in electrical conductivity were observed in relation to soil depth and the doses of coffee wastewater applied (Figure 5.14). However, when comparing the results obtained for electrical conductivity of the soil in this study with that of another author, it must be considered that this study used distilled water in the ratio of 1:2.5 (soil to distilled water), while other studies used saturated soil extract, which usually has a ratio of 1:1 (soil to water), producing values about four to five times higher than those obtained in this study.

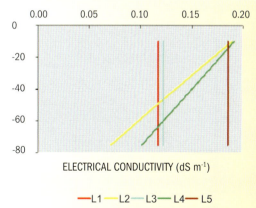

ELECTRICAL CONDUCTIVITY (dS m⁻¹)

—L1 —L2 —L3 —L4 —L5

Figure 5.14 Electrical conductivity in the soil, in relation to soil depth, before and after increasing doses of coffee wastewater: L1, L2, L3, L4, and L5 (66.4, 99.6, 132.8, 166.0, and 199.2 grams per plant of potassium).

Figure 5.15 The effect of applying large doses of coffee processing wastewater, equivalent to the application of 199.2 g per plant of potassium, on the coffee plant.

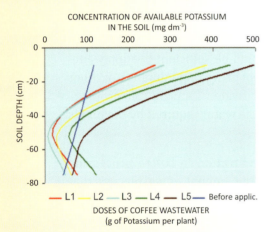

Figure 5.16 Available concentration of potassium in the soil, in relation to soil depth, before and after increasing doses of coffee wastewater: L1, L2, L3, L4, and L5 (66.4, 99.6, 132.8, 166.0, and 199.2 grams per plant of potassium).

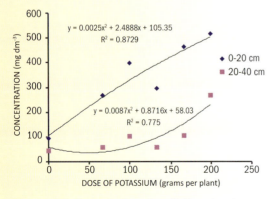

Figure 5.17 Concentration of available potassium in layers at 0–20 cm and 20–40 cm of soil depth, in relation to increasing doses of coffee wastewater, which supplies potassium to the plants.

The increase in electrical conductivity in the soil is primarily attributed to the large quantity of potassium added to the soil with the application of coffee wastewater. The application of the highest dose level (L5) resulted in the addition of sufficient ions to cause significant leaching, predominantly cationic (K+, Na+, Ca 2+, and Mg 2+), in the soil profile.

In applying coffee processing wastewater that included 199.2 grams per plant of potassium to a coffee crop, extreme dryness could be observed in the coffee plants, as shown in Figure 5.15[60]. This can be attributed to the large quantity of ions present in the wastewater. The presence of these ions leads to a decrease in the osmotic potential of the soil, which then compromises the absorption of water by the coffee plant. In other words, the plant started to expend more energy to absorb water and nutrients.

Sodium and potassium salts are considered to be the greatest contributors to soil salinity, and in coffee processing wastewater, potassium is the most evident component according to the results obtained in various research projects[61].

With the application of increasing doses of coffee wastewater onto the coffee plant, intense potassium leaching was observed in the deepest soil layers (60 cm), most notably in treatments where the plants received the highest dose of coffee wastewater (Figure 5.16)[62].

Figure 5.17 illustrates the curves of available potassium in the layers of soil at depths of 0–20 cm and 20–40 cm, in relation to the dose of potassium applied via the coffee wastewater.

In Brazil, foliar analyses of macro- and micronutrient concentrations were conducted on coffee plants that had received limestone, conventional nitrogenated fertilizer (urea), phosphate (simple superphosphate), and fertilizer with Zn, Cu, and B, through foliar coffee wastewater applications over a period of two months (the second two weeks of May through the first two weeks of July), corresponding to the production of coffee wastewater. The tables at right show inverse relationships with the applied doses of coffee wastewater in terms of the foliar concentrations of potassium (Figure 5.18), calcium (Figure 5.19), and magnesium (Figure 5.20). The concentration of potassium increased with the dosage until dose L3, after which it dropped. Inversely, the concentration of calcium and magnesium decreased with the dosage until L3 and increased thereafter. The decrease in potassium concentration leads to an increase in the concentration of calcium and magnesium in the coffee leaf.

Besides providing nutrients, the application of coffee wastewater leads to greater absorption of some macro- and micronutrients by the coffee plants, while some macronutrients leach into the soil.

Since an excess of available potassium in the soil can lead to a lack of calcium and, principally, magnesium in the leaves of the coffee tree, it is recommended that these macronutrients be added in areas where coffee wastewater is applied.

Considering that only the highest dosage applied was close to the amount of water necessary to irrigate a coffee field, and because of the chemical effects the larger dosages have on the soil, it is recommended that the dosage of coffee wastewater not be applied at levels established for irrigation.

The productivity results of coffee plants subjected to the previously cited treatments, based on the potassium contained in the wastewater (Figure 5.21), indicate an increase in productivity with the smallest application dosage, followed by a decrease with higher dosages[63].

Studies also verified the tendency of coffee plants to decrease in productivity with an increase in the level of wastewater applied to the plants, compared to plants that received inorganic fertilization[64].

Observations:

- Although wastewater contains some nutrients, it is recommended to supplement cultivated soil with chemical fertilizers that are specific to the crop and that do not include sources of potassium.
- Soil chemistry should be monitored over time in order to correct discrepancies in fertility.

C. How to Apply

Fertigation with wastewater can be done with furrows, sprinklers, drippers, or sludge sprinklers

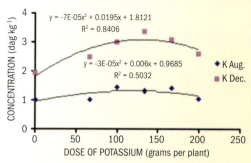

Figure 5.18 Concentration of potassium in coffee plant leaves with application of increasing dosages of coffee wastewater, in relation to the supply of potassium to the plants.

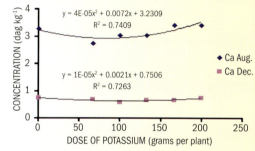

Figure 5.19 Concentration of calcium in coffee plant leaves with application of increasing dosages of coffee wastewater, in relation to the supply of potassium to the plants.

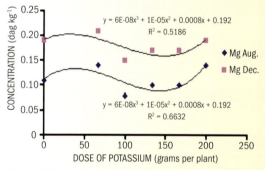

Figure 5.20 Concentration of magnesium in coffee plant leaves with application of increasing dosages of coffee wastewater, in relation to the supply of potassium to the plants.

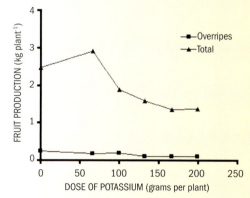

Figure 5.21 Productivity of coffee plants fertigated with different dosages of coffee wastewater, in relation to the supply of potassium to the plants.

Figure 5.22 View of a drip application system in a coffee plantation.

(portable fertilization tanks). The application method should be selected according to the type of crop, its susceptibility to disease, and the water infiltration capacity of the soil.

When applied with sprinklers, sugars contained in the wastewater, mostly those coming from the demucilaging process, can create ideal conditions for pests and diseases that attack the leaves of the coffee plant.

The effects of fertigation of coffee plants with sprinklers were evaluated for a period of 80 days. The fertigation was performed using wastewater from both hydraulic separation and pulping of coffee fruit, with a dose equivalent to the application of 166 kg ha^{-1} of potassium; a portion of the fertigated leaves was then washed, and another portion of the leaves was left unwashed[65]. Table 5.7 shows the results of the chemical analysis of the coffee leaves before and after being subjected to fertigation with coffee wastewater.

After applying coffee wastewater to the canopy of the plants, there was an increase in the concentration of macro- and micronutrients in the leaves, both when the application was followed by washing the leaves, and when it was not. The plants with unwashed leaves had higher concentrations of P, K, Fe, Zn, Cu, and Mn; this is most likely due to the longer period of time the plant tissues were exposed to the coffee wastewater.

Table 5.7 Results of foliar analysis before and after application of wastewater from coffee fruit pulping.

Treatments		Concentration of Nutrients in the Leaves							
		P	K	Ca	Mg	Fe	Zn	Cu	Mn
		dag kg^{-1}				Mg kg^{-1}			
Coffee wastewater + "clean" water	Before	0.113	1.52	2.05	0.23	141.00	5.10	13.85	131.50
	After	0.180	1.99	3.51	0.36	314.00	8.20	23.80	281.50
Coffee wastewater	Before	0.140	1.80	1.85	0.23	114.60	6.25	14.95	159.75
	After	0.240	2.77	2.95	0.29	435.15	10.10	26.80	346.50

Source: Moreira et al. (2004).[66]

After the period of fertigation, a visual analysis was conducted of pest attacks and fungus growth on the leaves. There was no negative effect from the treatments on the sanitary conditions of the plants, indicating that the fertigation of the plants with coffee wastewater using sprinklers did not cause any phytosanitary problems for the coffee plants, at least in the short term. Yet to minimize these risks, after applying wastewater from the demucilaging process it is recommended that the leaves of the crop be washed with irrigation water (from either surface or underground sources) for at least 20 minutes after finishing the wastewater application.

To avoid problems caused by the sprinkler application method, wastewater should be applied using a localized method such as drippers (Figure 5.22) or micro-sprinklers, which are effective at minimizing the risks of both pest attacks and environmental damage. However, such localized systems necessitate preliminary and primary treatment processes. Suspended solids must be removed prior to fertigation to prevent clogging of tubes. In drip systems, the concentration of suspended solids should be less than 50 mg L^{-1} to minimize the risks of clogging[67]. The use of organic filters has been recommended to reduce the concentration of suspended solids in the wastewater from swine production and from the hydraulic separation and pulping of coffee fruit[68].

After applying wastewater from coffee fruit pulping using a drip irrigation system, a biofilm formed inside and outside the drip, which caused a partial or total clogging of the emitters[69]. Figure 5.23 illustrates biofilm accumulation inside and outside the emitters, which were partially (a, b) and completely (c, d) clogged.

Note in Figure 5.23 that the structural cause of the loss of capacity was the deposition of organic solids, contributing to the development of biofilm. The formation of biofilm was the result of the interaction of the suspended and dissolved solids with the bacteria formed by mucilage, primarily mesophyllic aerobic bacteria from the untreated coffee processing wastewater, and enterobacteria from filtered coffee processing wastewater that has been passed through an organic filter made of coffee parchment.

With the application of coffee wastewater, there was a 100% reduction in the coefficient of distribution uniformity (CDU) after 36 hours, while with the filtered coffee wastewater this reduction occurred after 144 hours. An organic filter slowed the clogging of the drippers; however, it did not prevent the development of biological film on the tubing and on the emitters.

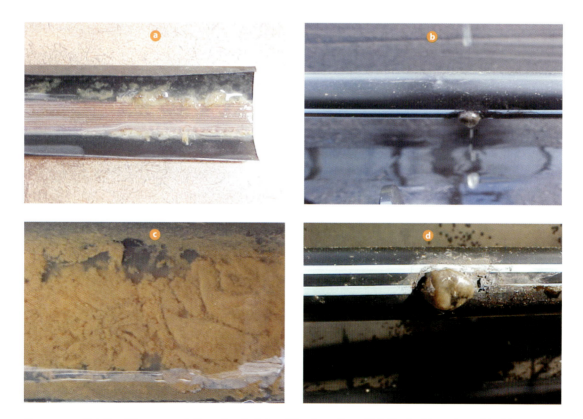

Figure 5.23 Details of the accumulation of material and the formation of biofilm inside and outside drippers after the application of wastewater from coffee fruit pulping. The photos show partially clogged (a, b) and completely clogged drippers (c, d).

5.5.5.2 Treatment in constructed wetland systems

The treatment of wastewater in constructed wetland systems has been practiced in Europe since the 1960s, with good results. The mechanisms involved in the treatment include filtration, microbial degradation of organic material, absorption of nutrients, and adsorption into the soil, among others.

Vegetation growing in these areas plays an important role in treating wastewater by extracting a large part of the macro- and micronutrients available, thus preventing their accumulation and resalinization of the medium, and preventing contamination of surface and ground waters. The vegetation in wetland areas enables the development of biologically active films, which, in turn, cause the degradation of organic compounds in solution and suspension in the wastewater.

Natural wetland systems with developed vegetation can be employed in the treatment of wastewater; however, the quality of the substrate and the water should be constantly monitored to maintain the efficiency of the system and to minimize the risk of groundwater contamination.

Monitoring the quality of surface and ground waters for a period of six months, it was verified that natural wetland systems with cattail have a high capacity for reducing BOD, COD, N, P, and K in coffee processing wastewater with 276 kg ha^{-1} d^{-1} of BOD[70]. During the monitoring period, the data obtained showed average removal rates of 90% of BOD, 87% of COD, 70% of nitrate, 75% of ammonium, 86% of phosphorus, and 89% of potassium. It is important to emphasize that releasing effluent into these wetland systems, particularly when the coffee pulper machine was at peak production, degraded the quality of the receiving watercourse and aquifers downstream from the wetlands, and that the dilution provided by the insurgent phreatic aquifer mitigated the environmental conditions for the plants. For this reason, wetland systems should be constructed as closed systems with impermeable tanks made of concrete or canvas to minimize the risk of ground water contamination and avoid environmental damage.

There are also legal restrictions to the use of natural wetland systems as a means of treating wastewater. From a regulatory standpoint, wetlands are usually considered receiving bodies of water and, consequently, wastewater should be dispersed into these ecosystems according to the standards set by local environmental legislation. For this reason, wetland systems to treat wastewater should be constructed of tanks for cultivating aquatic vegetation.

Constructed wetland systems are canals that have an impermeable coating, are filled with a support material (generally gravel), and contain aquatic macrophyte vegetation where effluents are dispersed. The flow can be either surface or subsurface. Organic material is removed by physical (filtration and sedimentation), chemical (oxidation and adsorption), and biological (biodegradation and phytoaccumulation) means. Generally, the wastewater bodies are made by excavating earth, covering the excavation with an impermeable material and filling it with gravel. Constructed wetland ponds can be standalone or connected either in series or in parallel. They can contain a monoculture of only one macrophtye, or an array of species[71].

Water treatment results from the integration of physical, chemical, and biological interactions in the wastewater bodies that occur because of the presence of the support medium, bacterial communities, and macrophtyes. Among the mechanisms for removing organic material, bacteria are the key players since they can decompose organic material by anaerobic, anoxic, and aerobic processes.

Since the waters of constructed wetlands are predominantly anaerobic, aerobic conditions are achieved by supplying oxygen via the roots of macrophytes[72]. Oxygen is moved internally through the macrophyte to the lower parts of the plant via aerenchyma. Macrophytes are able to supply an oxygen surplus, producing more than they consume through tissue respiration and rhizosphere oxygenation. The oxygen transported to the rhizosphere contributes to oxidative processes in the medium. Oxidation conditions, together with the anoxic conditions that are present, stimulate the aerobic decomposition of organic material, the growth of nitrifying bacteria, and the inactivation of compounds that would be toxic to the plant roots[73]. In addition to supplying oxygen to the system, macrophytes absorb macronutrients (nitrogen and phosphorus) and micronutrients (including metals).

A. Types of Constructed Wetland Systems

The two basic types of constructed wetlands systems (CWS) for treating wastewater are surface flow and subsurface flow. Subsurface flow can be subdivided into horizontal and vertical flow. The most common CWS is subsurface flow, in which the wastewater flows by gravity horizontally through the support medium in the bed, entering into contact with the facultative microorganisms that live in association with the support medium and with the roots of the emerging macrophytes[74].

B. Vegetation Cultivated in Wetland Systems

As with all plant-soil wastewater treatment systems, the choice of vegetation species is of fundamental importance to the success of the wetland system.

The most appropriate species for use in this type of treatment are emergent plants (Figure 5.24), which are characterized by having roots in a saturated substrate while maintaining a portion above water for the purpose of photosynthesis, thus producing a large quantity of biomass. These characteristics give emergent plants a high capacity to purify wastewater, making them ideal for use in wetland systems that treat wastewater.

In general, the emergent plant species most used in wetland areas are *Phragmites* (reeds), *Scirpus* (grassweeds or bulrush), and *Typha* (cattail).

Studies were conducted by the Department of Agricultural Engineering at UVF using CWSs to treat coffee processing wastewater to determine which plant species were better adapted to this water or to define recommendations for alterations to the chemical

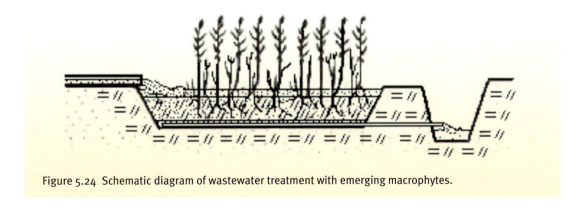

Figure 5.24 Schematic diagram of wastewater treatment with emerging macrophytes.

composition of the wastewater that would better allow for its unfiltered treatment. One study examined the macrophyte species *Typha* in terms of its efficiency in treating coffee processing wastewater with an organic load of 850 kg ha^{-1} d^{-1}, corresponding to a daily depth of 20 mm. The study concluded that after three days of application the initial part of the system (the first 3 m) presented signs of foliar "burning" and plant senescence (Figure 5.25). Possible factors attributed to this were the high absolute concentration of potassium; a high concentration of potassium in relation to other nutrients; high levels of salts in the coffee wastewater and in the medium, shown by the average electrical conductivity of the coffee wastewater equal to 1.34 dS m^{-1}; the low pH values of the wastewater and, consequently, reduced alkalinity of the medium; the presence of phenolic compounds; the heavy organic load applied; and the high evapotranspirometric rates in the system[75].

Figure 5.25 Effects of applying unfiltered wastewater from the hydraulic separation and pulping of coffee fruit to cattail cultivated in a constructed wetland system.

In this study, two conditions contributed to an elevated concentration of salts in the medium: the volume of applied coffee wastewater was high when compared to what would be applied for irrigating agricultural fields, and the rate of outflow from the wetland system was insufficient due to the high evapotranspirometric rates of the cattail. To evaluate the resistance of the cattail in relation to salinity, the organic/inorganic load applied was varied by diluting the coffee wastewater with a non-wastewater source so that the average electrical conductivity measured 0.3 dS m^{-1}, corresponding to an average application rate of an organic load of 400 kg ha^{-1} d^{-1}. In the system, the hydraulic retention time was 10 days, while the cattail was cultivated for a period of 35 days. During the diluted coffee wastewater application phase, the cattail presented signs of senescence and all dead plants were replaced. After 35 days of coffee wastewater application, the remaining vegetation species initially planted also died and the replanted plants did not grow. During the cultivation period and wastewater application period (35 days), four samples of the wastewater being treated were taken at points throughout the CWS (at distances of 0, 3, 6, 9, 12, and 15 m from the CWS entry). The concentrations of BOD and COD in the coffee wastewater showed a tendency to decrease along the treatment system (figures 5.26 and 5.27); however, the average efficiency of the system was relatively low, at 36% for BOD and 57% for COD. On average, the removal of total solids, total nitrogen, phosphorus, and potassium were 54%, 60%, 82%, and 66% respectively.

From the results obtained, the following conclusions were drawn:

- The CWS planted with cattail was moderately effective in removing the organic material in coffee wastewater.
- Rates of untreated coffee wastewater application equal to or greater than 400 kg ha^{-1} d^{-1} of BOD are inadequate for treatment in CWSs cultivated with cattail.
- Removal of nutrients was limited by death of the cattail.
- The effect of inhibitory substances on the development of the cattail plants indicated the need to remove these substances through primary treatment before release into the CWS.

Steps to increase the survival time of vegetation in CWSs treating coffee wastewater include: adding bivalent elements that counteract the effect of potassium, adding nitrogen and phosphorus-based nutrients, correcting the pH and pre-treating the coffee wastewater to remove phenolic compounds, reducing the organic load applied, and increasing the hydraulic retention time in the system.

Another study, completed by the Department of Agricultural Engineering at UFV, observed CWSs cultivated with *Typha* species (cattail) and *Alternanthera philoxeroides* (alligator weed) submitted to different experimental conditions and the application of different organic loads in the treatment of coffee processing wastewater." The study also looked at the agricultural performance and nutrient extraction of the plants in different operational conditions of the system. The experiment comprised nine CWSs with horizontal subsurface flow wherein the first 0.75 m was planted with alligator weed, and the last 0.75 m was planted in cattail. Each unit measured 0.4 m high, 0.5 m wide, and 1.5 m long, with gravel as a support medium (Figure 5.28).

Based on the results obtained, it was concluded that, in general, an increase in the surface load of the coffee processing wastewater in the system led to a decrease in the efficiency of the CWS in the removal of pollutants. Hydraulic retention times over 100 hours led to increased efficiency in the removal of COD; however, even this did not enable the effluent to achieve levels that would allow for its dispersion into waterways per the environmental legislation of the state of Minas Gerais, Brazil, where the studies were conducted.

The efficiency of the removal of organic matter and phenolic compounds from the coffee processing wastewater increased when calcium oxide was added to correct the pH to approximately 7.0, and when N and P were added so that the ratio of BOD/N/P was 100/5/1. In these conditions, the removals of BOD, COD, and phenolic compounds were 63%, 85%, and 65% respectively, and the alternanthera was capable of extracting approximately 4.6%, 28.8%, and 9.1% of the N, P, and K respectively, applied via the coffee processing wastewater.

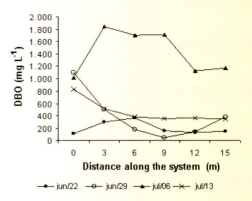

Figure 5.26 DOB concentrations in the coffee wastewater, along the CWS.

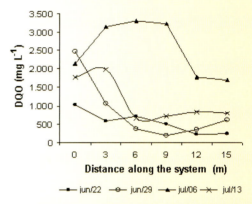

Figure 5.27 COD concentrations in the coffee wastewater, along the CWS.

Figure 5.28 Constructed wetland systems with horizontal subsurface flow after the planting of alternanthera and cattail (a), and a more detailed look at the alternanthera growing in the support medium (b).

6

Coffee Drying

Flávio Meira Borém

Ednilton Tavares de Andrade

Eder Pedroza Isquierdo

6.1 Introduction

Coffee is generally harvested when moisture content of the fruit is relatively high—between 30% and 65% (wb), depending on the state of maturation—and it is therefore subject to rapid deterioration. Because of this, harvested coffee must be dried before storage. While other food preservation methods are available (cryopreservation, chilling, and controlled atmosphere, among others), drying remains the most common method used for coffee. Of the various post-harvest stages (processing, drying, storage, milling, and transport), drying is the most important stage in terms of energy consumption, overall post-harvest costs, and quality preservation. Although much information is available regarding costs and energy consumption, little is known about the metabolic changes that occur during drying for the various processing methods, and how these changes affect final product quality.

6.2 Principles of Drying

Drying is necessary to maintain product quality during storage[1], and it can be defined as the simultaneous process of energy and mass transfer between the product and the drying air. In most cases, this consists of the removal of excess water in the product by evaporation through the forced convection of heated air. Various factors can influence coffee as it dries: drying method, temperature and relative humidity of the drying air, air flow, and drying time. Lack of control over these factors can compromise final product quality.

During drying, the reduction in moisture content is due to the movement of water from the interior to the periphery of the fruit. This movement is caused by water vapor pressure differences between the product surface and the surrounding air. To undergo drying, the partial pressure of the water vapor on the surface of a product (Pvg) must be greater than the partial pressure of the water vapor in the drying air (Pv), (Figure 6.1).

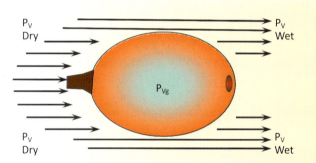

Figure 6.1 Schematic of water and air movement during drying.

The dimensioning, optimization, and determination of commercial viability of drying systems can be determined through mathematical simulation. The simulation, which is based on the successive drying of thin layers of product, uses a mathematical model that represents the loss of water during the drying period. For further reading on mathematical drying models, consult specialized literature listed in the bibliography[2].

6.2.1 Theoretical Basis

The water migration that occurs in the interior of coffee is not fully understood. Various thermal and physical mechanisms have been proposed to describe the transport of water within products that are porous, capillary, and hygroscopic. Some researchers affirm that the movement of water inside drying grain products is possibly caused either by liquid diffusion or vapor diffusion, or a combination of both mechanisms, with one or the other being predominant during different stages of drying (Figure 6.2). However, more recent theories suggest that in certain drying stages,

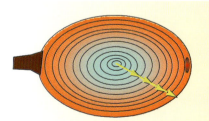

Figure 6.2 Isomoistures of coffee from interior to periphery during drying.

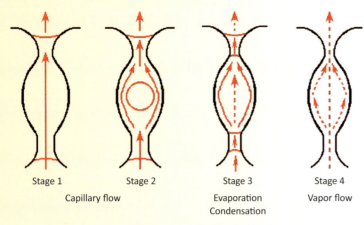

Stage 1	Stage 2	Stage 3	Stage 4
Capillary flow		Evaporation Condensation	Vapor flow

Figure 6.3 Water movement during the drying of a porous capillary material (Silva, 2000).

water movement is predominantly determined by liquid diffusion.

The drying process, under constant temperature, relative humidity, and air velocity, can be divided into two periods. The first period is defined by a constant rate of drying, the second by a falling rate. During the constant rate drying period, the product temperature remains the same as that of the saturated drying air, and the exchange of heat and mass offset each other. The energy lost by the air, in the form of sensible heat, is compensated by the latent heat of vaporization. The internal movement of water to the product surface does not affect the drying rate since it is equal to or greater than the maximum rate of evaporation on the product surface. In this stage, only free water evaporates. In other words, the constant rate drying period of an agricultural product is defined as the period in which the moisture content of the product is sufficient to maintain water upon its surface while constantly exposed to the same ambient conditions.

The free water removed in this stage is adsorbed water that is weakly retained and absorbed water retained by capillarity. There is a decrease in the diameter of the pores and capillaries, and a consequent decrease in the volume of the product approximately equal to the evaporated water. This period can be observed in the first hours of drying unripe and ripe coffee with high initial moisture levels. For most agricultural grains, including coffee, the constant rate drying period is very short or even nonexistent. This is because in most cases there is internal resistance to moisture transport through the seed, and this resistance means that the evaporation rate at the surface of the seed is higher than the flow rate of water to the surface.

During the falling rate drying period, the internal transport rate of water is less than the rate of evaporation. Therefore, the transfer of heat from the air to the product is not compensated by a transfer of water vapor from the product to the air. As a result the temperature of the product increases, rising to meet the drying air temperature. There are four distinct stages of grain drying within the two main drying periods. The first stage is equivalent to the period of constant rate drying, and the last three stages are part of the falling rate drying period, with each stage delineated by the facility of movement of wa-

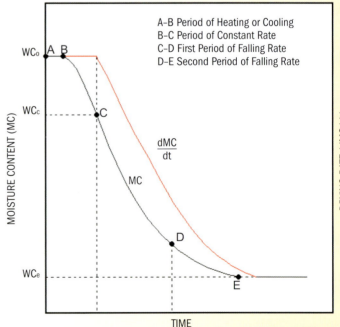

A–B Period of Heating or Cooling
B–C Period of Constant Rate
C–D First Period of Falling Rate
D–E Second Period of Falling Rate

Figure 6.4 Curve of the variation in moisture content (MC) and the drying rate (dMC/dt) with time

ter through the porous capillaries of the product, as illustrated in Figure 6.3. At the point where the product passes from the constant rate drying period to the falling rate drying period, the moisture content of the product is termed "critical moisture content" (Figure 6.4).

Since most agricultural products are harvested with levels below critical moisture content, most theories and mathematical drying models have been developed to better predict variations in the moisture content and temperatures of grains and agricultural products for the falling rate drying period.

6.3 Drying Methods

6.3.1 Natural Drying

Natural drying occurs when coffee fruit are still on the plant. (Figure 6.5). The fruit are heated through solar radiation while air movement occurs by the natural action of wind. Although adequate moisture content for storage can be reached through these conditions, drying time cannot be controlled as it depends on climatic variations. Natural drying also brings a higher risk of fungus attacks, which decrease product quality. Conditions for natural drying vary from region to region, are difficult to accurately predict, and are impossible to control. On top of this, the fruit frequently fall to the ground, increasing the risk of fermentation and the formation of defects that decrease the value of the final product. Although not the ideal method for coffee drying, natural drying may occur when the harvest is late and at the end of the harvest. Also, this may be the only economically viable means of drying coffee in less-developed regions.

Figure 6.5 Natural drying of coffee.

Figure 6.6 Coffee drying on patios (courtesy of Grupo Sete Cachoeiras, Três Pontas, Minas Gerais Brazil).

6.3.2 Patio Drying

In patio drying (Figure 6.6), the wet product is exposed to the sun on flat surfaces and rotated either manually, using animal labor such as horses or mules, or mechanically.

The principle advantage of patio drying is low energy cost, as solar radiation is used to heat the coffee and remove water. Patio drying is the traditional method of coffee drying. In favorable environmental conditions and with proper care, it can result in a quality product. It is also the most environmentally conservative method as it does not require the burning of fuels. Drying coffee in thin layers through exposure to the sun leads to a grayish green coloration that is considered to be favorable, while mechanical drying in a dark atmosphere results in a yellow-green color that is considered undesirable.[3]

Large areas, increased labor, and longer drying times are required when patios are the sole means of coffee drying. These conditions can potentially expose the coffee to climatic variations that can, in turn, elevate the risk of contamination and fermentation, reducing final product quality. These climatic conditions vary from region to region as well as within the same region, depending on the time of year; for the most part they are unpredictable. The risks of deterioration increase when the harvest is not properly planned and when adequate infrastructure is not available. One common sign of a poorly planned harvest and/or lack of adequate infrastructure or technology is the accumulation of too much coffee on the patio As the quantity of coffee on the patio grows, the drying process becomes more difficult to control given the wide ranges of moisture content and heterogenous lots in the drying coffee. These conditions can lead to the development of the sour bean defect and an unfavorable bean appearance.

The average time required to complete coffee drying on the patio varies with the characteristics of the coffee and the climate, but generally falls between 15 and 20 days for dry process coffees, though drying times can reach as much as 30 days in wet regions. For parchment coffee, drying times are usually between 8 and 12 days. Drying times are influenced not only by processing method and environmental conditions, but also by patio type (earth, brick, concrete, asphalt, suspended bed) as well as by the care taken when drying, especially in regard to layer thickness and the number of times the coffee is rotated throughout the day. Studies examining the influence of layer thickness on coffee quality for both dry process and wet process coffees have shown significant quality loss when layer thickness was 8 cm for dry process coffees and 4 cm for parchment coffees.[4]

Producers should work with homogeneous lots, taking into consideration the point in the season when the coffee is harvested as well as the maturation state and moisture content of the coffee being dried. Doing so will lead to more uniformity and increase quality in the final product.

Patios can be constructed of various materials. The choice of surface type should take into consideration coffee quality preservation, implementation costs, ease of use, and total drying time. Features of a good patio include good water absorption, low heat retention, and high durability, resulting in shorter drying times and a higher quality product. In addition to patios, coffee can be dried on suspended or raised beds, in greenhouses, and using hybrid patios.

Certain patio types have disadvantages. Because of their tendency to absorb energy, asphalt surfaces can lead to high coffee mass temperatures. Surfaces that are highly impermeable, such as ground covered with canvas, risk water accumulation. Suspended beds generally require longer drying times[5]. The following section describes the basic characteristics of patio types commonly used in coffee production.

Figure 6.7 Dirt patio.

6.3.2.1 Dirt Patios

Despite the fact that they are still commonly found on many small coffee farms and in less-developed regions, dirt patios (Figure 6.7) are not recommended for coffee drying. Their common use is due to low implementation cost and simple setup, consisting basically of clearing an area and leveling the earth to create a flat surface. Besides frequently rendering low quality coffee, dirt patios do not meet basic hygienic standards for good production processes.

In the 1950s and 1960s, some authors went as far as recommending dirt patios with firmly compacted surfaces, a 2% surface inclination, and lined with manure when cement or brick were not available[6]. Clearly this recommendation is no longer accepted for hygienic coffee production or for achieving a high quality product. Studies with Arabica and Canephora coffees dried on different surface types showed that coffees dried on non-paved surfaces resulted in poor appearance and poor sensorial qualities for both dry process and wet process coffees[7].

6.3.2.2 Concrete Patios

Concrete is the material of choice for patio drying. Compared to dirt patios, concrete patios yield faster drying times and higher quality coffee[8]. However, few concrete patios are properly constructed to recommended specifications. Concrete patios constructed in slabs (Figure 6.8) are subject to cracking, concrete erosion, weed growth, and coffee accumulation that can lead to fermentation and compromise the quality of various lots during drying[9]. To avoid these problems and optimize concrete patio design (Figure 6.9), it is advised to follow the recommendations described in Chapter 12, Post-Harvest Coffee Facilities.

6.3.2.3 Thin Layer Asphalt Patios

The high costs of both concrete and asphalt have made them less financially viable for coffee drying. Fortunately, lower-cost alternatives, such as thin layer asphalt (also called slurry seal asphalt), have been used with good results[10]. Thin layer asphalt patios (Figure 6.10), sometimes referred to as "asphalt mud" patios, are different from asphalt patios in that only a very thin layer of asphalt is used, normally around 5 mm (Figure 6.11), which reduces the overheating risks of standard asphalt. Thin layer asphalt patios can be thought of as an impermeable layer over a dirt patio.

There are many questions as to how thin layer asphalt patios affect coffee quality. Recent studies, using both natural and parchment coffees dried in thin layers and rotated frequently, show that no smell or off-flavors were transferred to the coffee from the asphalt.

Figure 6.8 Concrete patio constructed in slabs.

Figure 6.9 High resistance concrete patio.

Figure 6.10 Thin layer asphalt patio.

Figure 6.11 Close-up of a thin layer asphalt patio showing thickness and compacted sub-base.

6.3.2.4 Suspended Beds

Suspended beds, originally called "portable aerial patios," were invented by Geronymo L.C. Souza in 1888.[11] They can be made by joining various wire mesh rectangular boxes to form a 3.0 m x 1.5 m surface, then mounting these boxes onto a wooden frame 0.8 m high to form the bed subsurface. This subsurface can then be covered with shade fabric, bamboo mat, or other covering material (Figure 6.12).

Figure 6.12 Suspended bed.

Figure 6.13 Suspended bed with plastic covering (courtesy of Fazenda Ponto Alegre, Cabo Verde, Minas Gerais, Brazil).

Figure 6.14 Greenhouse for coffee drying (courtesy of Fazenda Ponto Alegre, Cabo Verde, Minas, Gerais Brazil).

Some authors credit suspended beds with producing higher coffee quality for both fully washed and pulped natural coffees compared to traditional patio drying[12]. However, since coffee quality is dependent on so many factors, such as the origin of the raw material (cultivar and production location), layer thickness when drying, and care taken in processing, generalizations about the best type of drying surface for maximizing quality should be avoided.

Studies of drying times in suspended beds have been contradictory since they are dependent on the unique characteristics of the local environment. However, when compared to concrete and thin layer asphalt in the same drying conditions, suspended beds require a longer drying period[13].

Suspended beds should be placed in well ventilated, sunny locations, avoiding places with high humidity. In these locations the principal advantage of suspended beds is that they can produce coffee that is free of defects, in contrast with lots dried on flatlands that are lower and much more humid. The main disadvantage of suspended beds is that the loading and unloading of the coffee onto the beds is more difficult and less ergonomic than with traditional patios. Plastic coverings can be used to cover the product during rain and also to cover coffee at night, after half-dry, so that dew does not form on the coffee (Figure 6.13).

6.3.2.5 Greenhouses

Greenhouses (Figure 6.14) are an alternative drying method in regions with high relative humidity, especially at night and in the early morning when dew can form and rehumidify the dried coffee.

Studies have shown that the air temperature inside a greenhouse remains above the external temperature while decreasing relative humidity. Greenhouses decrease drying times, thus increasing processing capacity, and also minimize the undesirable effects of rain and dew[14]. Special care should be taken when using a greenhouse given the risk of excessively high temperatures inside. Some preventive steps include monitoring the thermodynamic characteristics of the air and using exhaust systems to reduce temperatures that rise above those recommended for coffee drying.

6.3.2.6 Hybrid Patio Dryer

The hybrid patio dryer (Figure 6.15) is a system that allows for complete coffee drying on the patio, even in regions with heavy fog and precipitation during the drying period. This system combines traditional patio drying, when drying conditions are favorable, with air heated by a furnace when moisture levels are higher, such as during a fog or at night. The patio is modified by putting air ducts on the patio floor, thus allowing hot air to be distributed to the coffee mass that is layered on top of the ducts. This drying system is relatively inexpensive to implement considering that it requires smaller patios and shorter drying times[15]. In less-than-ideal drying conditions, it can also result in higher coffee quality compared to traditional patios.

6.3.2.7 Patio Management for Dry Process (Natural) Coffees

To maintain quality, it is imperative that coffee pass through the hydraulic separator and be spread out to dry on the day it is harvested.

At the beginning of the patio drying process, the patio surface may become completely wet due to excess water on the coffee coming from the hydraulic separator. To avoid this, the water should be drained when the coffee is still in the cart or wagon, before it is spread over the patio. In addition to free surface water, the moisture content of the coffee fruit is also high. Although this water is easily removed when the coffee is exposed to the sun, its presence facilitates the rapid development of fungi and fermentation, especially in the lower coffee layer that is in direct contact with the patio.

In the first few days of drying, both ripe and over-ripe natural coffees can be easily pulped unintentionally, with minimal pressure, through the action of walking, raking, or machine traffic on the coffee. This can result in non-uniform drying since the unintentional parchment coffee will dry more quickly than the intact natural coffee. Because of this the coffee should be spread out in a single-fruit-high layer of around 14 L m^{-2} during the first few days of drying, exposing all coffee directly to the sun (Figure 6.16). While the coffee is still wet and easily pulped, rotating should be avoided. This will guarantee a higher quality lot with a more uniform appearance. It will also permit a drastic reduction in coffee moisture content while avoiding fermentation and reducing the presence of parchment coffee mixed in with natural coffee pods. However, this process should only be performed using thin layers at the beginning of the drying process in the presence of abundant sunlight, low relative humidity, and good, natural ventilation.

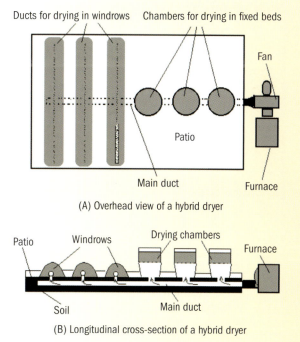

Figure 6.15 Overhead view and longitudinal cross-section of a hybrid dryer, with options for drying using a fixed-bed or in windrows.

Figure 6.16 Beginning of the drying of natural coffees in thin layers.

Figure 6.17 Manually rotating the coffee on the patio.

Figure 6.18 Rotating natural coffee with a tractor.

After the coffee fruit wither, generally after two days of sun exposure but varying with drying conditions, the coffee should be continually rotated either manually or with the help of proper equipment. More frequent rotation will result in more uniform drying, as well as a higher quality appearance of the final product. Natural coffees should be rotated at least 12 times a day.

Manual rotation is done with wood or metal rakes and workers should always walk in line with their shadow (Figure 6.17). The use of tractors (Figure 6.18) or animals to rotate drying coffee has advantages and disadvantages. While labor is decreased, the weight of the tractor or animal can pulp and/or hull the coffee, exposing the endosperm and potentially compromising the quality of the final product. With tractors this problem can be mitigated by using deflectors in front of the wheels.

The mixing of lots is quite common, with producers at times mixing lots with up to five-day differences in harvest time. This mixing leads to coffees with wide variations in moisture content, resulting in lots with disuniform color and, in the worst cases, the development of sour defects. It can also lead to the coffee being stored with high moisture content. For this reason, daily harvest volumes should be planned so that the coffee can be processed in homogeneous lots. If harvest lots from different days must be mixed, it should be done with great care. The first priority should be maintaining homogeneous moisture content and the second mixing similar quality fruit.

After the first days of drying, when the coffee is partially dry, the thickness of the layer can gradually be increased up to a maximum of 5 cm for ripe coffee and 10 cm for floaters (Figure 6.19).

When moisture content is below 30%, the coffee has reached half-dry. At this stage, at around 3:00 p.m. every day, the coffee should either be mounded or put into thick rows that run in line with the patio's declivity, then covered with rags or canvas. This will conserve and distribute the heat absorbed during the day, ensuring better uniformity and a redistribution of moisture throughout the coffee mass. The next morning the coffee should be uncovered and spread out over the patio. This process should be repeated every evening until the coffee reaches a moisture content of 11%, the ideal level for coffee storage.

Coffee that has reached half-dry can also complete the drying process in mechanical dryers. To maximize the quality of natural coffees, producers should keep the coffee on the patio until half-dry. Depending on the climatic characteristics of the region, this can occur within three days (in dry regions) or five days (in more humid regions). This period of patio drying, besides permitting slower drying, is also more economical. The drying process can then be finished in mechanical dryers using heated air. The drying process used on the patio for coffee that will go to the mechanical dryers is no different from coffee that will finish drying through one of the patio methods described here. A big advantage of combined patio and mechanical drying is that it can reduce the patio area needed by up to 50%.

Figure 6.19 View of patio with natural coffee with appropriate layer thickness for the end of the drying process.

Another way to complete the drying process on patios is using "volcanoes" (Figure 6.20). The volcano system is used to finalize the drying process in lots with moisture content of less than 20%. In this phase the coffee is already being mounded and covered at the end of each afternoon. However, on subsequent mornings, instead of spreading the coffee evenly across the patio, the coffee is kept in mounds. After the coffee on the surface of a mound is warmed by the sun, it is scraped off and put around the mound in a circle 30 cm away. This is repeated every 30 minutes or so, depending on the intensity of the sunlight, until the entire

Figure 6.20 Drying coffee using "volcanos."

mound has been taken down, usually around midday. The process is then reversed in the afternoon and the coffee is mounded again, taking the top heated layer from the circular row every 30 minutes. The process of re-mounding the coffee should be finished by around 3:00 p.m.; the mound should then be covered to retain heat. This operation is repeated until the coffee reaches 11% moisture content.

While techniques for drying ripe and overripe coffee fruit have steadily evolved over the years, processes for drying unripe coffee fruit have changed drastically in recent years.

6.3.2.8 Patio Management for Unripe Coffee Fruit

The traditional recommendation for drying unripe coffee fruit was to arrange the coffee in rows 20 cm high, rotate the coffee periodically until the fruit skin was completely darkened, then spread the coffee out following dry process drying norms. Thick rows were recommended as this was thought to homogenize the unripe fruit and prevent overheating from solar exposure, since high temperatures are the main cause of the black-green defect.

However, rows of fruit 20 cm high with high moisture content levels can lead to fermentation. This, in turn, not only leads to higher temperatures that cause the black-green defect, but also to the deterioration of the coffee fruit. In other words, this technique often caused what it set out to prevent.

New processes for drying unripe coffee fruit avoid these problems and yield a better product than the traditional method. The free water on the coffee fruit surface is easily

Figure 6.21 Drying unripe coffee in thin layers at the onset of the drying process.

Figure 6.22 Removing the canvas from coffee that was covered for the evening.

evaporated, which inhibits the coffee from reaching elevated temperatures, since the transfer of heat and mass can then offset each other. Thus it is now recommended to first spread the coffee in thin layers (Figure 6.21) and constantly rotate the coffee until it reaches half-dry to prevent the blackening of the skin. From this point on, rows 15 to 20 cm high should be formed to slow the drying rate and impede overheating, preventing the formation of the black-green defect. The rows are periodically moved and turned over to ensure that drying is slow and uniform. These procedures have yielded unripe coffees with better appearance and quality than traditional methods.

6.3.2.9 Parchment Coffee Patio Management

Parchment coffee that has been wet processed should be spread out in a one-bean-high layer of roughly 7 L m^{-2}, permitting rapid removal of free water from the parchment surface as well as dehydration of any remaining mucilage. The coffee should not be covered during its first night of drying. Starting on the second night it should be covered with a canvas to avoid exposure to evening fog and dew. The coffee should be maintained in thin layers and covered at night until half-dry is reached. For parchment coffee, half-dry occurs at 25% moisture content. At this point the coffee should be put into piles at night and covered. Figure 6.22 shows the canvas being removed from coffee that was covered for the evening.

During the day, the coffee should be frequently rotated—16 or more times—and maintained in thin layers. Tools used to rotate the coffee should be flat and light to avoid cracking the parchment and inflicting damage to the endosperm. In places where the parchment is cracked, evaporation rates can be higher and this may also damage the endosperm. Damage to the endosperm can accelerate the bean whitening that may occur in the warehouse if the coffee is stored with high moisture content levels.

As drying progresses, the endosperm will contract and no longer adhere to the inside of the parchment that envelops it. When this occurs, layer thickness should increase progressively every day during daytime drying. In most regions, under optimal conditions, layer thickness should be doubled from its initial bean-high thickness on the second full day of drying, with the next increase again doubling the thickness, typically on the third day. If drying is to be completed on the patio, a third doubling should be done on the fifth full day of drying. The layer thickness achieved on the fifth day should be maintained until drying is complete.

After half-dry is reached, the coffee should be piled into mounds or windrows and covered at around 3:00 p.m., continuing the process used with natural coffees described above, until reaching 11% moisture content (Figures 6.23 and 6.24). As with natural coffees, the final drying of parchment coffees can be done using "volcanoes."

As with natural coffee, combined drying (patio/mechanical) can be used with the coffee remaining on the patio until half-dry is reached. However, unlike natural coffee, during mechanical drying the rates of water removal from parchment coffee are much higher when high temperatures are used as well as when coffee moisture content is above 40%. Studies have shown that parchment coffee should remain on the patio at least three days under adequate conditions. This reduces the initial moisture content of the coffee going to the dryers, thus minimizing the negative effects that high temperatures and drying rates may incur. The highest frequency of "specialty" quality coffees (defined per SCAA standards as coffee that has no defects, a distinctive character in the cup, and scores 80 or higher on the SCAA cupping form) was obtained with higher pre-drying times[16].

Figure 6.23 Forming windrows of parchment coffee after half-dry.

However, there are two key situations where reaching half-dry on the patio is not possible. In cases where patio space is not sufficient, producers often move the coffee to the dryers after only one or two days, before half-dry is reached. Also, when environmental conditions inhibit coffee drying (e.g., rain, heavy fog, etc.), the coffee should go directly from wet process milling to the dryers. In both of these situations, producers should pay even more attention when drying to preserve coffee quality.

Figure 6.24 Canvas covering parchment coffee in the late afternoon.

TABLE 6.1 Frequency (%) of SCAA cupping scores for parchment coffee subjected to different periods of pre-drying.

Rep.	Drying Period (days)									
	1					3				
	>86	>80 <85.9	>78 <79.9	>70 <77.9	Phenolic Beverage	>86	>80 <85.9	>78 <79.9	>70 <77.9	Phenolic Beverage
I	11	11	5	73	-	45	22	5	28	-
II	0	22	5	68	5	5	55	0	40	0
III	0	11	0	84	5	0	50	17	33	0

Table adapted to SCAA scoring from the Official Brazilian Classification (Classificaçao Oficial Brasileira).

6.3.3 Drying at High Temperatures

High temperature drying is necessary in regions where coffee cannot be completely dried on the patio and for large properties with high harvest volumes. On these large properties, although climatic conditions may be favorable for complete patio drying, the large area necessary to completely dry all of the crop on the patio is a limiting factor. The combination of patio and mechanical drying is best in these cases.

"High-temperature drying," defined as using forced air at temperatures at least 10 °C above ambient temperatures, involves high air temperatures and high air flows, resulting in shorter drying times. The air flow used depends of the type of dryer and can vary from 10 to 100 m^3 min^{-1} ton^{-1}. High temperature drying is used in the horizontal, vertical, and fixed-bed dryers commonly used in coffee processing.

Technicians and producers often state that drying coffee in mechanical dryers at high temperatures results in astringent flavors or disuniform and mottled beans. Studies have confirmed that mechanical drying does alter the color of coffee beans, making lots disuniform in color as well as leaving individual beans mottled[17]. Overdrying coffee to levels lower than 11% can fade the beans, and drying at temperatures above 80 °C can overheat the beans and turn them grey. This observation reinforces the fact that coffee quality can be greatly compromised during the drying stage of coffee processing when drying is not properly conducted.

Maintaining coffee quality during high temperature drying requires controlling both product temperatures and controlling the drying rate. As the coffee mass loses water, its temperature will increase and reach the drying air temperature. This, combined with the fact that often the coffee mass being dried is heterogeneous in terms of maturation, makes coffee drying one of the most complex operations in agricultural product processing. Fruit at different moisture content and maturation stages have different heating capacities for a given quantity of energy supplied. As such, for each maturation stage the fruit will obtain different temperatures after a set period of time. This is why it is imperative to separate and, if possible, pulp the fruit to enable working with homogeneous lots.

There are various mechanicals dryers currently on the market that will enable a producer to obtain a quality final product. However, whether due to lack of capital or lack of equipment that meets small volume processing needs, patio drying still predominates.

The mechanical dryers most commonly used for coffee are quite large, with a capacity of 5,000 to 15,000 liters. The most common types of dryers found in Brazil are rotary and vertical dryers. On small properties, fixed-bed dryers are common, as they are practical for drying small quantities of coffee.

Although concurrent flow dryers are highly efficient and can produce a quality final product, they are still not well known by producers and there are few companies making and selling them. Concurrent flow dryers use less energy per kilogram of evaporated water than traditional dryers when drying coffee[18].

6.25 Rotary dryer equipped with a fan, furnace, and loading bin (courtesy of Pinhalense S.A. Máquinas Agrícolas).

6.3.3.1 Rotary Dryers

Rotary dryers, or horizontal dryers, (Figure 6.25) are also commonly called pre-dryers since they can receive coffee that still has high moisture content. A rotary dryer consists of a horizontal, tubular cylinder that rotates on a central axis at a velocity of 2.5 to 3.0 rpm; air flows radially and continuously moves the coffee around the inside of the cylinder. When properly operated, rotary dryers can result in uniform drying; they are easy to load and unload, and enable lower drying times. However, they require special attention to temperature and air distribution to guarantee a higher quality coffee.

Rotary dryers can be fitted with hoppers or resting chambers where the coffee can rest. These resting periods promote the homogenization of the coffee, economize energy, and increase the machine's drying capacity.

Loading can be performed manually or mechanically using hoppers or silos that are usually located above the dryers. Discharge doors are designed to minimize operation time. Rotary dryers can operate using a direct or indirect furnace, or with LPG burners.

Heated air is moved by a fan through the plenum, which also functions as a duct to evenly distribute the air through to the end of the dryer (Figure 6.26). The hot air passes through perforations in the duct and moves radially around the cylinder of the dryer. The rotation of the dryer causes continuous movement of the coffee. The air carrying the moisture removed from the coffee exits through perforations in the external walls of the drying cylinder.

The double function of the plenum can result in disuniform distribution of air along the dryer. Studies have shown the existence of temperature and moisture gradients in rotary dryers when drying natural coffee. These gradients, especially the temperature gradient, were associated with a reduction in coffee quality during drying since coffee mass temperatures as high as 70 °C were recorded when the temperature of the drying air was 90 °C[19]. In more recent rotary dryer models this problem has been minimized by the installation of air deflectors at different points along the duct.

Rotary dryers can be used in three ways:

1. Pre-dryer: When used as a pre-dryer, the rotary dryer receives coffee with high moisture content directly from the field or from the hydraulic separator. It should be operated initially without heated air for at least 30 minutes, then heat should be added until the interior coffee mass reaches a temperature of 30 °C. This temperature should be maintained until the coffee reaches half-dry. Once half-dry is reached, the coffee can either be maintained in the rotary dryer for continuous drying, or sent to vertical dryers to complete the drying process. This type of drying can compromise final coffee quality and should be used only in cases of extreme necessity and when an experienced technician can supervise the drying process.

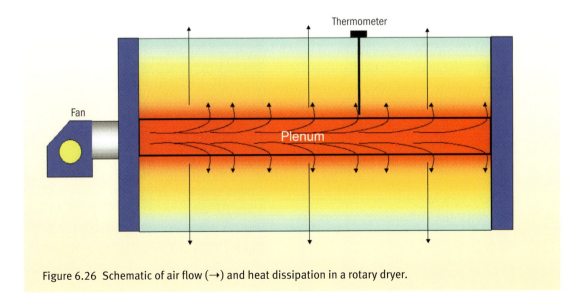

Figure 6.26 Schematic of air flow (→) and heat dissipation in a rotary dryer.

2. Continuous dryer: After pre-drying, the rotary dryer can also be used to complete the drying process. Instead of transferring the coffee to a vertical dryer as described above, it can simply be left inside the rotary dryer. After half-dry, the coffee mass temperature can be increased up to 45 °C for natural coffees and 40 °C for parchment coffees. For specialty coffees, the maximum temperature for both natural and parchment coffee is 40 °C.

   Using the rotary dryer as a continuous dryer is very inefficient. When coffee reaches half-dry, its volume contracts, leaving a large empty space at the top of the dryer. This gap results in heat loss and reduced drying efficiency.

3. Complementary dryer to pre-drying on the patio: This is typically referred to as combined drying. The coffee is taken to the dryer after reaching half-dry on the patio. The dryer should be loaded with lots of homogeneous moisture content, maturation state, and product quality. When loading, 10 to 15 cm of space should be left at the top to allow the coffee to move around the dryer freely. Initially, the dryer should operate without heat for a period of one to three hours, allowing for the homogenization of the coffee and partial removal of the weakly bound water. Next, the air should be heated, initially to a maximum of 60 °C for commodity grade natural coffees and 50 °C for specialty grade natural coffees, all parchment coffees, and unripe coffees. Air temperature should then be increased to a maximum of 90 °C for commercial grade natural coffees and 60 °C for all other coffees. However, the temperature of the coffee mass should not rise above 45 °C for commodity grade natural coffees, 40 °C for both specialty grade natural coffees and all parchment coffees, and 35 °C for unripe coffees.

Whenever possible, the supply of energy to the coffee should be interrupted for the evening, allowing for the redistribution of water in the interior of the product. This leads to more uniformity, higher quality and lower electrical and combustible energy consumption. The next day, the process should be restarted. In the first hours the dryer should operate with natural air, followed by heated air. Drying with resting intervals should be conducted until moisture content is between 12.5% and 13.0%. At this point the coffee can be discharged and, while still hot, put into resting bins, permitting it to attain 11% moisture content.

6.3.3.2 Vertical Dryers

In vertical dryers (Figure 6.27) coffee is moved by gravity in vertical columns constructed of perforated metallic plates. Normally these dryers are used to complete the drying process, either when rotary dryers have been used as pre-dryers or with coffee from the patio. However, since these dryers need elevators to constantly move the coffee, they are not adequate for drying natural coffees with high moisture content as the elevators can easily pulp the ripe fruit. Vertical dryers are also called cross-flow dryers since the coffee mass moves perpendicular to the air flow (Figure 6.28).

The coffee inside a vertical dryer falls at a constant velocity, forming a column. Since the cof-

Figure 6.27 Vertical dryer.

fee nearest the plenum receives hotter and dryer air than the coffee near the exit, moisture content and temperature gradients may form. Moisture gradients can be minimized by increasing or reversing air flow and by reducing drying air temperature[20]. Exhaust air from the dryer should be vented out of the facility to remove the humid air from the drying atmosphere.

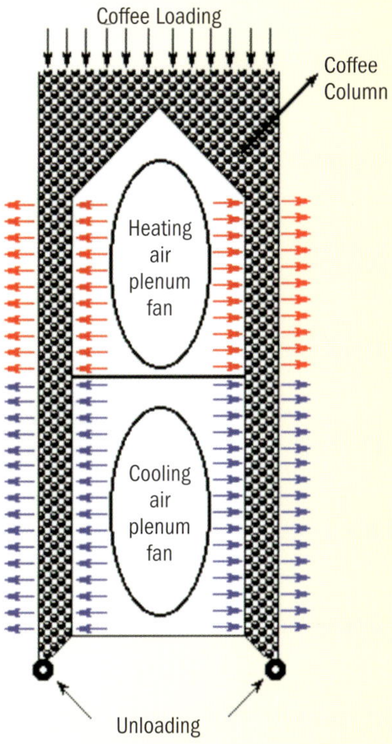

Vertical dryers have a resting chamber located in the upper part of the dryer where the coffee is loaded. However, these chambers are rarely used properly since producers do not verify that they are constantly full and leveled.

Despite their high drying capacity, low initial cost, and ease of use, vertical dryers use large amounts of energy, disuniformly dry the coffee, and can often result in low product quality. To mitigate their impact on product quality, the maximum recommended temperatures in vertical dryers are 70 °C drying air temperature and 45 °C coffee mass temperature for all coffees[21].

6.3.3.3 Fixed-Bed Dryers

In fixed-bed dryers (Figure 6.29) the product remains stationary in the drying chamber while air is forced mechanically through a ventilator, passing through the product layer and reducing its moisture content. After drying with heated air, the product is cooled in the same dryer, removing the heat source and running just the fan.

Fixed bed dryers have low initial and operational costs. However, they also have lower drying capacities and allow for the formation of a moisture content gradient in the coffee layer, which can compromise final product quality. This drying system is simple, relatively cheap, and compatible with the investment capacity of many small coffee producers. Besides drying coffee, fixed bed dryers can also be used for drying ears of corn, bean stalks, hay, and other agricultural products.

The fixed-bed dryer can be used both for pre-drying and for the complete drying of coffee. The recommended drying air temperature is 50 °C with a maximum layer thickness of 50 cm. The coffee mass should be inverted every three hours to ensure more uniform moisture content and temperature in the product[22]. Approximately 23 hours are needed to completely dry a coffee with 60% initial moisture content, at an air temperature of 50 °C, with a layer thickness of 40 cm, and with the product turned over every three hours.

Figure 6.28 Schematic of product flow in a cross-flow dryer.

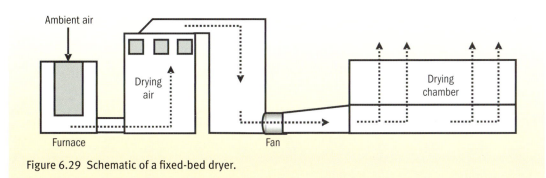

Figure 6.29 Schematic of a fixed-bed dryer.

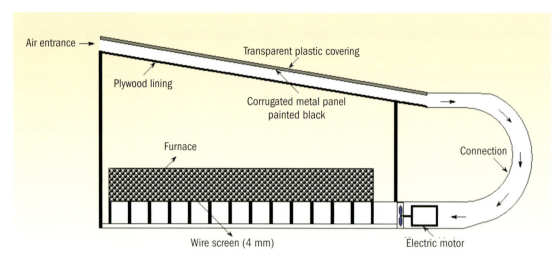

Figure 6.30 Fixed bed dryer with solar heater.

The furnace type commonly used with these dryers is direct fire with descending flow, where all of the combustible gases are mixed with the ambient air and suctioned by the fan. A cylindrical cyclone between the furnace and the fan impedes the injection of ash and sparks into the drying chamber, thus avoiding possible fires. Care must be taken when using the furnace, especially when burning dry firewood, to minimize smoke contamination of the coffee[23].

6.3.4 Drying Using a Solar Collector

Solar collectors (Figure 6.30) are not commonly used to heat the drying air of agricultural products. Not only are commercial systems of this type hard to find, but due to the large area needed to capture sufficient solar radiation to increase the ambient air temperature to the desired level, these systems are not very economical. However, the heating of drying air with solar energy can be complemented with other sources of energy in periods of low sunlight levels. Solar heating can result in low operational costs if it is done with the necessary care in managing the drying process.

6.3.5 Low Temperature Drying

Low temperature drying occurs when heated forced air is from 5 to 10 °C above the ambient air temperature. In general, with this type of drying, the product may lose water until thermal and hygroscopic equilibrium between the product and the drying air is reached.

Drying at low temperatures, while a slow process, uses less energy than high temperature drying. This drying system, if properly planned and managed, can be economical and technically efficient[24].

The most recent and common example of low temperature drying are the drying silos commonly used for corn and other grains. In this case the silos can be filled all at once or in layers. The air flows upward and the product is completely dried when the drying air front reaches the top layer.

Choosing the optimal air flow is critical to the success of low temperature drying systems. Inadequate air flow will increase drying times, which can compromise final product quality. However, excessive air flow, while reducing drying times, can result in increased energy consumption and result in higher operational cost.

Air flow recommendations for drying, also called minimum air flows, are generally expressed in air flow per unit of volume of product. They are determined by defining the minimum amount of air necessary to permit drying of the complete coffee mass without deterioration in the upper layers[25].

6.4 The Impact of Drying on Coffee Quality

After fungal attacks and undesirable fermentation, high temperatures and drying rates are considered to be the principle factors that reduce coffee quality in post-harvest.

In regions with hot and humid climates, more care should be taken to avoid physical and sensorial defects and taints. In these regions, the main objective is to avoid the production of fermented or phenolic coffee as well as defective coffee such as blacks, sours, and moldy beans. On the other hand, when the quality focus is achieving or maintaining the flavor potential of a coffee, more care should be taken in climates that are dry and hot. In this case, thermic damage caused by high drying air temperatures and high drying rates can be a barrier to quality.

To produce high-quality coffee, drying should be slow enough to prevent damage from the stresses caused by high drying rates, and fast enough to prevent fermentation. However, many producers who lack adequate drying infrastructure will attempt to dry coffee as quickly as possible, causing serious quality problems. Poor drying technique, especially using mechanical dryers, frequently results in beverages with high astringency and off-flavors, both for dry and wet process coffees.

Drying can be expedited either by increasing the temperature of the drying air or increasing the drying rate. In rotary dryers, for example, the temperature of the coffee mass can reach levels that are not recommended for the production of specialty coffees. This can occur when the temperature of the coffee mass is controlled using thermometers with probes that are only 10 cm long. Since rotary dryers have a radius up to 90 cm and the air cools as it moves outward through the coffee, the average temperature reading using this thermometer is lower than the temperature further inward; e.g., while the thermometer reads 40 °C, the inner temperature can be up to 60 °C. For this reason, it is recommended that the temperature at the halfway point of the dryer radius not exceed 40 °C[26], as measured by a thermometer of sufficient probe length to accurately read the temperature.

Table 6.2 presents sensorial analysis results for coffee samples dried in rotary dryers with different temperature control points in the drying mass.

The effect of temperature can also be perceived in the alteration of the color of the coffee[28]. This happens principally in mechanical dryers when drying is poorly managed and the coffee develops heterogeneous coloration. If drying temperatures exceed 80 °C, the beans will become overheated and grayish in color. Upon re-absorbing humidity they will whiten more intensely and irregularly.

Besides the direct effect of temperature, the interaction between temperature and drying rate will affect product quality. Elevated temperatures and high drying rates can rupture the internal structure

Table 6.2 Sensorial analysis of coffee dried in a rotary dryer with different temperature control points in the drying mass.

Depth		Moisture Content (%wb)		Sensorial Analysis
		Initial	Final	
100 mm	40 °C	42.26	10.71	Hard[1]
	45 °C	42.26	10.31	Hard
350 mm	40 °C	42.67	10.30	Soft[2]
	45 °C	42.67	10.01	Hard

Source:[27] [1] hard cup: coffee that has a flavor that is acrid, astringent, or rough, yet without the presence of off-flavors; [2] soft cup: coffee that has an agreeable aroma and flavor that is mild and sweet

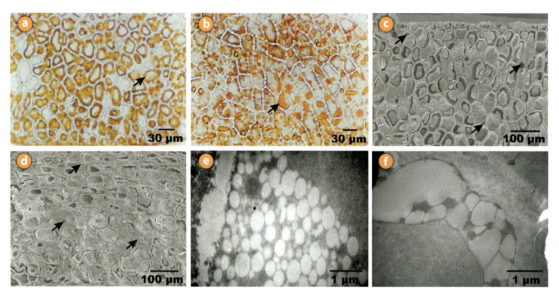

Figure 6.31 **ⓐ** and **ⓑ** Light microscopy of a histochemical test of a coffee endosperm showing **ⓐ** that the oil corpuscles are intact inside the cells (see arrows), **ⓑ** large oil droplets in the intercellular spaces indicating a plasmatic membrane has ruptured. **ⓒ**–**ⓕ** Electronic microscopy of a coffee endosperm: **ⓒ** Scanning electron micrograph of the coffee endosperm cells after drying at 40 °C. Note that the material in the cell interior has been preserved and the intercellular space is empty (see arrows). **ⓓ** Scanning electronic micrograph of a coffee endosperm dried at 60 °C. Note that the endosperm cells are completely full of cellular matter and that the intercellular spaces are obstructed (see arrows), indicating that the cells have ruptured. **ⓔ** Transmission electron micrograph of coffee endosperm cells dried at 40 °C. A large number of vesicles are still intact. **ⓕ** Transmission electron micrograph of coffee endosperm cells dried at 60 °C. The oil vesicles have coalesced and collapsed[29].

of the bean, exposing oils and other components to the action of oxygen and compromising coffee quality. Figure 6.31 illustrates the effects of drying temperature on the integrity of the endosperm.

Histochemical studies have shown that when parchment coffee is dried at an air temperature of 40 °C, the oil corpuscles remain compartmentalized and are uniformly distributed throughout the internal perimeter of the plasmatic membrane. In contrast, in the endosperm of beans dried at 60 °C, the oil corpuscles are fused into large droplets in the intercellular space, indicating a rupture of the plasmatic membrane of the oil vesicles. In micrographs obtained using an electronic sweeping microscope it was possible to observe that when beans were dried at 40 °C, the internal content of the cells remained intact and full of cellular material, and that the space between the plasmatic membrane and the cellular cavity was empty. However, in beans dried at 60 °C, the cellular structure ruptured, resulting in a total filling of the cellular cavity, indicating an overflow of part of the protoplasm. The results of these analyses, conducted with transmission electron microscopy, corroborate all of the observations previously described. When drying is performed at 40 °C, the vesicles in the interior of the endosperm cells are not compromised, while at 60 °C they rupture and coalesce[30].

The cell membranes are damaged mainly at moisture content levels between 30% and 20% (wb) when the coffee is dried at 60 °C[31]. Based on these results, studies were performed using high temperatures at the beginning of the drying, followed by lower temperatures, in an effort to decrease drying exposure time without damaging the coffee. However, the initial high temperatures caused sensorial coffee quality loss independent of drying and processing methods. A coffee mass temperature above 40 °C, independent of the point in the drying when it occurs, causes physiological damage, principally in coffees processed using the wet method. This physiological damage results in a lower quality beverage (Tables 6.3 and 6.4)[32][33].

Beans with poorly structured, disorganized, and damaged membranes due to high temperatures leach a higher quantity of solutes and have higher electrical conductivity[35]. During drying, physiological alterations can occur that compromise cup quality. Various studies have been conducted to find a correlation between maintaining the physiological quality of the coffee seed during drying and the sensorial quality of the resulting beverage[36][37][38][39].

A combination of low air temperature and low relative humidity favors an increase in the drying rate of agricultural products while avoiding the thermic damages that can come from using high temperatures[41]. However, for a given temperature, the isolated effect of the drying rate on the chemical, physiological, and sensorial aspects of coffee has not been thoroughly studied.

Studies conducted on natural coffee showed that an increase in the drying rate negatively affected cup quality (Table 6.5). Although this effect is not isolated and is dependent on the temperature of the drying air, even at temperatures between 35 °C and 40 °C, the negative effects of the increased drying rate on coffee quality are more pronounced. For these temperatures, an increase in the drying rate reduced the physiological quality of the seeds and led to elevated incidents of cellular disorganization. The loss of selective permeability by the cell membranes permits the chemical components, previously contained, to now enter into contact with hydrolytic and oxidative enzymes, setting into motion a series of chemical and biochemical reactions that can negatively impact the sensorial quality of the beverage[42].

Drying natural coffee at low temperatures and lower drying rates (11.35 g kg^{-1} h^{-1}) results in higher cup quality even when compared to coffee fully dried in the sun. Therefore, mechanical drying of coffee, when properly done, can produce natural coffees equal to and even superior to coffees that are sun-dried. In many cases, the negative effects of drying can be minimized using intermittent drying or resting periods.

Intermittent drying is a process whereby the drying of coffee with high moisture content is interrupted for hours or even days; drying is resumed after this resting period. This drying/

Table 6.3 Average final sensorial analysis scores, in terms of drying treatment.

Drying Treatment	Final Score[1]
Patio	80.35 a
60 °C / 40 °C	79.05 b
60 °C	77.64 c

Source[34]: [1]Methods followed by the same letters are not statistically different per the Scott-Knott test, at 5% probability. Sensorial analysis completed in accordance with the Specialty Coffee Association of America (SCAA) method.

Table 6.4 Average final sensory analysis scores for different drying treatments and processing methods.

Processing Method	Drying Treatment [1]			
	Patio	50/40 °C	60/40 °C	40/60 °C
Wet	79.54 aA	79.08 aA	78.67 aA	75.96 bA
Dry	79.33 aA	76.29 bB	75.54 bB	75.88 bA

Source: Oliveira (2010).
[1]Averages followed by distinct letters differ by 1% probability per the Tukey test. Lower-case letters apply to the rows, upper-case to the column. Sensorial analysis was completed according to the method defined by the Specialty Coffee Association of America (SCAA).

Table 6.5 Sensorial analysis of natural coffee dried in the sun, with air heated to 35 °C, 40 °C, and 45 °C, and at different drying rates (DR).

T_{db} (°C)	DR (g kg^{-1} h^{-1})	Sensorial Analysis[1]
35	11.34	83.94 a*
	13.35	82.87 ab
	14.59	80.37 b
40	15.89	82.68 a
	21.22	81.00 a
	23.14	82.31 a
45	23.19	79.25 a
	29.82	83.00 b
	53.12	79.06 a
Sun drying		80.25
Average		81.47
CV (%)		1.86

[1]Averages followed by the same lower-case letters are statistically the same (P<0.05), per the Tukey test.
*Averages differ statistically (p>0.05), in relation to coffee dried completely in the sun, per the Dunnett test.

resting cycle can be repeated as needed. This process is used to dry various agricultural crops as it reduces the effective drying time, or the time the product is exposed to heated drying air, and it can increase the quality of the final product[43].

Intermittent drying of pulped natural coffee, alternating 12 hours of drying with air heated to 50 °C with 12 hours of rest, reduced the effective drying time by 24.56% compared to continuous drying[44]. Studies of intermittent drying have not encountered significant effects on cup quality[45]. These results are very favorable to coffee producers. When an infrastructure is insufficient for timely drying, it is possible to interrupt the drying of coffee with higher moisture levels and store the partially dried coffee in appropriate silos or bins, freeing up dryers for lots with even higher moisture content that are subject to deterioration. Drying is completed a few days later by putting the rested coffee back into the dryers. This process allows for energy savings, increases coffee uniformity, and, if done properly, does not compromise final cup quality.[46]

7

Energy Used in Coffee Drying

Jadir Nogueira da Silva

Roberto Precci Lopes

Evandro de Castro Melo

7.1 Introduction

Regardless of the coffee species or processing method, coffee must be dried before it can be stored. Depending on the type of drying system employed, management practices, and the efficiency of the equipment used, well-managed drying can contribute to an overall decrease in post-harvest costs through energy savings. If drying is not completed properly, more steps must be taken to minimize product quality loss and to complete the drying process.

This chapter will show the importance of optimal energy use in coffee drying, the principal sources of energy, and the equipment used to heat the drying air. It is hoped that with this knowledge, producers will be able to minimize costs without losing sight of final product quality.

7.2 Energy for Water Evaporation in Coffee Seeds

Drying is a basic operation with the objective of partial elimination of the water contained in coffee seeds until adequate moisture content is reached for storage or commercialization. For coffee, the drying operation is very important because of the high moisture content of the harvested fruit, a characteristic that also demands more energy in the drying process.

In determining energy consumption for drying systems, an important property of coffee is the enthalpy of vaporization, also called latent heat of vaporization, which represents the quantity of energy necessary to evaporate a unit of mass of water contained in the coffee product. Equation 7.1 can be used to determine this value for the three main processing methods[1]:

$$L = (2500.49 - 2.43\ T) \times [1 + A \times \exp(-B\ MC^C] \qquad (7.1)$$

where: L = latent heat of vaporization of water of the product (kJ kg^{-1})
   T = temperature (°C)
   MC = moisture content of the product (db)

and: A, B, C = dimensionless parameters that depend on the product, per Table 7.1

Table 7.1 Parameters A, B, and C in the equation of the latent heat of vaporization for water contained in coffee

Processing Method	Parameters of the equation for the latent heat of vaporization of the water contained in coffee		
	A	B	C
Dry Process	7.7866×10^6	19.6621	0.0499
Wet Process – Pulped Natural	1.7665×10^7	20.6416	0.0464
Wet Process – Fully Washed	1.8377×10^7	20.1732	0.0363

Source: Afonso Júnior (2001)

Analyzing the equation, it can be seen that the energy necessary to evaporate the water in coffee increases when moisture content and temperature decrease, independent of the processing method. Table 7.2 shows that seeds with low moisture content require more energy to dry than seeds with higher moisture content.

Using the same equation, the values for the latent heat of vaporization of the water in natural (dry process) coffees are always higher than in the other two processing methods. Because of the presence of the skin and mucilage, natural coffees have the highest volume of product and the lowest diffusivity of water, factors most likely responsible for the

Table 7.2 Latent heat of vaporization values for natural coffees at two temperatures and moisture content levels.

Temperature	Moisture Content	Latent Heat of Vaporization for Natural Coffees
(°C)	(% wb)	(kJ kg⁻¹)
25	11	2,858.5
	68	2,503.3
65	11	2,744.8
	68	2,403.6

Source: adapted from Afonso Júnior (2001)[2]

substantial increase in energy needed[2]. Comparing the values of the latent heat of vaporization between the pulped natural and fully washed processing methods, no major differences are observed in the data rendered by equation 7.1. However, comparing both wet methods to the dry method, studies have shown that pulp removal leads to a 30% decrease in energy consumption[3].

Because of its high moisture content at harvest, coffee requires long drying times and high specific energy consumption, which justifies the study of efficient drying methods. To make a technically appropriate decision when choosing a drying unit, it is indispensable to know the unit's operational characteristics, system capacity, energy efficiency, and, principally, its effect on final product quality.

7.3 Specific Energy Consumption and Drying System Efficiency

Specific energy consumption of a drying system is the total quantity of energy used in evaporating a unit of mass of water contained in the product being dried. In coffee dryers this energy comes from the burning of a fuel to heat the drying air and from the electrical energy used to drive the fans and transport the coffee.

Not all of the energy supplied to the drying air is used to remove water from the coffee. Some of the energy is lost in the form of sensible heat in the exhaust air, in the heating of the coffee, and in the body of the dryer to the surrounding environment through conduction, radiation, and convection. In fact, studies have shown that 50% of the energy produced by burning fuels is not used to evaporate water from the coffee product[4]. These measures can be taken to reduce specific energy consumption:

- Recycle part of the exhaust air. This is possible when the relative humidity of the exhaust air is lower than the recommended equilibrium moisture content.
- Use the sensible heat of the heated coffee through dry-aeration or through intermittent drying with resting periods.
- Thermally insulate the points where the heat is dissipating from the furnace, ducts, and dryer.

Studies have shown that 50% of the energy produced by burning fuels is not used to evaporate water from the coffee product

Table 7.3 shows the specific energy consumed by various types of dryers. The range of values—from 4.8 to 22.3 mJ per kg of water evaporated—can be attributed to dryer type, fuel type (liquefied petroleum gas or firewood), and the drying air heating system (direct or indirect). Other factors, such as moisture content, product type (presence or lack of pulp), drying and ambient air temperatures, and air flow rate, can also affect these values. The percentage of electrical energy consumed is also variable, and normally ranges from 2.3% to 10.4% for rotary dryers, 11.0% to 16.6% for cross-flow dryers, and 2.5% to 3.3% for mixed-flow dryers.

The amount of energy consumed in drying is directly related to the efficiency of the drying equipment. The coffee producer should monitor not only the parameters of drying that affect coffee quality, but also those that affect drying efficiency. Various factors can affect the performance of a drying system, such as initial and final product moisture content, air temperature and relative humidity, physical properties of the product, any resistance to air flow, exhaust rate of drying air, furnace type, fuel characteristics, and other factors cited above.

Table 7.3 Specific Energy Consumption (SEC) for various types of coffee dryers.

Dryer Type	SEC MJ kg^{-1} water	Fuel Type	Source
Rotary	8.2 – 8.5	LPG	Reinato et al. (2003)[5]
	6.2 – 7.8	LPG	Cardoso Sobrinho (2001)[6]
	12.3 – 15.9	Firewood	Reinato et al. (2003)[7]
	9.5 – 17.3	Firewood	Cardoso Sobrinho (2001)[8]
Fixed Bed	7.7 – 13.1	Firewood (direct fire)	Lacerda Filho (1986)[9]
	16.8	Firewood (indirect fire)	Campos (1998)[10]
Intermittent Cross-flow	10.6 – 20.0	Firewood	Pinto Filho (1994)[11]
	10.0 – 22.3	Firewood	Cardoso Sobrinho (2001)[12]
Intermittent Concurrent-flow	4.8 – 5.7	Firewood	Osório (1982)[13]
	5.4 – 10.8	Firewood	Lacerda Filho (1986)[14]
Intermittent Counter-flow	6.5 – 8.3	Firewood	Silva (1991)[15]
Intermittent Mixed-flow	5.7 – 6.1	Firewood	Pinto (1993)[16]

The efficiency of a dryer is determined by the ratio of the quantity of energy necessary to remove water from the coffee to the actual energy used by the dryer to complete the drying process. The dryer efficiency calculation uses a specific methodology that is beyond the purview of this chapter[*]; however, the following example can assist a coffee producer in evaluating dryer performance.

Example: An intermittent high-temperature dryer is used to dry 2,830 kg of fully washed coffee that was pre-dried on the patio. The initial moisture content (MC_i) of the product going into the dryer is 24%, and it takes 12 hours to achieve a final moisture content (MC_f) of 14%. During the drying, 25 kg h^{-1} of firewood is used and the electrical energy consumption to move the heating air and the coffee is 25 kWh. Knowing that the calorie potential of the firewood used is 12,044 kJ kg^{-1}, and assuming that the latent heat of vaporization of the water in the coffee is 2,500 kJ kg^{-1}, it is possible to determine: a) The specific energy consumption and b) dryer efficiency

Weight loss related to difference in moisture content (WL):

$$WL = (MC_i - MC_f)/(100 - MC_f) \times 100 \qquad (7.2)$$

$$WL = \frac{24 - 14}{100 - 14} \times 100 = 11.6\%$$

Water removed during drying: 11.6% of 2,830 kg = 328.3 kg

---

[*] For more on this subject, see Silva (2000)[17], "Secagem e Armazenagem de Produtos Agrícolas".

A. Specific energy consumption

Given that 25 kWh = 90,000 kJ, and using equation 7.3, specific energy consumption (SEC) can be determined.

$$SEC = \frac{\text{Energy used by the dryer}}{\text{Quantity of water removed}} \qquad (7.3)$$

$$SEC = \frac{90,000 + 12 \times 25 \times 12,044}{328.3}$$

$$SEC = 11,280 \text{ kJ kg}^{-1}$$

Though the specific energy consumption here is elevated, it does fall within the values cited in Table 7.3. The value is high because the drying system used high temperatures while the coffee had low initial moisture content. In other words, there was a low quantity of water available for evaporation compared to the high quantity of energy made available during the drying process.

B. Drying system efficiency

Drying system efficiency can be obtained by looking at the ratio of the quantity of energy necessary to evaporate water from the coffee to the quantity of energy necessary to heat the drying air and move the product (equation 7.4).

$$\eta = \frac{\text{Energy necessary to evaporate water in the product}}{\text{Energy used by the dryer}} \times 100 \qquad (7.4)$$

$$\eta = \frac{328.3 \times 2,500}{90,000 + (12 \times 25 \times 12,044)} \times 100$$

$$\eta = 22.4\%$$

The low system efficiency does not necessarily mean that the equipment consumes a lot of energy. It could mean that the energy available is not being properly used. For example, it could be that the dryer is not fully loaded or that the initial moisture content of the product is low. Drying is a dynamic process with parameters that vary during the drying process as the product's moisture content levels decrease. Certain dryers may be more efficient for given moisture content levels. Coffee producers should be aware of all of these factors and control them to maximize equipment efficiency.

The efficiency and power requirements of a dryer are relevant parameters for the analysis of energy consumption. The total energy required (hph ton^{-1}) can be determined by dividing the total power consumed in horsepower by the nominal capacity (tons/hour) of the dryer. This renders the energy consumed by the equipment per ton of product. For dryers this power corresponds to the power of the fan motor and the system for discharging and transporting the coffee (screw conveyor and elevator). Considering all of the forms of energy used, the total specific energy consumption can be expressed in tons or in cubic meters of product (kJ ton^{-1} or kJ m^{-3}). These values can be very useful in comparing different dryer brands. The lower the number, the more economical the product.

C. Fuel consumption

One step a producer should take in preparation for harvest should be to ensure that there is sufficient fuel to dry all of the coffee. Estimates of fuel consumption depend on furnace efficiency, dryer exhaust rates, the heat potential of a particular fuel type, and the desired drying air temperature. The following example illustrates how to estimate fuel consumption (FC) for drying coffee using charcoal as the fuel source.

Ambient air temperature (T_1): 27° C
Drying air temperature (T_2): 60° C
Furnace efficiency (η): 90%
Exhaust Rate (Q): 70 m^3 min^{-1}
Mass density of the ambient air (ρ_{ar}): 1.069 kg m^{-3}
Average specific heat of the air (c_p): 1.005 kJ kg^{-1}
Lower heating value of the fuel (LHV): 30,067.9 kJ kg^{-1}.

Fuel consumption can be determined using equation 7.5:

$$FC = 60 \, \frac{\rho_{ar} \times Q_{ar} \times C_p(t_2 - t_1)}{\eta \times LHV} \tag{7.5}$$

Applying this equation to the example:

$$FC = 60 \, \frac{1.069 \times 70 \times 1.005(60 - 27)}{0.90 \times 30,067.9} = 5.5 \text{ kg } h^{-1}$$

With this information as well as drying time, quantity of product harvested, and price per kWh of fuel, it is easy to determine the quantity of charcoal needed and the total energy cost for drying.

7.4 Energy for Drying Coffee

The availability of energy for coffee drying can be a problem for coffee producers, given the scarcity and high cost of natural energy resources. The rising cost of these resources combined with the recent electrical energy crisis shows the necessity of energy conservation for commercial, residential, industrial, and agricultural activities.

In modern coffee production, coffee is dried with equipment that utilizes combustible fuels and electrical energy. Energy costs can be reduced by using efficient and technically appropriate equipment operated properly by well-trained workers. However, energy inputs (fuel, electricity, and equipment that uses the energy) are not always used with energy savings in mind. A widespread example of this can be seen in the firewood furnaces commonly used to dry coffee (Figure 7.1). Many are in a precarious state, are poorly dimensioned, and permit excessive energy loss. The majority of these furnaces do not have a control mechanism for the fuel used, and in most cases are poorly operated. This can lead to in-

Figure 7.1 A typical wood furnace used for drying coffee.

complete combustion, product contamination by combustion residues present in the drying air, and difficulties in maintaining constant air temperature during drying.

To effectively use energy during drying, it is necessary to have knowledge of a particular fuel's characteristics, the factors that affect its combustion, and the characteristics and correct operation of the equipment used.

7.4.1 Fuels

A fuel is a product that, under favorable temperature and pressure conditions, is able to produce energy in the form of light and heat as a consequence of the chemical reaction of its components—carbon, hydrogen, and sulfur—with the oxygen in the air.

A fuel is characterized by its physicochemical properties and can be either natural or synthetic. The principle characteristic of a fuel is its calorific value. Also known as heat value or energy content, calorific value is the quantity of energy released during the complete combustion of a unit of fuel mass or volume. There are two basic ways to measure this value. When the measurement takes into consideration the latent heat of condensation for all of the water vapor formed by the combustion of the hydrogen present in the fuel, its calorific value is the higher heating value (HHV), also called gross energy or gross calorific value. When latent heat of vaporization is not considered, the calorific value is the lower heating value (LHV), also called the net calorific value. In coffee drying, because the water vapor does not condense and all of the product resulting from the combustion of hydrogen remains in a vaporous state, it is common practice to use the lower heating value[18].

In experiments, the higher heating value is measured using a calorimeter. However, if a calorimeter is not available, a good approximation can be achieved using the elemental composition of the fuel (in percent by weight dry basis) and the enthalpy of the reaction of the combustible elements with oxygen.

Dulong's formula:

$$HHV = 33774\ C + 141744 \left(H_2 - \frac{O_2}{8} \right) + 9238\ S \tag{7.6}$$

where: HHV = higher heating value, kJ kg^{-1} of dry fuel
C = fraction of carbon in the fuel (kg of carbon kg^{-1} of dry fuel)
H_2 = fraction of hydrogen in the fuel (kg of hydrogen kg^{-1} of dry fuel)
O_2 = fraction of oxygen in the fuel (kg of oxygen kg^{-1} of dry fuel)
S = fraction of sulfur in the fuel (kg of sulfur kg^{-1} of dry fuel)

The LHV is determined analytically using equation 7.7[19], subtracting from the HHV the enthalpy of water vaporization of the water vapor formed in the reaction of hydrogen with oxygen.

$$LHV = HHV - 2440\ (9H_2) \tag{7.7}$$

where: LHV = lower heating value (kJ kg^{-1} of dry fuel)
2440 = Enthalpy of vaporization of water, kJ kg^{-1} of water at a reference temperature of 25 °C
$9H_2$ = Portion of water vapor formed by hydrogen combustion.

One of the parameters with the largest influence over the heat value of a fuel is its moisture content. Moisture increases the energy necessary in pre-ignition and decreases the heat released by combustion. In calculating thermal efficiency of a fuel with higher moisture content, the lower heating value (equation 7.7) should be corrected for the elementary composition in the wet basis since there is still moisture present in the fuel (equation 7.8[20]).

$$LHV_{wf} = HHV (1- MC_{wb}) - 2440 [9H_2 (1- MC_{wb}) + MC_{wb}] \qquad (7.8)$$

where: LHV_{wf} = lower heating value (kJ kg^{-1} of wet fuel)

MC_{wb} = fraction of water in the wet fuel (kg kg^{-1})

Ash load, the residual solids resulting from combustion, is an undesirable characteristic of fuels. When ash proportions are elevated, the ash can obstruct the flow of combustion air in the grates, increasing the amount non-converted carbon in the ash and causing high temperature ash fusion. High ash content can also increase the cost of fuel transport, cause corrosion, and increase losses related to the sensible heat of the ash that is removed from the equipment.[21] Since combustion takes place at high temperatures, it is important to know the characteristics of ash to avoid these potential drawbacks.

7.4.2 Combustion

Combustion is a series of exothermic chemical reactions between oxygen and combustible elements, such as carbon, hydrogen, and sulfur, that are present in the combustible material[22]. The maximum release of thermal energy is achieved when the oxidizable elements—those that react with oxygen—are converted into substances or compounds that are no longer oxidizable. The energy released in this reaction, known as the enthalpy of reaction, or combustion, is an extremely useful parameter.

Good combustion should release all of the chemical energy of the combustible material such that the absorption of this energy by surrounding surfaces is uniform, stable, and continuous. There should be minimal loss from incomplete combustion due to lack of air, wet combustible material, an inefficient turbulence process, or inadequate fuel-oxidizer mixing. In most cases oxygen is supplied by atmospheric air. Air used for combustion is called *combustion air*; the combustion air and combustible material are the reagents of the reaction. Combustion results in a gaseous mixture of CO_2, CO, SO_2, H_2, water vapor, N_2, and large amounts of O_2.

The exact composition and concentration of this mixture depend on the type of combustible material used and the quality of the combustion. The N_2 present in the combustible material and in the oxidizing air does not react in the process. Ash is a solid residual waste of combustion along with any residual carbon, or soot, present in the ash or in the gases. Figure 7.2 shows different reagents and products commonly encountered in combustion.

Combustion begins with biomass decomposition through the action of heat, and occurs in the following phases: drying, pyrolysis/gasification, ignition of volatile substances, and combustion of fixed carbon. Upon contact with the heat of the furnace, wood goes through a drying process that starts on surface areas and moves inward. The duration of this drying process varies with the size of the wood piece. During the drying phase the temperature

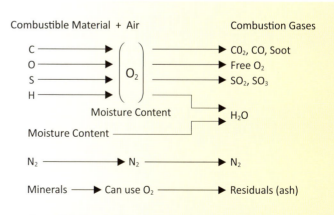

Figure 7.2 Reagents and combustion products.

remains lower because of the evaporation of water in the combustible material. After drying, the temperature rises and volatiles are released. The processes of evaporation, pyrolysis, and gasification all absorb heat from the furnace. Heat is generated by the burning of both fixed carbon and volatiles. If the level of volatiles is high, combustion produces a long flame (Figure 7.3 a and b)[23]. While the volatiles burn in a gaseous state, fixed carbon burns in a solid state when the temperature reaches 400–550 °C. It should be noted that while these phases generally occur in this order, there is much overlap and at any point during combustion they may occur simultaneously. An example of this is the interference of the water vapor released in the first phase with the distillation that occurs in the second phase; this interference causes oscillations in temperature within the combustion chamber.

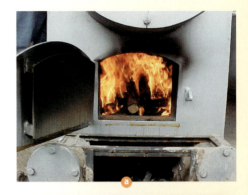

Figure 7.3 **a** Combustion of fuel in the furnace of an internal boiler. **b** Detail view showing the burning of volatiles.

The entire combustion process should follow fundamental principles that ensure fuel efficiency. For this it is necessary both to control factors that interfere with the quality of combustion and to facilitate the three elements of the "combustion triangle": temperature, combustible material, and oxygen. These elements must be combined in the correct quantities and at the right times to promote complete and efficient combustion. In practice, observing the following three conditions will ensure facilitation of the combustion triangle:

1. *Temperature equal to or greater than ignition temperature:* the combustion chamber should provide adequate conditions for the combustible material to attain ignition temperature so that the combustion is self-sustaining. For example, the minimum temperature of wood should be 300 °C. Combustion will not occur at a lower temperature, even if sufficient air is present.

2. *Adequate air mixture or turbulence with the combustible material:* the combustion air and the combustible material must be in close contact. To ensure this contact, a solid combustible material, such as a biomass, should be appropriately broken down and combustible liquids should be pulverized. Besides helping to ensure complete combustion, this has the benefit of reducing the amount of furnace space needed.

3. *Sufficient time and space for the combustion reaction to occur:* adequate time should be provided for all of the combustible material in the furnace to be consumed and transformed into combustion gases.

These factors are known as the "three Ts" of combustion: temperature, turbulence, and time. The mechanisms responsible for the control of combustion should be constantly monitored to ensure that the above conditions are met for the duration of the combustion.

"Theoretical air" is the quantity of air that must be supplied for combustion to achieve the complete burning of carbon, hydrogen, sulfur, and any other elements present in the combustible material that are subject to oxidation. Theoretical air is calculated based on the elemental composition analysis of the combustible material. In reality, the quantity of theoretical air is not sufficient to achieve complete combustion; the actual quantity of air needed is higher and is therefore called "excess air." This value is greater than theoretical air by 30%–60% for combustible solids, 10%–30% for combustible gases, and 5%–20% for combustible gases[24].

During combustion, care must be taken with high levels of excess air to prevent[25]

- the slowing of the combustion reaction,
- a reduction in the efficiency of the combustion system,
- incandescent particles or unburned material,
- the need for a more powerful fan.

Conversely, an air quantity that is less than theoretical air can cause incomplete combustion, resulting in the formation of carbon monoxide (CO), where a portion of the carbon bonds with oxygen, instead of the carbon dioxide (CO_2) that forms in complete combustion. Lower air quantities can also result in the formation of soot, or carbon particulate suspended in the combustion gases. Soot is wasted fuel that can also cause environmental damage.

For farms that use water vapor or hot water as a heat exchange method, boiler efficiency is an important factor that should be constantly monitored. Principle energy losses occur in the combustion gases exhausted through the boiler chimney. To reduce these losses, combustion can be controlled by measuring CO_2 and O_2 levels in the combustion gases using portable gas analyzers. Some commercial readers also provide the temperature of the gases, an important parameter for the quantification of enthalpy in the exhaust air. The quality of the exhaust air can also be monitored by measuring its soot content. The comparison standards used to evaluate combustion efficiency are also useful for measuring acceptable emission standards.

Independent of fuel type, the gases that result from incomplete combustion contain residual toxins that can damage the coffee, the environment, and health. The compounds considered most dangerous if in direct contact with drying coffee are sulfur oxide, nitrogen oxide, polycyclic aromatic hydrocarbons, condensed pyrolytic volatiles (smoke), carbon particles (soot), and ash. Concentrations of these compounds vary by fuel type. In humid environments sulfur oxides form acidic compounds that can corrode drying equipment; when emitted into the atmosphere, sulfur oxides contribute to acid rain. Nitrogen oxides can react with the protein in agricultural products and produce carcinogenic nitrosamines. As such, in situations where it is impossible to control combustion efficiency, and therefore drying air quality cannot be guaranteed, hot air generators equipped with heat exchangers are recommended.

7.5 Biomass Energy

The drawbacks of fossil fuels—increased environmental pollution, risk of scarcity, frequent price increases—have driven the study of various alternative energy sources for coffee drying. However, for supplying energy on a large scale, the majority of these alternatives present limitations such as low energy density, lack of availability in high demand periods, the advanced technology required for capitation and conversion, or their use is not economical for the majority of coffee producers. Biomass, because of its widespread availability, is considered the best alternative energy for agricultural use, especially firewood, wood charcoal, and agricultural byproducts from coffee tree trimmings.

7.5.1 Firewood

Firewood is wood that has been cut into pieces of adequate size to be used as fuel or transformed into charcoal. It is probably the oldest form of energy used by humans, and for many developing countries it is the principle source of energy.

New technologies for converting firewood into combustible liquids, solids, and gases with high aggregated value have recently gained great interest worldwide and receive large investments for further research and development. Combustion, the direct burning of the wood, is the most traditional use of firewood; however, gasification and pyrolysis are thermochemical processes that have received more attention in recent years[26].

Native forests have been the principle source of firewood for many agricultural industries. However, predatory and intensive extraction from colonial to current times have contributed to scarcities and high prices, making supply and efficiency urgent issues.

Native forests are largely being replaced by reforested areas as main sources of firewood. By planting small forests on the property for self-consumption, farms are able to take advantage of areas not suitable for agriculture and ranching. Eucalyptus is the principle tree used in reforestation[27]. It is a tree of Australian origin comprising more than 600 species, many of which were developed in and adapted to the climate of Brazil. In fact, the fastest growing Eucalyptus species is found in Brazil.

During the production of commercial firewood, the trunks and thin branches of the trees are rejected and constitute forestry waste. What's more, industries that use wood for non-energetic purposes, such as sawmills and furniture manufacturers, generate industrial byproducts in the form of log ends, wood chips, and sawdust of different sizes and densities. All of these wood industry byproducts can be used as combustible material.

The mass density of firewood varies from 250 to 450 kg m^{-3}, depending on the wood species and its moisture content[28]. The lower the moisture content, the better the combustion and the higher the calorific potential, as illustrated in Table 7.4. Table 7.5 shows the lower heating value of several species of wood. The elemental composition of wood (db) is: 47.5% carbon, 6.0% hydrogen, 44.0% oxygen, 1.0% nitrogen, and 1.5% ash[29]. An immediate analysis (an analysis of easily obtained information) of the elemental composition of wood rendered the following: 70.0% to 75.0% volatile materials, 20.0% to 27.0% fixed carbon, and 0.5% to 2.0% ash[30].

The following are some of the advantages and disadvantages of using firewood as an energy source[31].

Advantages:

- Firewood remains the cheapest available fuel, both per unit of mass and unit of energy.
- Skilled labor is not required.
- Combustion gases have low ash load and sulfur levels.
- It can be stored outdoors.

Table 7.4 Lower heating value of wood* as a function of water content.

Water content (% wb)	Lower heating value (kJ kg^{-1})
0	19,880
10	17,644
20	15,412
30	13,180
40	10,947
50	8,715
60	6,483

* Species not distinguished

Table 7.5 Lower heating value of select wood species (dry).

Species	LHV (kJ kg^{-1})	Source
Cedar	18,092	1
Cypress	21,474	1
Oak	19,527	1
Eucalyptus	19,255	2
Pine	20,511	2

Source: 1 Diniz (1981)[32]. 2 Mitre (1982)[33].

Disadvantages:

- Firewood is labor intensive, which increases final product cost, especially in producing countries and regions with higher labor costs.
- Planning is required for proper use.
- Large storage space is required.
- Supply is volatile because of large quantities used and growing demand.
- Maneuverability is low compared to fossil, liquid, and gas fuels.
- Heat potential is low compared to fossil fuels.

When used in coffee drying, firewood should be of good quality, of adequate size for burning in the combustion chamber, preferably split, and should have less than 30% moisture content. The presence of smoke indicates deficient combustion, which can be caused by excess moisture in the wood or low-quality wood. In cases where incomplete combustion is inevitable, heat exchange furnaces are recommended, especially for drying parchment coffee.

7.5.2 Charcoal

Brazil is the largest charcoal producer in the world, responsible for more than 25% of global production[34]. In spite of this fact, charcoal is not commonly used for drying agricultural products in Brazil. This is due to the low-tech, lengthy, and labor intensive process commonly used in its production. Other impediments to using charcoal include the lack of low-cost technology available for its use in small furnaces and the lack of knowledge that it can be used as an effective means of drying agricultural products.

Charcoal is created by the carbonization of wood through the pyrolysis process. Heat is applied to wood in the absence of oxygen, causing components of wood to be released in the form of elemental gases (non-condensable gases) and pyroligneous liquid (condensed combustible volatiles), yielding the solid matter called charcoal. Figure 7.4 lists the percentages in dry weight for *Eucalyptus grandis* carbonized at a temperature of 500 °C [35].

The quality of charcoal obtained depends on the species and moisture content of the wood, the size of the charcoal pieces, and the method of carbonization. Dense wood is preferred as it produces denser charcoal. Moisture content should be under 25%. Quality charcoal is dense, with low friability, and has uniform granulometry. Chemically, it should have a high percentage of fixed carbon and a low ash load.

The heat potential of charcoal is associated with the carbonization temperature. When carbonized at 400 °C, the LHV is 31.047 kJ kg^{-1} [36]. Since charcoal has low friability, it produces fine particles that constitute from 10% to 20% of the total product because of breakage during production, transport, storage, and screening. Compacting these fine particles using ligands produces briquettes with high calorific potential. Using these briquettes in coffee drying furnaces designed especially for them results in high autonomy and combustion with high energy production.

Bulk density is an important parameter in handling charcoal. Numerically, this parameter is expressed as kilograms of charcoal per cubic meter (kg m^{-3}).

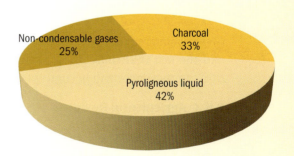

Figure 7.4 Percentages in dry weight for *Eucalyptus grandis* carbonized at a temperature of 500 °C.

Figure 7.5 Oven container for production of charcoal. **a** Filling the container with wood. **b** Closure of the oven. **c** Charcoal obtained from a test.

Bulk density varies with the granulometry of the charcoal because of the empty spaces created when it fills the volume in which it is measured. In general, the bulk density of charcoal varies between 200 and 300 kg m^{-3} [37].

There are various oven types for carbonizing wood. Conventional ovens used to produce charcoal have a gravimetric rendering between 25% and 33%. These ovens generally exhaust smoke directly, contaminating the work atmosphere and the environment. However, new technology for producing charcoal on rural properties is available. (Figure 7.5). This technology carbonizes wood in 8 to 10 hours and minimizes pollution by burning the gases in a small furnace. The energy produced by this combustion can then be reused to dry the wood that will be converted into charcoal. This technology allows coffee producers that use charcoal furnaces to produce their own charcoal in an ecological and efficient manner[38].

Though little utilized in the industry, charcoal is an appropriate fuel for coffee drying since it

- comes from the pyrolysis of wood, so most phenolic compounds are eliminated during the carbonization process;
- has higher calorific potential than firewood;
- is easy to handle and store;
- has a low level of ash, fine particles, and sulfur;
- can be purchased in a wide array of sizes;
- can be moved by gravity, permitting the use of constant feed systems;
- allows for easy control of combustion and heating air;
- emits low levels of smoke and odors when burned.

Given the increased use of charcoal burning ovens, the development of more modern ovens, and the fact that charcoal is a renewable energy source, various researchers believe that this energy source will become as important as firewood for heating air for coffee drying[39].

7.5.3 Agricultural Residues

Agricultural residues from commercially grown crops such as rice, corn, and coffee can be used as a source of energy for drying. The quantity of residue generated in harvesting and pre-processing these crops (straw, hulls, cobs, etc.) is generally sufficient as a source of energy for drying the crop.

Although coffee hulls can provide an energy source for drying air, they can also be used as an excellent, potassium-rich organic fertilizer (Figure 7.6). However, the energy required to transport and distribute the hulls should be taken into consideration. Surplus coffee hulls can be used as an energy source.

Table 7.6 shows the mass density and calorific potential of several agricultural byproducts that can be used as an energy source for drying coffee; Table. 7.7 shows the amount of byproduct generated per kilogram for some standard agricultural products. In terms of composition and calorific potential, agricultural byproducts are thermically and chemically equivalent to wood. Because of this there is current focus on their use as fuel, although because of their low density the energy they produce is less concentrated.

Figure 7.6 Coffee hull residue that can be used to generate energy or as a source of potassium for crops.

7.5.4 Forestry and Wood Industry Byproducts

In coffee regions near logging operations, the use of logging byproducts should be considered if the species of harvested trees is suitable for use as a fuel. The byproduct from logging industries and sawmills has high energy potential, as each cubic meter of wood produced generates an average of eight cubic meters of waste[40]. Furthermore, sawdust generated by wood milling represents 30% of the total amount of wood milled[41].

Sawdust is found in large volumes around sawmills and is often an environmental pollutant, principally to the rivers and waterways near the mills. Because of its high energy potential, this material should not be discarded. In fact, the HHV of sawdust with 12% moisture content is 13.822 kJ kg[-1]. Both agricultural and forestry byproducts are very important as fuels when they are produced industrially as a source for briquette production. In the industrial sector, briquettes are used principally for steam production and are a substitute for firewood, combustible oil, and natural gas. In commercial and service sectors, briquettes are chiefly used in ovens as a substitute for firewood, natural gas, LPG, and electricity[42]. Briquettes are still not commonly used as a fuel source for coffee drying; however, they are a

Table 7.6 Lower heating value of agricultural byproducts.

Byproduct	Moisture content (%wb)	Mass density (kg m^{-3})	Lower heating value (kJ kg^{-1})
Rice hulls	12	140	12,977
Peanut shells	12	150	12,977
Wheat chaff	20	160	13,395
Corncobs	13	220	17,598
Coffee hulls	13	250	15,488
Sawdust	40	300	8,372
Sugarcane bagasse	50	150	7,535

Source: Lopes et al. (2001)[43].

Table 7.7 Byproduct production of principal agricultural products.

Material (1 kg)	kg of byproduct
Sugarcane	0.30
Soy	2.55
Corn	1.50
Rice	1.22
Wheat	1.50
Beans	2.00
Cassava	0.15
Sisal	8.50
Coffee	0.40

Source: Batista (1981)[44].

viable substitute for fossil fuels and charcoal. The main advantages of briquettes in a rural setting are

- high energy density, which facilitates storage and transport;
- increase in energy efficiency due to compressed form, which increases the heat potential of the biomass;
- greater ease of use than other byproducts;
- decreased dirt buildup on equipment;
- alleviation of dependency on a single type of raw material;
- flexibility of briquette composition. (When byproducts with higher heat potential are in short supply, other materials of lower heat potential can be used in higher quantities to make the briquettes, thus maintaining their energy efficiency.)

Briquette production requires moisture content control of the material used. In general, the moisture content should be 12% to 15% (wb). Higher moisture content can compromise the burning efficiency of the briquettes. The briquettes are densified, or compacted, using high pressure, which causes an increase in temperature that in turn causes a "plastification" of the particles through lignin, which acts as an agglomerant. The ideal dimension of briquettes for ovens and boilers is 75 to 100 mm diameter by 250 to 350 mm width[45].

One disadvantage of briquettes, which stems from their provenance from agriculture and forestry, is that they often contain a large amount of dirt and soil. This can result in high levels of silicates in the ashes, which can lead to ash fusion and clogging as well as air flow obstruction.

7.5.5 Technologies to Heat Air for Coffee Drying Using Biomass

Furnaces and gasifiers are the principle devices used for burning biomass to heat drying air. The following considers these options.

7.5.5.1 Furnaces that Directly Heat the Drying Air (Direct-fired Furnaces)

In furnaces that directly heat the drying air, gases resulting from combustion are mixed with ambient air and are infused by fan directly into the product mass. The mixture of part of the oxidizing gas with the combustion gas is undesirable when the combustion process is incomplete, as this generates contaminating compounds. Since direct-fired furnaces directly apply the thermic energy of the combustion gases when combustion is complete, they are more efficient than indirect-fired furnaces. However, these furnaces require a cyclone to force the fly ash, formed mainly from carbon, into a spiral movement, separating it from the gas flow by centrifugal force. Depending on the type of fuel used (firewood, agricultural byproducts) and the characteristics of the fuel (moisture content, size), direct-fired furnaces can contaminate the drying product with smoke and/or particulate that infiltrates the drying air.

A direct-fired charcoal furnace has been developed that includes a cyclone and a fuel deposit (Figure 7.7). Because of its reduced dimensions it costs less than traditional firewood burning furnaces[46]. Studies of this furnace have concluded the following:[47]

- It does not have excess air flow that could impede proper combustion.
- It permits a gravity-powered charcoal flow into the combustion chamber that is continuous and uniform.
- It allows for stable burning and can operate continuously without labor as long as the hopper is full.

- Combustion occurs in a continuous and regular manner (Figure 7.8) in the combustion chamber and the temperature at its center reaches 1,320 °C.
- Continuous and regular fuel burning in the combustion chamber allows for a stable drying air temperature.
- The average fuel consumption for an air flow rate of 70 m^3 min^{-1} and a temperature of 60 °C is 5.6 kg h^{-1}, with thermic efficiency of 91.0%. With the size of the deposit used, the functional autonomy of the furnace was 12 hours without refueling.
- The largest thermic load volume is 567 kW m^{-3} with charcoal consumption of 14.4 kg h^{-1} and a heating air temperature of 120 °C when air flow is at its maximum.
- Temperature control of the heating air is easily obtained.
- The furnace is sensitive to temperature variations in the ambient air.
- Its operation requires less labor compared to direct heat furnaces that use firewood.
- It can be used in drying systems of varying capacities.

Figure 7.7 Charcoal furnace for direct heating of the drying air, with fuel deposit, cyclone, and fan.

When this furnace type was used to dry parchment coffee in a hybrid system that used sun drying when possible and heated air at night and during periods of inclement weather, no cup quality alterations were noted compared to a direct-heat charcoal system[48].

In the 1980s, a firewood-burning downflow furnace with direct heating was developed; this furnace was widely accepted for drying natural coffees. However, this type of furnace (Figure 7.9) is still not commonly used despite the fact that it is easily built, economical, and efficient[49].

7.5.5.2 Furnaces that Indirectly Heat the Drying Air (Indirect-fired Furnaces)

In furnaces that use indirect heat to dry coffee, combustion gases are run through a heat exchanger. This type of furnace loses thermic energy through the chimney and the heat exchanger, resulting in lower energy efficiency compared to direct-fired furnaces.

Indirect-fired furnaces are optimal for agricultural products that require controlled and moderate drying temperatures. One common type of indirect-fired furnace uses a shell

Figure 7.8 Charcoal burning in the combustion chamber in the interior of a gravity fed furnace.

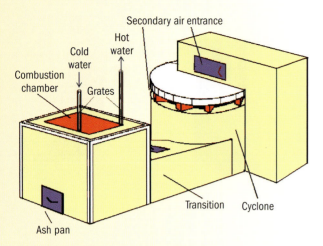

Figure 7.9 Firewood-burning downflow furnace for direct heating of drying air.

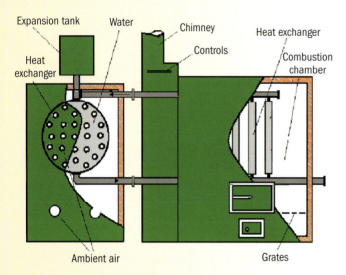

Figure 7.10 Indirect-fired furnace with automatic maximum temperature control.

Figure 7.11 ⓐ Furnace for indirect heating of the drying air. ⓑ Air to air heat exchanger tubes.

and tube heat exchanger, where a hot fluid receives the energy of the gases in the combustion chamber (Figure 7.10). Cold air enters through the tubes in the heat exchanger and is heated by the fluid that circulates in the shell until reaching a maximum temperature determined by equilibrium with the boiling temperature of the circulating fluid. The advantages of this furnace type are durability and uncontaminated drying air, which is especially pertinent to drying parchment coffee as any smoke flavors acquired by the coffee will greatly reduce its value and limit its marketability. When burning firewood, this furnace type uses 16 kg h^{-1} of wood for heating 80 m^3 min^{-1} of air to a temperature of 60 °C[50].

Industrial dryers usually heat the drying air indirectly using air-to-air–type heat exchangers (Figure 7.11). This type of furnace has low efficiency, around 30%, due to high loss of enthalpy in the gases that escape through the chimney. Another disadvantage of this system is the physical strain put on the heat exchanger tubes exposed to high temperatures. With use, it is common for perforations to form in the tubes, allowing smoke to enter into the drying air, which compromises coffee quality.

When agricultural crop residues are used for fuel, indirect-fired furnaces should be constructed with inclined grates that prevent the fuel from sticking or jamming and allow for constant burning. An indirect-fired furnace was developed specifically for burning coffee milling byproducts (Figure 7.12); this furnace achieved a thermal efficiency of 54%. The furnace operated with air flow of 65 m^3 min^{-1}, heating the ambient air from 23 °C to 92 °C, with fuel consumption of 36 kg h^{-1}[51].

7.5.5.3 Firewood-Powered Boilers: Using Steam to Dry Coffee

Recently, boilers have been introduced to heat drying air. A boiler is a heat exchanger that produces steam under pressures higher than atmospheric pressure, in most cases using the energy released in the combustion of a solid, liquid, or gas fuel. Modern boilers (Figure 7.13) are made up of diverse equipment precisely integrated to maximize thermal return[52].

Boilers operate as follows: water receives heat through contact with a heated surface. As its temperature increases, the water changes from liquid to vapor under a determined pressure that is higher than that of the atmosphere. The enthalpy of the vapor is used to transfer thermic energy to the drying air through water-to-air heat exchangers.

All boilers are composed of three basic parts: furnace, water tank, and steam chamber. Ducts and pipes for discharging gases (chimney) are separate structures that are added to the body of the boiler.

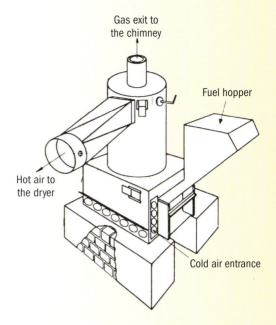

Figure 7.12 Indirect fired furnace for burning coffee milling byproducts.

The furnace is the part of the boiler where fuel is burned to produce vapor. If charcoal or firewood is used, the furnace will comprise grates on which the fuel is burned, a combustion chamber for the flames to develop, and an ash pit through which the combustion air flows.

Water tanks and a steam chamber, the internal part of the boiler, are made up of various airtight metal containers such as tubings, cylindrical housings, etc. that are interconnected and that have adequate resistance to the pressure of the steam. The interiors of these containers, whatever their connected form, hold the water that will be converted into steam. Their external surfaces almost entirely contact the combustion flames or gases. The lower part of this structure is the water tank, while the limited space between the water and the upper part of the structure is the steam chamber.

There is moisture in the exiting steam from smaller boilers because of the contact of steam with the water surface in the steam chamber. By definition, saturated steam is considered a mix of liquid water and dry steam. The quality of a wet steam is determined by the mass of dry steam contained in one kilogram of wet steam[53]. Tables for enthalpy of vaporization are elaborated for dry vapor and should be corrected when determining boiler efficiency.

The smoke ducts and chimney are used to remove the combustion products after they have transferred part of their enthalpy to the water vapor via the heating surface. The chimney also serves to increase the discharge velocity of the gases, producing a natural draft that increases air flow in the combustion chamber. This promotes air flow into the furnace, which accelerates combustion. These gases can also be removed using fans in a process called "artificial draft." When these fans are placed at the base of the chimney, the removal is called induction.

Figure 7.13 Boiler.

Attempts to manipulate the flow of hot combustion gases inside the boiler gave rise to the fire tube boiler, a great advancement in that it allowed for an increase in the heated surface area exposed to water and thus a more uniform distribution of steam generated by the water mass (Figure 7.14). Fire tube boilers can be classified into two basic types, reverse flame and direct flame, according to how the internal combustion gases flow. In a reverse flame fire tube boiler, combustion gases circulate in one direction through the furnace and combustion chamber before doubling back, often through interior tubes, to exit the boiler through the chimney. In a direct flame fire tube boiler, the gases flow in one direction, following a direct path to the chimney[54].

The following advantages and disadvantages have been cited for fire tube boilers[55][56].

7.14 Fire tube boiler under assembly, showing ⓐ fire tubes, chimney, and ⓑ inlet for cumbustion air.

Advantages:

- They are easy to build and require little masonry.
- They supply variable levels of steam.
- They provide a possible use for treated wastewater.
- They are relatively low-cost.
- Tubes are easy to replace.
- Maintenance is low-cost.

Disadvantages:

- They start slowly.
- Pressure is limited.
- They release a large amount of soot when burning solid fuels. This soot deposits on surfaces and aggravates environmental problems.
- Water flow circulation may be insufficient for optimal efficiency.

Research has shown that on some farms firewood-powered boilers use 50% less fuel than conventional dryers such as rotary dryers[57]. These systems feature a line of steam that feeds various compact heat exchangers (Figure 7.15) that in turn supply hot air to the dryers; they bring the advantages of improved drying air quality, fuel and labor savings, lower fire risk, lower pollution levels in the work atmosphere, and uniformity of drying air temperature. Principle disadvantages are high investment cost for the initial installation, and higher maintenance costs.

Another study was conducted on a drying system that had the option of direct or indirect heat for drying both natural and parchment coffee[58]. The system consisted of a small horizontal fire-tube boiler with a capacity of 100 kg of steam per hour. The thermic potential of the heat exchanger was 103.6 W. The combustion chamber

7.15 Steam lines and heat exchangers used for drying coffee at Martins Soares, Minas Gerais, Brazil.

accepted either firewood or charcoal (Figure 7.16).

When the indirect heating option was chosen, saturated water vapor was used as a thermic fluid for the transfer of energy to the air via a heat exchanger. Eucalyptus wood was used as the energy source for the furnace's internal boiler, which operated with a natural draft. Charcoal was used for the external furnace, and the boiler was operated with induced airflow. Thermic efficiency consuming 146 kg h^{-1} of charcoal was 45.2% heating 94.1 m^3 min^{-1} of air to 62.1 °C. Thermic efficiency

7.16 System for direct and indirect heating of air for drying high quality coffee, showing **a** hot air exhaust, **b** heat exchanger, **c** refueling reserve, **d** external furnace, **e** chimney, **f** water vapor, **g** boiler with internal firewood furnace.

using 177 kg h^{-1} firewood was 86.5% in heating 95 m^3 min^{-1} of air to a temperature of 55.5 °C. The lower efficiency of the charcoal can be attributed to high enthalpy loss via exhaust gases because of induced airflow as well as direct passage of the gases to the exhaust ducts[59]. The higher efficiency obtained from firewood can be attributed to the natural draft and air flow that caused combustion gases to spend more time in the boiler, the fact that gases were returned to the interior of the boiler, and the fact that the fuel was burned in the boiler's internal combustion chamber. The lower thermic efficiency of charcoal in this system could be balanced by labor savings, and if the enthalpy of the escaping gases could be used, it would allow for three small-scale dryers to be used at the same time, two with indirect heating and one with direct heating. This would make the average charcoal use of each dryer around 5.0 kg h^{-1}.

7.5.5.4 Biomass Gasifiers

Gasification, one of the technologies used to convert solid biomass into combustible gases, converts materials made of carbon, hydrogen, and oxygen into a combustible gas containing carbon monoxide (CO) and hydrogen (H2), key energy components.

In the last two decades, biomass gasification has been studied as an alternative way to convert the chemical energy in agricultural and logging industry byproducts such as hulls, straw, wood chips, sawdust, etc. Gasification has stirred interest from various sectors as a "green" alternative that takes advantage of byproducts that could otherwise cause environmental and societal problems where they are generated and that demand financial and material resources for proper disposal (e.g., sewage sludge). Gasification provides a way to add value to these byproduct materials, decrease their potential environmental impact, and make them into renewable fuels. The most recent studies in this area center around co-gasification, the gasification of two biomasses mixed together[60].

Gases generated in the process of gasification provide a clean energy source that can be used directly in dryers and thus have a high efficiency similar to that obtained with direct fire[61].

Gasification has stirred interest from various sectors as a "green" alternative that takes advantage of byproducts that could otherwise cause environmental and societal problems where they are generated. It provides a way to add value to these byproduct materials, decrease their potential environmental impact, and make them into renewable fuels.

During gasification, various physical and chemical reactions occur. The physical reaction involves heat and mass transfer; chemical reactions are redox reactions. Gasification is a heterogeneous process since it involves both solid and gas states.

When studying gasification, it is necessary to define the parameters involved in the process, such as the equivalence relation, specific gasification rate, and primary air flow. The typical equivalence relation for biomass gasification is between 0.2 and 0.4, depending on biomass characteristics such as granulometry, composition, density, and humidity. The technique of gasification offers the following advantages[62]:

1. Thermic efficiency is high, varying between 60% and 90%, depending on the system implemented.
2. Energy produced is clean.
3. Dried coffee is not contaminated by smoke or gases.
4. The energy level can be controlled and consequently the gasification rate can be easily monitored and controlled.

Gasification, however, presents the following disadvantages:

1. The biomass must be clean and free of dirt or other elements that can compromise the gasification process.
2. When a biomass with high ash load is used in a down draft (co-current fixed bed) gasifier, the potential fusion of the ashes can alter the performance of the gasifier.
3. If not completely burned, the tar formed during the gasification process can limit the uses of the gases produced.

7.17 Mixed flow gasifier used in coffee drying.

A mixed flow gasifier using wood chips was developed for coffee drying (Figure 7.17)[63]. An afterburner was added to the gasifier to burn the gas produced and to generate hot air. The gasifier was used to dry parchment coffee in a dryer with airflow of 46.3 m³ min⁻¹ and drying air temperature of 60 °C. The amount of fuel used varied between 15.3 and 18.8 kg h⁻¹. The system proved to be viable for coffee drying and there was no product contamination by smoke or fly ash.

Recent studies show that an upflow (countercurrent) gasification reactor can also be used to generate hot air for coffee drying[64]. When the gasification reactor was connected to a combustion chamber for the gases produced, the hot air generated was clean and thus appropriate for coffee drying. However, traces of combustible gases were found in the combustion chamber, meaning that the combustion chamber needed to be improved. Figure 7.18 shows an upflow gasification reactor. Figure 7.19, shows a down-flow gasifier that uses eucalyptus as its fuel source[65]. In this down-flow gasifier, the gases are generated in a reduction zone that is below the oxidation and pyrolysis zone. This results in gases that are free of phenolic compounds, which allows them to be used both for generating heat and for generating potential energy, such as with an internal combustion motor.

Figure 7.20 shows a flowchart of a gasification/combustion system, with entries for the biomass and the ambient air, and exits for the ash and the heated air.

7.6 Drying Coffee with LPG

LPG, or liquid petroleum gas, has only recently become a fuel source for drying coffee; in fact, until recently its use was prohibited for coffee drying in Brazil. Use of LPG has increased with the stabilization of the Brazilian Real relative to the U.S. Dollar in the 1990s and 2000s. In this period, many farms converted their firewood dryers to LPG since its consistency guaranteed better product quality capable of satisfying an increasingly demanding market. However, high cost has made LPG economically prohibitive.

7.18 Upflow gasification reactor; detail shows burning of gases in attached combustion chamber.

The use of LPG in drying facilitates easy, automated temperature control, eliminating the need for a machine operator to monitor constant temperature. Importantly, LPG also leads to a cleaner work environment as it does not produce smoke, dust, or gas, require firewood handling, or create fuel spills, etc., all key aspects in maintaining a clean service area free of contamination.

LPG burners establish the positioning and type of flame, and maintain continuous or intermittent ignition, depending on the drying needs.

There are numerous other advantages to using LPG for coffee drying:

7.19 Downflow gasification reactor; eucalyptus wood fuel in background.

- Complete combustion prevents carbon formation in deflectors (generally installed in the exit of LPG burners) and hot air ducts.
- LPG has high calorific potential (47,440 kJ kg^{-1}) compared to firewood (19,000 kJ kg^{-1}).
- There are no sparks in the fly ash, decreasing the risk of fire in the dryers.
- The coffee is not exposed to soot and smoke that can degrade product quality.

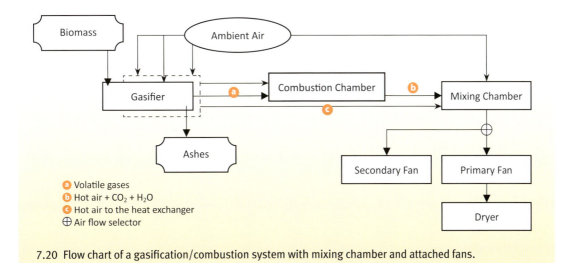

ⓐ Volatile gases
ⓑ Hot air + CO_2 + H_2O
ⓒ Hot air to the heat exchanger
⊕ Air flow selector

7.20 Flow chart of a gasification/combustion system with mixing chamber and attached fans.

- The absence of carbon and lack of fuel pumping reduces system maintenance in terms of replacing filters and cleaning jets. (Fuel moves as a result of pressure differences.)
- Equipment lasts longer due to the absence of corroding combustion gases and sulfur.
- There is minimal duct obstruction where the fuel flows.
- System automation is simple.
- Initial setup requires less time.
- Fuel flow can be finely adjusted, resulting in better temperature control.
- Storage space for fuel is minimal.
- Labor requirements are lower.

On the downside, LPG prices are tied to oil prices and therefore to the dollar. Also, there may not be a convenient or constant source for LPG near a farm, especially in times of high demand.

Many studies have been conducted comparing the cost of LPG to other fuels. Given big differences in exchange rates of the U.S. dollar and the Brazilian real as well as variable coffee market prices, the studies concluded that firewood was a better fuel source. Currently, coffee producers that use LPG to dry coffee are largely abandoning this fuel and moving back to more traditional fuel sources such as firewood and other lower-cost biomass fuels.

A study compared the cost of drying natural coffee using eucalyptus firewood and LPG[66]. Cross-flow dryers with firewood-fueled indirect-fire furnaces were used as well as other dryers in which the drying air was heated by steam generated in a firewood-powered boiler via hot-air-to-gas heat exchangers. The study also examined the use of horizontal rotary dryers with indirect heat using firewood furnaces. Isolating fuel costs, the cost of LPG was around three times that of firewood, while there was no quality difference in the product among the various energy options cited above.

7.7 Solar Energy

Solar energy drying is still the most common method of drying coffee worldwide. Various methods of drying with solar energy have been developed to increase drying efficiency and product quality. Independent of the technology adopted, drying capacity, drying time, and the quality of the final product are dependent on climatic conditions. These conditions vary with weather conditions and from region to region, making solar energy drying very unpredictable, with high risk of losing product or compromising product quality where weather conditions are unfavorable. Solar drying also requires intensive labor and large areas set aside for drying, and it leads to difficulties in coordinating the harvest with the drying. Further, a coffee producer who plans to use solar energy as a main energy source should, independent of the system adopted, have an auxiliary energy source, such as biomass, on hand. This will help to prevent product deterioration when climatic conditions are not favorable for drying.

The principal technologies used to dry coffee with solar energy are conventional patios, suspended beds, solar collectors, and solar rotary dryers. More recently, hybrid patios have come into use; also known as patio dryers, hybrid patios use both solar energy and heated air drying from combusting biomass.

A common alternative to using only solar energy for coffee drying is an integrated system that employs solar pre-drying combined with mechanical finishing. Coffee is dried by the sun until it reaches half-dry with moisture content between 35% and 40%, after which drying is finished in mechanical dryers. This option reduces drying time and allows for energy savings in heated air drying.

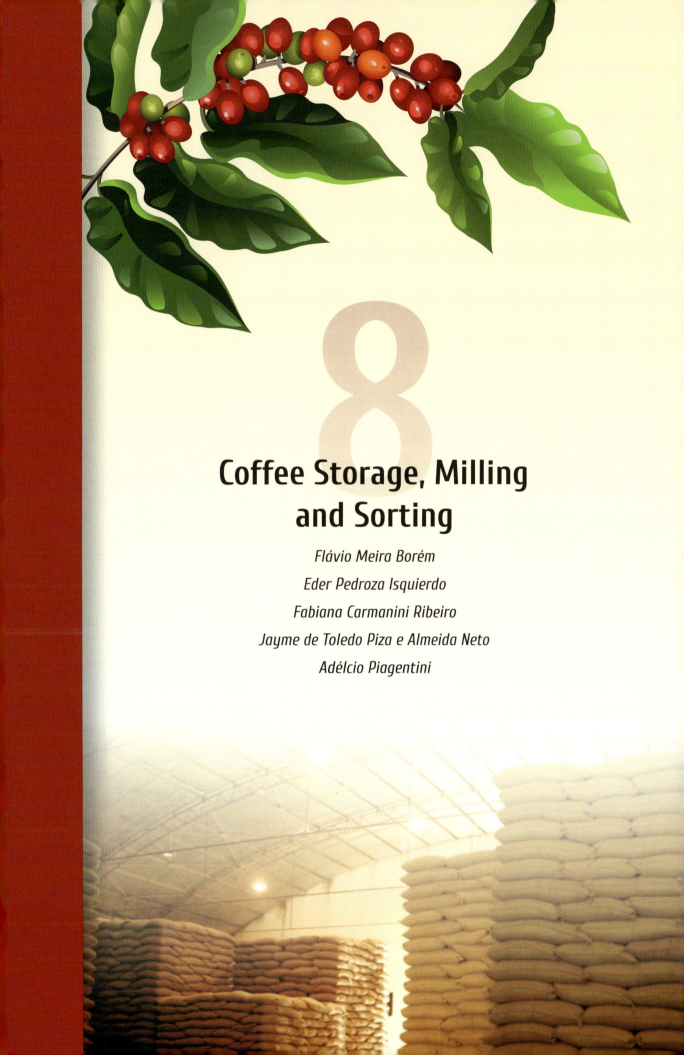

8

Coffee Storage, Milling and Sorting

Flávio Meira Borém

Eder Pedroza Isquierdo

Fabiana Carmanini Ribeiro

Jayme de Toledo Piza e Almeida Neto

Adélcio Piagentini

8.1 Introduction

The ultimate goal of coffee storage is to maintain product quality during the period between harvest and sale, allowing for adequate coffee distribution in order to supply different markets. For this to happen, a well-structured system of properly constructed warehouses and silos must exist on the farms as well as at cooperatives, mills, and distribution centers. Besides guaranteeing proper storage, these structures also provide the equipment necessary for milling, sorting, size grading, and classifying coffee. During storage, various changes can occur that jeopardize coffee quality. In addition to fungi and insect attacks, the metabolism of dried coffee—whether stored as dried pods, parchment, or green coffee—results in changes to the color, flavor, and aroma of the coffee. Furthermore, factors such as temperature, relative humidity of the ambient air, CO_2/O_2 concentrations, amount of light exposure, initial coffee quality, moisture content, state of maturation, and type of storage, among others, will determine how well coffee quality is preserved during storage.

8.2 Storage Network

A good storage network is an indispensable element of the coffee production chain. Such a network comprises conventional warehouses and bulk grain storage structures designed to receive processed coffee, preserve it in perfect technical conditions, and redistribute it at a future date. Due to the ability of a storage network to stabilize prices and guarantee supply, it is a key element in the economic management of an agricultural activity.

Furthermore, both product storage and movement should be managed through an efficient logistical system designed to reduce the interval between production and final delivery, making it possible for consumers to access products and services where and when they want and in the physical condition they desire, through channels of distribution and marketing[1].

A storage system is composed of storage units that are classified according to their location and stage in the supply chain. They can be classified as farm units, collective units, and terminal units.

Figure 8.1 Wooden coffee silos for coffee storage on farms.

Figure 8.2 Storage of coffee in bags.

Farm unit warehouses or silos are owned by the farm and are generally used only by one producer. In general, wood coffee silos (Figure 8.1) are primarily used for storage of dried coffee pods (natural coffee) or parchment coffee in bulk and, to a lesser degree, for conventional storage in bags.

Collective units, such as cooperatives and coffee warehouses, are located near farms or production centers and serve various users. Terminal units are located in industrial sites or ports. Both of these types of storage facilities store milled coffee, typically in standard coffee bags (Figure 8.2). Recently, exporters have forgone the bags and increased bulk storage of green coffee in order to reduce operational costs. A storage unit that is technically equipped, conveniently located, and has an efficient logistical system can make the production chain more economical.

8.3 Storage Methods

Storage methods can be classified according to product packaging and handling. Conventional storage refers to coffee packaged in jute bags. Bulk storage describes coffee that is handled and stored without bagging. Whichever method is chosen, the warehouse location should be clean and well ventilated, and it should receive direct sunlight.

8.3.1 Conventional Storage

Conventional storage of coffee in jute bags is the most common method. Facilities for conventional storage are relatively simple masonry and metal structures that are technically equipped to maintain the physical and sensorial quality of the coffee, guaranteeing its long-term preservation (Figure 8.3).

In addition to jute bags, green coffee can be packaged in "big bags" (Figure 8.4). These polypropylene fiber bags can hold 1,200 kg and allow for mechanized handling, which reduces loss during loading and unloading, minimizes manual labor, and eliminates the need for jute bags. However, these bags reduce static storage capacity to a maximum height of three bags because of their instability. This number can be higher, up to six bags, if physical support is provided.

Figure 8.3 Coffee warehouses.

Hermetic storage systems that allow for atmospheric modification and control are viable alternatives for preserving coffee quality, principally for higher-quality specialty coffees whose higher sale price can justify the increased storage expense[2].

GrainPro™ bags have recently come on the market in response to demand for improved green coffee packaging. These plastic bags are impermeable to CO_2 and are normally used as liners in jute bags[3]. They are commonly used for coffee that will be transported long distances or for prolonged periods, such as with maritime transport. These bags are more expensive, have a limited capacity of 60 kg, and do not provide much resistance to rugged wear and tear, often arriving in importing countries damaged and perforated[4].

Figure 8.4 Coffee stored in "big bags."

Hermetically sealed big bags can be used to store coffee in a controlled or modified atmosphere to preserve coffee quality for prolonged periods. These bags are less expensive and easier to handle than the 60 kg bags[5].

Conventional storage has several advantages, including the ability to keep small lots separate, use any quantity of bags to form blends, and remove bags that show deterioration. It is adaptable to the needs of small-scale markets, it lowers initial facility costs, and it facilitates easier access to stored lots for inspection than in a bulk storage system.

However, conventional storage has several disadvantages, one of which is the high operational cost to reposition bags due to the intense labor this requires. Other disadvantages include less flexibility in the operations of loading, unloading, and product movement; less adaptability to process automation; the inability to apply thermometry or aeration in order to control temperature; and larger area requirements compared to bulk storage.

Another limitation of conventional storage is the product's greater exposure to ambient air, which can result in problems caused by moisture exchange. Coffee can lose moisture to drier air, reducing the weight of the stored coffee. It can also absorb moisture in more humid times of the year, accelerating the processes of whitening and product degradation.

Nonetheless, the green coffee commerce is strongly rooted in the use of the conventional jute bags in terms of tradition and units of sale. This limits the use of other types of handling and storage. Jute bags remain the best option in cases where producers want to maintain the traceability of small lots of their coffee throughout the supply chain.

Conventional warehouses should be designed and built to create secure long-term storage conditions, with good ventilation, impermeable floors, controlled lighting, and adequate headroom. Temperature, moisture levels, and light strongly influence product quality and are important factors in determining the suitability of a facility for storing coffee. More specifically, controlling temperature and relative humidity, and maintaining low light levels are key to ensuring proper storage conditions. In warehouses located in hot, dry regions, small changes in temperature and relative humidity can be made using specialized equipment that

Figure 8.5 Rodent control in coffee storage facilities.

employs an adiabatic cooling system, which reduces temperature by increasing the ratio of air humidity to the enthalpy constant. This equipment cannot be used in humid regions, where dehumidifiers are recommended to ensure adequate storage conditions. Extractor fans can also be installed to facilitate air movement.

Other procedures can be used to further safeguard the coffee: protection from mice (Figure 8.5); higher foundations and impermeable flooring with proper ground drainage; isolation of the storage area from coffee processing areas to avoid infiltration of dust and humidity; keeping both the interior and exterior of the storage facility clean; and protecting the first two rows of coffee from humidity by wrapping them or preventing direct contact with the floor.

When opting for conventional jute bag storage, some technical building criteria should be followed for optimal coffee protection during storage and to facilitate operations inside the warehouse:

- Bay doors should
  - be installed on tracks and permit vehicle transit.
  - be located to enable straight, longitudinal, or transverse aisles across the warehouse. This will facilitate transporting, stacking, removing, loading, and unloading the coffee as well as allow for vehicle transit within the warehouse.
  - have awnings that are high enough for safe work conditions during coffee loading and unloading.
- Headroom should be at least 6 m, allowing easy access for workers on top of a stack and minimizing the effects of ambient heat inside the storage facility throughout the day.
- Internal walls should be coated to prevent accumulation of dust and microorganisms.
- Sides of the structure should allow for good ventilation while being secured with screens to prevent rodents and birds from entering the facility.
- The ground where the facility is located should be adequately drained before construction. Floors should be impermeable and made of concrete or asphalt.
- The design of the storage facility should maximize the use of the area, with a carefully constructed floor plan for the various stacks, blocks, and primary and secondary aisles.
- A fire prevention system must be installed.

Figure 8.6 Division of storage facility into aisles, blocks, and cells.

Figure 8.7 Identified primary aisle of a conventional storage facility.

Figure 8.8 Numbering of blocks in conventional storage facilities.

Conventional storage facilities use specific terminologies that facilitate the various handling and product depositing operations. The warehouse area includes all of the property occupied by the various structures and can be divided into usable and non-usable areas. Usable area refers to the area that can be used to store product, while non-usable area is for product receiving, transit aisles inside the storage facility, etc. The usable area usually represents 80% of the total area[6]. In order to efficiently organize and then locate the coffee lots, storage facilities are usually arranged according to the architecture of the building, specifically the roof and the configuration of doors and supporting columns.

Traditional warehouses have a pitched roof and are divided into sectors. Although the denomination of these sectors does not follow a fixed rule and it is usually up to each company to develop their own nomenclature for sector division, coffee warehouses are generally divided into sectors, gangways, blocks, cells, and stacks.

Gangways, or aisles, are pathways that divide the storage facility longitudinally or transversely, depending on the location of the bay doors. They should allow for movement within the facility and access to all stored lots without necessitating the removal of stacks or parts of a lot. Aisles can be primary or secondary. Primary aisles (Figures 8.6 and 8.7) are generally located in a longitudinal or transverse direction within the storage facility, with the ability to connect two bay doors that are in direct alignment. Secondary aisles separate the blocks, or even different lots within the same block, and are connected to a primary aisle. Secondary aisles are temporary, being created or eliminated when necessary.

The aisles should have sufficient width to allow for transit of forklifts, mobile conveyor belts, and in some cases even trucks, so as to facilitate the loading, unloading, and free movement of coffee. The aisles should also be clearly marked and easily identified (Figure 8.7).

In traditional warehouses, block location is determined by the location of the supporting columns of the roof and crossbeams, as well as by the distribution of the longitudinal and transverse aisles. In more modern facilities with larger open spaces, the division of the warehouse into blocks is based on the best use of the usable area rather than on the architecture of the building. How the blocks are numbered (Figure 8.8) varies with the design of the storage facility. In narrow facilities with only one longitudinal aisle, even and odd numbers should be given to the different sectors.

Figure 8.9 Identification of lots deposited in coffee warehouses.

Figure 8.10 Stack of coffee bags.

Figure 8.11 Protection of the lower layers of the stack with a tarp to avoid humidification of the coffee during longer storage periods.

Wide storage facilities can have more than one longitudinal aisle. This results in various blocks lining the aisles in different directions.

Lots can be formed according to coffee type and producer, thus preserving the identity of the deposited material. They can also be formed according to pre-established standards based on coffee quality, independent of the depositor and without producer identification.

The lots are arranged in blocks with specific numeration according to the criteria of the cooperative or warehouse (Figure 8.9). From an administrative perspective, the lot is the registration unit for the storage facility. It will have a defined address and its history must be registered from the moment of arrival at the facility until its departure.

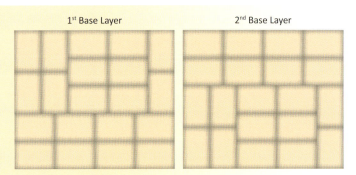

1ˢᵗ Base Layer   2ⁿᵈ Base Layer

Figure 8.12 Base layer of 20 bags (90 x 60 cm), occupying an area of 3.0 x 3.2 m.

Lots are deposited in stacks, created by the superimposition of successive layers (rows) of bags (Figure 8.10). To guarantee their stability, layers should be created by cross-stacking, or interlacing the bags, such that the abutment of two coffee bags in a row does not line up between successive layers.

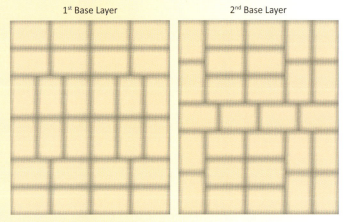

1ˢᵗ Base Layer   2ⁿᵈ Base Layer

Figure 8.13 Base layer of 28 bags (90 x 60 cm), occupying an area of 3.6 x 4.2 m.

Stacks usually take the form of a rectangle or square, depending on the type of base layer used. In some cases, "doubling" can occur, meaning that one lot is superimposed on top of another within the same stack. When this happens, it is necessary to correctly identify the bags that belong to each lot.

The base layer is the first layer of bags on the floor or platform. The number and arrangement of bags in the base layer define the geometric shape of the stack and the maximum quantity of bags that can be stacked. When stacking is done directly on the floor and the coffee will be stored for long periods, the first two layers of bags should be fully wrapped in a plastic tarp to avoid humidification of the product in the bottom layers (Figure 8.11).

Adequate securing of the stack guarantees its support and stability and can be achieved by rotating the pattern of each layer 90° or 180° or by superimposing rows with different layouts such that a bag never matches the position of the bag directly below it (Figures 8.12 and 8.13). In coffee storage, it is common to use base layers of 20 or 28 bags. In addition to securing the stack, the worker responsible must be able to stabilize the stack by correctly aligning it along a plumb line, while ensuring the correct number of layers is used. In order to maintain safe work conditions, stacks of no more than 25 layers of coffee in jute bags are recommended.

Figure 8.14 Cut separating two stacks located in the same section.

More than one stack can be assembled in the same block, forming cells. Cells should be arranged in the blocks to allow for transit along the longitudinal and transverse aisles. Stacks should be placed at least .7 m from the walls for easy access and to mitigate humidity problems.

Stacks within the same cell are separated by a cut (Figure 8.14). A cut is a small gap that divides a cell into two or more stacks. This division is done to make the best use of the space and better facilitate the removal of parts of a lot without affecting the stability of the stack or of adjacent stacks. Using cuts reduces costs and optimizes operations within the storage facility.

Figure 8.15 Wood coffee silos.

Figure 8.16 Equipment for loading coffee into containers.

8.3.2 Bulk Storage

In bulk storage, coffee is packaged and handled without bagging. This is the predominant form of storage for both natural and parchment coffees on the farm. Wood coffee silos, traditionally used for this purpose (Figure 8.15), are usually multi-compartment structures made of wood with either flat or angled bases. Flat-based silos have greater capacity but require labor to be completely unloaded.

Silos should be made of wood to guarantee greater thermal isolation and to avoid humidification of the coffee. Nonetheless, in certain regions or times of year, humidification can occur in the coffee located in the upper part of wooden silos. In these cases the coffee should be covered with cloths or tarps.

Storing green coffee beans in bulk has become more popular with large exporters who want to reduce handling and transportation costs. Some warehouses and cooperatives are already using the bulk system in combination with the conventional system. In this case, the coffee arrives at the storage facility in jute bags and is stored conventionally, then placed in bulk containers prior to export (Figure 8.16).

Bulk storage offers some advantages over conventional storage, including faster loading and unloading, lower transportation costs, greater temperature control for the stored product when equipped with an aeration system and thermometry, more efficient use of the storage area, and lower human resource requirements due to the significant use of mechanized processes. However, construction of bulk storage silos requires greater initial investment, and with bulk storage it is difficult to individualize smaller lots and to remove damaged or deteriorated beans. Another disadvantage of bulk coffee storage is the challenge of precisely measuring and controlling inventory.

8.4 Procedures for Coffee Storage Facilities

The procedures and nomenclature described below are common to coffee warehouses, with variations among some companies due to location, internal norms, and sales volumes.

8.4.1 Nomenclature

A. Human Resources

- **Manager**: the administrator responsible for the warehouses. The manager is responsible for the physical integrity of both mobile and immobile assets of the storage facility as well as for the deposited merchandise. In addition, this person should supervise the services offered, ensuring they are completed as required. The manager must coordinate all activities involving product reception, storage, hulling, sorting, blending, and dispatch, among other activities in line with the administration of each warehouse[7].
- **Warehouse foreman**: coordinates all logistics and operations of the warehouse, identifying locations for unloading, stacking, etc.

- **Checker**: subordinate to the warehouse foreman and responsible for the completion of all routine activities in the warehouse, such as reception, hulling, and dispatch, while adhering to specific protocols for operations.
- **Weigher**: worker responsible for the control of the scale.
- **Reception and dispatch administrators**: employees responsible for the issuance of receipts, bills of lading, or other legally appropriate documentation, as well the order of the loading and unloading of all products.
- **Controllers**: workers responsible for the verification of loading, unloading, and removal operations, among others. Controllers confirm inventory movement, generating service instructions (see below) or other necessary documents.
- **Sampler**: worker responsible for collecting and sending samples for classification.
- **Stacker**: worker responsible for the proper stacking of coffee in bags and, depending on the size of the lot, selecting the best type of base layer to maximize the area of the storage facility, with worker safety in mind.

Figure 8.17 Scale, reception and dispatch.

- **Loader**: operator who handles the manual movement of coffee bags, loading and unloading of trucks, and manual transport for the stacking and removal of lots.

B. Departments

- **Entrance gate**: department responsible for the registration of vehicles and identification of people passing through the storage facilities. This is the first point of contact of the coffee with the warehouse. In this department, all movement activities are registered and a database is maintained for future movement activity. In addition, front gate workers issue a ticket for access to the scale. This registration process is necessary since all product movement within the warehouse can only be completed with the appropriate fiscal and internal documentation.
- **Scale**: department that registers the weight of the trucks (Figure 8.21), then verifies, records, and videos (when equipped) everything about the load from the moment it is weighed.
- **Reception and Dispatch**: department that receives and confirms all necessary documentation for transport within the facility (Figure 8.17). For coffee to be weighed, this department verifies information on fiscal receipts as well as the ticket issued by the entrance gate.
- **Accounting**: department responsible for the control of physical inventory, lot verification, and reconciliation of inventory and producer supply figures with fiscal and accounting documentation. Accounting is also responsible for the fiscal control of inventory, auditing of all documentation related to the movement of coffee, storage security, and monitoring all processes that prepare the coffee for analysis, among other activities.

C. Codes and nomenclature used in the movement of coffee

- **EDN** (Entrance Document Number): documentation necessary to enter and deposit coffee in the storage facility, issued by Reception and Dispatch.
- **SI** (Service Instruction): internal document that outlines the service to be carried out, including information such as number, quantity, weight, and location of lots. The SI is issued by the Classification Department for the processes of coffee hulling, blending, sorting, and bagging.
- **LO** (Loading Order): internal document issued at the time of coffee dispatch, specifying the lot number, quantity of bags, weight, and lot location.
- **LT** (Loads Transfer): document used in the transfer of loads from affiliate facilities to the main storage facility. The LT should be completed at the lot's place of origin with information about the lot that will be transferred (quantity of bags, weight, and standards), and it must be presented at the time of dispatch to the main facility.

8.4.2 Description of operations

The different operations carried out by conventionally equipped storage facilities—receiving, storing, and dispatching of coffee bags—can be seen in figures 8.18 and 8.21. However, it should be noted that these procedures will vary per the internal procedures of each company, the structure of storage facilities, and whether bulk processes are used.

Figure 8.18 Flowchart of operations conducted when coffee enters the storage facility (courtesy of Cooparaíso).

8.4.2.1 Unloading green coffee

Before transporting coffee, the producer should either have a fiscal receipt or a receipt issued by the warehouse that is responsible for the coffee while in transit (Figure 8.18a). When the vehicle arrives at the storage facility, it proceeds to the entrance gate where the vehicle, driver, and coffee origin are registered (Figure 8.18b). Once authorized to enter, the entrance gate issues a temporary entrance ticket with a sequential number, then the vehicle proceeds to the scale (Figure 8.18c).

Before weighing, the driver gives the fiscal receipt and the temporary ticket to the person responsible for reception and dispatch in exchange for the issuance of an Entrance Document Number (EDN) (Figure 8.18d). This document should include the name of the producer, name of the farm, number of bags, type of coffee (natural, parchment, or green), type of packaging, etc.

Before weighing and unloading, a pre-sample is taken by perforating the exposed bags on the sides of the shipment on the truck. This pre-sample is expedited to the classification department, which issues a Service Instruction (SI) for unloading the coffee based on the warehouse's classification standards. This reduces operating costs by ensuring the coffee is placed in the warehouse to facilitate future movement for operations such as removal and blending.

After pre-sampling and issuance of the EDN, the coffee is taken to the scale where the weigher takes the initial weight (gross weight) (Figure 8.18d). The vehicle then proceeds to the unloading facility according to the classification standard listed on the pre-sample card (Figure 8.18e). After unloading, the vehicle returns to the scale for final weighing in order to obtain the net weight of the coffee. The net weight of the coffee is entered on the EDN, which, along with the quantity of bags, yields the producer's balance (Figure 8.18f). When unloading lots of fewer than five bags, weighing should be done in the storage area, using scales with a capacity of 300 kg. The emission of sample protocols, inspection reports, and transaction reviews (Figure 8.22g, Figure 8.22h, Figure 8.22i) are internal operations that are verified by appropriate departmental authorities.

At the time of unloading, the checker takes the EDN, indicating where the coffee should be deposited, and confirms the number of bags, type of coffee, packaging, and the condition of the stack where the coffee is to be deposited. Another sample should be taken, this time an official sample that will serve as the reference sample for this lot. These samples will also help identify possible quality differences in the received coffee. When differences are found, the coffee is separated into different lots. Samples are packaged in paper bags or other appropriate receptacles with a capacity of 1 kg, and taken to the classification department. Unloading should be handled with utmost care, with verification of consistency between the entrance document and the quantity received in the stack.

The deposit is then entered on the entrance document and the daily inspection report. The stack must also be correctly labelled. For loads that are divided into many lots, the stacks should be placed on base layers that hold ten bags. Next, EDN are completed with the number of the lot, quantity of bags, unloading location (address or facility number, aisle, and block), vehicle license, etc.

Green coffee can be unloaded in big bags or in bulk. When unloading big bags, all bag volumes should be sampled to check for any differences in quality. In this case, samples are removed using a special double-tube spear sampler that extracts small amounts from deep within the big bag. A forklift should be used to unload big bags. When there are two or more types of coffee in a load, weighing should be done separately as there could be differences in quality.

When unloading a bulk shipment, the depositor should provide advance notice of the delivery since, in some cases, prior communication is required for bulk loads in order to prepare for receiving, blending, and sorting. Storage facilities that do not normally receive bulk shipments should plan ahead and take measures to facilitate the unloading of bulk coffee so as to reduce operational costs. When the bulk coffee needs to be bagged, the warehouse supervisor organizes the loading area to handle bagging using either silos or hoppers.

8.4.2.2 Unloading bagged dried pod coffee

The procedures for unloading dried pod coffee are the same as described in the previous section. However, it should be noted that the conditions of the area that will receive the lot (silo or stack) should be verified. The care taken to confirm the quantity received and to complete internal documentation should be the same as that taken when unloading green coffee.

8.4.2.3 Direct discharge

With direct discharge coffee, the coffee arrives at the warehouse already linked to a Service Instruction. Therefore it is not necessary to take an initial sample to be sent to the classification department; rather, the classification department simply stamps the EDN instead of issuing an SI.

Apart from this, unloading procedures are similar to those previously described. However, since the coffee arrives with a Service Instruction to be directly unloaded into the warehouse (bypassing the initial verification protocols), extreme care should be taken to verify quantities both when receiving bags of coffee or bulk coffee directly into the hopper.

Once unloaded, the EDN should be completed with the number of the lot, quantity of bags, and the unloading location marked "DIRECT DISCHARGE."

8.4.2.4 Hulling, blending, and sorting

A. Hulling

When dried pod coffee is received and the depositor authorizes it to be hulled, after unloading and sampling have been completed the coffee is sent to the hopper and then to the silo. The name of the depositor (producer) and the number of the lot are recorded, opening an SI no later than the following day. The hulling machine is first labeled with the appropriate coffee name/lot, then the coffee is hulled. The resulting green coffee is packaged in jute bags and the name of the producer is marked on the bags. The bags are then sent to the inventory storage facilities, according to the assigned SI number. After hulling, a sample is collected following the procedures outlined under the section on unloading.

B. Blending

Blending is carried out according to the SI issued by the Classification Department. In the warehouse, the lots listed in the SI are removed and sent to be stored in column stacks, which are non-interlaced stacks of jute bags or big bags without base layers. The coffee is organized in this way to facilitate the production of homogeneous blends in the blending operation.

The column stacks are assembled near the silo in the longitudinal direction of the warehouse. The bags are stacked to a height of 10 bags. The different lots that will be blended are stacked in alternating fashion in a longitudinal direction (Figure 8.19). The lots to be transported to the blending silo are removed in a transverse direction, following the stacking sequence, so that each load, with different parts of each lot, is taken to the silo in the blend specified in the SI, guaranteeing maximum homogenization (Figure 8.20). If the SI specifies more than 10,000 bags, the stack is composed of blocks of 5 base layers with a height of 25 bags.

After a specific period of homogenization in a blending silo with a capacity of 500 bags, the coffee is bagged and a 30 g sample is removed from each bag. Once the truck, cart, or

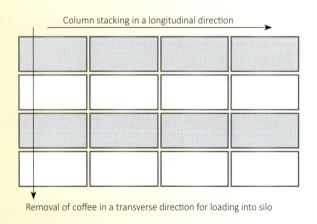

Column stacking in a longitudinal direction

Removal of coffee in a transverse direction for loading into silo

Figure 8.19 Column stack assembly for loading into silo.

container is loaded, or after the lot has been transported to the warehouse, a sample is sent to the Classification Department for final analysis of the shipment according to the standard desired and prescribed in the SI.

The loaded vehicle is directed to the scale and, if necessary, any weight difference is corrected.

When blending is done for coffee packaged in "big bags," the procedure is similar to that described for coffee in jute bags. However, the stack is made by separating the lots in the stack and unloading them into the hopper in smaller portions, according to the requirement in the SI.

C. Sorting, Classification, and Size Grading

Each cooperative or coffee warehouse has unique commercial standards. When the Sales Department makes a sale, the Classification Department should be informed of the specified internal standard and sales quantity. The Classification Department then issues an SI, defining how the blend should be made to meet the sales request. The Service Instruction is sent to the warehouse checker.

Figure 8.20 Blending silo for homogenization of coffee.

In the warehouse, the lots listed on the SI are removed from the stack in the specified proportion and transferred, in bulk, to a container to verify the weight of the blend. The blend is then sent to be sorted, classified, and graded. The procedures followed should be in line with the type of processing requested by the Classification Department, including basic dry milling, electronic selection, dry milling with ventilation, or complete dry milling that can include sorting, classification, and grading.

8.4.2.5 Transfers

The unloading of coffees coming from affiliate storage facilities is handled according to previous explanations for unloading green coffee, but with the checker working with the Transfer Loads documents, where the lots from affiliates, and their respective quantities, are recorded.

The checker verifies quantities, as well as possible separations, according to the classification of the transferred lots. The TL can refer to either a transfer from affiliates or to coffee arriving directly from the farm, in which case the same principles for normal unloading are used.

8.4.2.6 Removal

The process of removal occurs when the lot reported on the SI is not on top of the stack, and therefore lots that are not outlined on the Service Instruction must be temporarily relocated so that the lots on the SI can be accessed. Before temporary removal, the cut, stability, coffee standard/quality, and general stacking conditions of the stack should be verified to guarantee operational safety. Warehouses should also use the opportunity to replace any torn bags. After removal, the appropriate entries are made on the daily inspection report.

The key to temporary removal is to properly code lots when doubling occurs (more than one lot is in the same stack or base layer) so that lots can be identified quickly and correctly. In cases where the lots are removed and then taken to external storage facilities, TL's should be properly completed with the number of lots and their respective quantities.

8.4.2.7 Shipment/Embarkment

Coffee is shipped from the warehouse for various reasons, including direct sales, indirect sales, export sales, returns, transfers, or direct shipments.

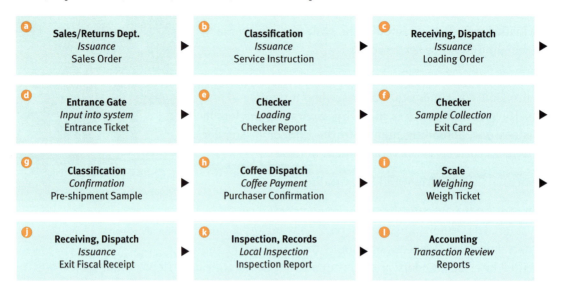

Figure 8.21 Flowchart of operations conducted when coffee leaves the storage facility (courtesy of Cooparaíso).

After completing a sales order (or in some cases a return credit), the Sales/Returns Department (Figure 8.21a) informs the Classification Department of the internal standard to be shipped. The Classification Department then specifies the lots and their respective proportions to make the blend in order to meet the standard specified in the sales contract. The Service Instruction, which outlines the shipment specifications, is then sent to the reception/dispatch department as well as the checker (Figure 8.21b).

The Reception and Dispatch Department, using the SI, issues a Loading Order, which is sent to the Entrance Gate to register the vehicle (license, model, and name of customer or destination) and issue a temporary entrance ticket. The empty vehicle is then weighed and its initial weight is recorded (Figure 8.21 c and d).

With the SI in hand, the checker will
- monitor the electronic scale, verifying weight in 10 bag increments,
- note the information on a weight monitoring form,
- confirm the transport load with the accountant (Figure 8.21e).

The way coffee is transported depends on its destination. Coffee can be packaged in jute bags or can be packed in bulk directly into containers, using container liners and injecting the coffee directly into the container.

Once shipment preparation is completed, all loading information should be entered into the Daily Inspection Report and on the Classification Card (Figure 8.21f). When silos are used in preparing the shipment, every bag should be punctured to obtain a sample that is then sent to the Classification Department prior to shipment (Figure 8.21g).

Direct shipments, where the loading takes place in the storage facility, can only be initiated with the SI in hand (remittance for shipment). The person responsible for shipping verifies the quantity in the stack with that listed on the SI, communicating to The Account and Warehouse Inspection Departments any discrepancies prior to service execution. Before shipping, the shipment is identified, as well as the general conditions of the transport vehicle, including bodywork, decking, lining, odor, cleanliness, oil stains, and humidity.

If, for whatever reason, the lot listed for shipment cannot be loaded, the Classification Department should be informed so that the SI can be updated. The Account and Warehouse Inspection Departments should also be informed in order to issue an Internal Communication.

After loading the vehicle, the driver waits for payment clearance and the fiscal receipt (Figure 8.21h). When the vehicle is exiting (Figure 8.21i), the entrance ticket must be shown along with this fiscal receipt in order to complete the exit registration (Figure 8.21j), including the quantity of bags and the receipt number, for internal control purposes. Once completed, the appropriate annotations are made on the Inspection Report, removing the lot from inventory (Figure 8.21k).

Figure 8.22 Simplified coffee milling flowchart.

8.5 Coffee Dry Milling

The purpose of coffee milling is to transform the dried coffee fruit pod and parchment coffee into green coffee that will then be further sorted, classified, and graded (Figure 8.22). This section explores the stages and associated equipment in the movement, processing, and storage of coffee.

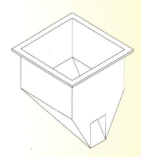

Figure 8.23 Hopper.

8.5.1 Auxiliary Equipment in Dry Milling

Auxiliary equipment is the equipment used for the accumulation, flow, and transport of coffee during dry milling, and whose operation does not alter the product composition. The principal function of this equipment is to ensure continual and uniform product flow during milling, sorting, and classifying, thus optimizing the performance of the primary machinery used in dry milling.

8.5.1.1 Accumulators

Depending on their form, size, and function, accumulators are called hoppers, wood coffee silos, bins, or silos.

A. Hoppers

A hopper (Figure 8.23) is a temporary accumulator that usually features a rectangular upper section and a lower section in the form of a pyramid. One side is straight and open so that the product may easily flow out of the hopper. Receiving hoppers, where product arrives for milling or sorting, are generally larger in volume than intermediary hoppers, which are mainly used to regulate product flow. The mouth of a receiving hopper is usually covered by a protective grate, which, in some cases, is thick enough to support the weight of a dump truck.

B. Wood Coffee Silos

A wood coffee silo is a large volume collection area or facility that is divided into compartments that hold up to 200 m³ of dried coffee pods or parchment coffee; it is used for holding coffee prior to dry milling. Generally a wood coffee silo is made with a flat base built directly onto the existing floor; this eliminates the need for a supporting structure and increases ca-

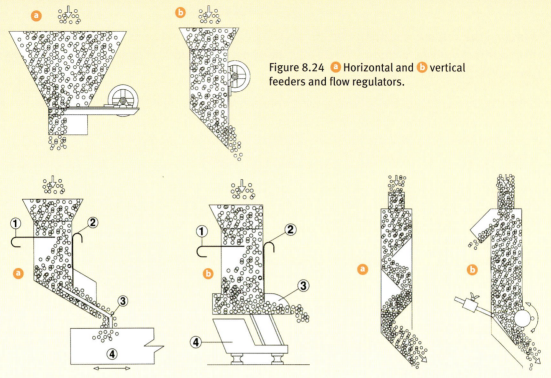

Figure 8.24 ⓐ Horizontal and ⓑ vertical feeders and flow regulators.

Figure 8.25 Drawer regulator with "stimulation." ⓐ The "stimulation" is produced by an oscillating rod (3) within the exit chute (4). One of the regulators (2) controls the flow and the other (1) simply opens or closes. ⓑ A linear vibrator (4) with frequency variation, besides stimulating the flow, helps to control the flow through the chute (3) together with the regulator (2) providing for more operational precision.

Figure 8.26 ⓐ Baffle and ⓑ tray seed ladders.

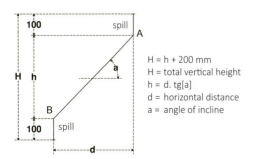

Figure 8.27 ⓐ Seed wheel and ⓑ horizontal auger.

Figure 8.28 Tube showing change in direction.

H = h + 200 mm
H = total vertical height
h = d. tg[a]
d = horizontal distance
a = angle of incline

Figure 8.29 Calculation of height (h) from point B to point A that is horizontal distance (d) from B.

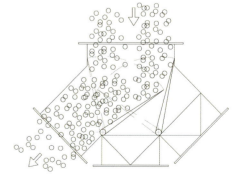

Figure 8.30 Distributor head with three flow options.

pacity. However, as mentioned, the downside to a flat-bottom surface is that it necessitates manual labor to move product in and out of the compartments, where a sloped bottom employs gravity to empty the silo.

C. Bins and Silos

Bins and silos are large-scale accumulators generally used for green coffee. These accumulators have a cylindrical or prismatic superior part, with a discharge funnel in the center. After milling, sets of bins and silos are used to separately store coffee that has already been classified so that subsequent operations are not duplicated.

8.5.2 Feeders and Flow Regulators

Feeders and flow regulators are machines that gather, dose, and position product flow, ensuring manageable volume and uniform flow. Figure 8.24 shows feeders and flow regulators in the form of (a) horizontal drawer and (b) vertical drawer. These registers are simply constructed and easy to regulate; however, dosage levels can often be thrown off when the product contains impurities or when moisture content is above the milling standard. Figure 8.25 shows two types of feeders with stimulated flow that allows for better dosing and flow control.

Figure 8.26 shows a baffle (a) and tray feeder (b). These feeders are ideal for positioning the product in that they provide a cascade of coffee that facilitates coffee flow. The baffle feeder spreads the product via inclined plates placed in alternating positions, and dosage is handled by separate equipment that precedes the baffle feeder on the line. Tray feeders have a "stimulated feed" caused by a rotating ribbed axel, as well as a "thief" in the feeder box that guarantees a maximum level of product and therefore a constant weight on the tray, which has a counterweight that can be regulated.

Seed wheels and horizontal augers, illustrated in Figures 8.27 (a) and (b), precisely control the flow of the product and seal the passage from one environment to another.

8.5.3 Transporters

8.5.3.1 Tubes and directional registers

Gravitational transport through tubes and gutters should be employed whenever possible when setting up grain transport infrastructures. Tubes should be set at a 45° angle, giving 100% declivity. In cases where clean and dry coffee enters the tube with velocity and an exit is open, an inclination of up to 35° can be used, giving a declivity of 70%.

Table 8.1 Transport capacity in m^3 h^{-1} for tubes of different diameters.

Tube diameter (mm)	Transport capacity (m^3 h^{-1})
100	25
150	50
200	80
250	100

Table 8.1 shows that the transport capacity m^3 h^{-1} by gravity depends on tube diameter. This capacity was calculated by taking into account coffee density (0.4 kg m^{-3} for natural coffee pods and 0.6 kg m^{-3} for green coffee) and an incline of 45°. When directional change is necessary within the tubes, a small vertical trajectory of 10 to 15 cm should be inserted to reaccelerate the product flow (Figure 8.28).

The solution to a common problem in planning tube layout is demonstrated in Figure 8.29. How does one calculate the height (h) of point A in relation to point B given the horizontal distance between the points (d) and the angle of inclination (a)?

Distributor heads, also called directional registers, may be inserted into the tubes to permit different flow options. The distributor head in Figure 8.30 allows for three flow options.

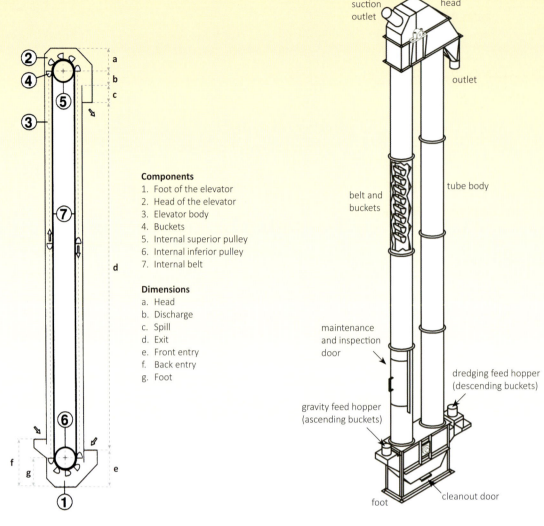

Components
1. Foot of the elevator
2. Head of the elevator
3. Elevator body
4. Buckets
5. Internal superior pulley
6. Internal inferior pulley
7. Internal belt

Dimensions
a. Head
b. Discharge
c. Spill
d. Exit
e. Front entry
f. Back entry
g. Foot

Figure 8.31 Schematic drawing of an elevator showing principal dimensions and components.

Figure 8.32 Coffee elevator. (Courtesy Pinhalense S.A. Máquinas Agrícolas)

8.5.3.2 Elevators

All vertical transport in dry milling is done using bucket elevators. Figure 8.31 shows the design of a bucket elevator, including its dimensions and principal parts. The internal pulleys are connected by a belt to which the buckets that receive the coffee are attached. Upon reaching the head of the pulley, the buckets release the product through a dispensing tube. High velocity discharge in elevators is done using a centrifuge, while discharge at lower velocities is performed by gravity.

When the coffee enters side A of the elevator and fills the buckets going up, the velocity of the belt can reach 1.5 m s^{-1}. When the coffee enters side B, where the buckets must carry the load at the foot of the elevator, the velocity should not go above 2.7 m s^{-1}. It should be emphasized that bucket elevators should only be used for coffees that have been milled.

Figure 8.32 diagrams a coffee elevator, naming its principal components and depicting the buckets going up and through a self-cleaning brush, an optional component on elevators. Note that the compartment at the foot of the elevator is telescopic and, together with the stretcher, it maintains a constant distance between the buckets. In general, coffee elevator capacities range from 6 to 150 m^3 h^{-1}; they vary in height depending on volume demands.

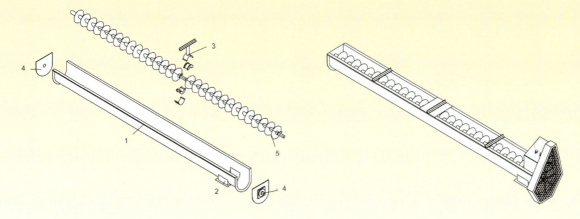

Figure 8.33 Exploded view of auger-style U-trough.

Figure 8.34 Auger conveyor with U-trough.

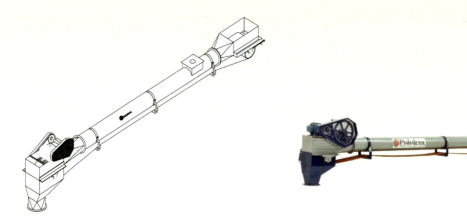

Figure 8.35 Schematic of a tubular belt conveyor. (Courtesy Pinhalense S.A. Máquinas Agrícolas)

Figure 8.36 Tubular belt conveyor. (Courtesy Pinhalense S.A. Máquinas Agrícolas)

8.5.3.3 Helicoid Conveyors

Figure 8.33 shows a helicoid conveyor (also known as an auger or screw conveyor) in an expanded view illustrating its various components. The rotation of the helical screw moves the coffee up the shaft.

Figure 8.34 shows a helicoid conveyor with an open trough. In general, the bodies of helicoid conveyors have open troughs and tubes designed for specific purposes. Helicoid conveyors used for coffee transport range in capacity from 1 to 50 m^3 h^{-1}, and their length is limited to 20 m. They allow for product entry and exit at any point and can be used as mixers to combine various products. They can also transport over an incline; however, their transport capacity decreases with an increase in inclination. They can be highly wasteful and are not self-cleaning.

8.5.3.4 Conveyor Belts

Figures 8.35 and 8.36 show a drawing and photo of a tubular belt conveyor, which range in capacity from 5 m^3 h^{-1} to 80 m^3 h^{-1} and are limited in length to 20 m.

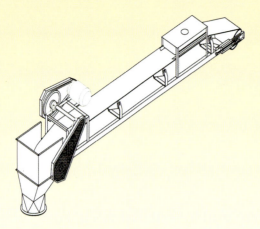

Figure 8.37 Inclined roller conveyor belt schematic (without "tripper"). (Courtesy Pinhalense S.A. Máquinas Agrícolas)

Figure 8.38 Inclined roller conveyor belt (without "tripper"). (Courtesy Pinhalense S.A. Máquinas Agrícolas)

Figure 8.39 Vibratory conveyor with inclined springs. (Courtesy Pinhalense S.A. Máquinas Agrícolas)

Natural or Parchment Coffee
↓
Receiving
↓
Cleaning
↓
Hulling
↓
Preliminary Selection
↓
Milled Coffee

Figure 8.40 Flowchart of coffee receiving, cleaning, and hulling.

Figures 8.37 and 8.38 present a diagram and photo, respectively, of an inclined roller conveyor belt called a mat top conveyor. This type of conveyor is capable of high capacities of between 40 m^3 h^{-1} and 300 m^3 h^{-1}. Intermediary exits can be added using a mobile discharge station called a "tripper," and the length of the belt can vary with project demands.

Figure 8.39 shows a vibrating chute with inclined springs (which is also used in the exit chamber of the wood coffee silo). There are two screens located after the exit that help eliminate impurities both in dried pod coffees and parchment coffees. Vibrating chutes range in capacity from 4 to 40 m^3 h^{-1} and are no longer than 15 m.

8.5.4 Coffee Cleaning, Destoning, and Hulling

Coffee cleaning, destoning, and hulling are the first steps in dry milling. These are the operations that remove impurities and transform dried coffee pods and parchment coffee into green coffee beans, leaving behind the coffee hulls and parchment. These operations are generally performed on the farm by the coffee producer to add value to the product. They also decrease the volume of material that must be transported, ensure a better yield estimate for the coffee, and keep the byproduct material on the farm where it can be utilized. A flowchart of these operations is shown in Figure 8.40.

In receiving, the various lots of coffee are identified and stored in wood coffee silos. Each lot should be properly identified to guarantee its traceability in terms of cultivar, location, age of plants, products used in crop management, and procedures used before dry milling. These identifications are essential for both the coffee producer to perfect their coffee, and to fulfill requirements for various quality certifications. Storing dried coffee pods or dried parchment coffee in wood coffee silos permits the producer to conserve product quality while waiting for the best moment to begin dry milling and sell the coffee.

Cleaning operations are carried out by a combined hulling unit machine called a cross-beater huller with oscillating screen (Figure 8.41). An illustration of how this equipment works, along with its various attachments, is shown in Figure 8.42.

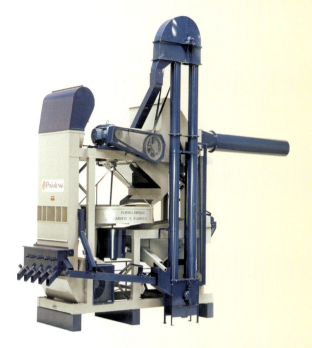

Figure 8.41 Cross-beater huller with oscillating screen. (Courtesy Pinhalense S.A. Máquinas Agrícolas)

Cleaning begins when the coffee stored in one of the hoppers (1) flows through register (1a) by means of an induced and controlled flow via a vibrating chute (2). The chute has two screens at the end, the first (2a) having holes smaller than a coffee bean and the second having holes larger than a coffee bean, thereby removing impurities. The elimination of impurities lighter than coffee is performed by the pneumatic column (4) at the entrance of the combined hulling unit fed by the bucket elevator (3). The removal of impurities heavier than coffee is done by a "destoner" in a process called "air float." In this process, air is blown over a perforated, vibrating, sloping table bed that separates the product based on density, with the less dense coffee floating and moving towards the lower part of the table. At the same time, the heavier impurities do not float but rather stay in contact with the surface of the table and are moved upward by the vibrations.

Hulling starts at the point where the elevator (6) receives clean coffee and feeds it to the huller (7), which is the heart of the milling process. Modern hullers have a rotor with blades that can be regulated, and an external perforated screen (Figure 8.43). The entrance (A) is an opening in one of the lateral discs (E), through which the coffee enters and is then pushed by the rotation of a rotor (C) and the inclination of the blades (D) against a perforated screen (B), which breaks the hulls and frees the beans. The product mass that moves through the perforated screen contains beans, hull, or parchment, and unhulled coffee. It passes through the pneumatic column (8, Figure 8.42) where the hull is aspirated away. The clean beans and coffee that was not hulled fall onto a circular oscillating screen (9) whose bottom is slightly inclined (9a) so that the two components are separated by centrifugal force. The hulled coffee, being more dense, exits through an opening (9d) in the edge of the circular screen. The less-dense whole (unhulled) coffee converges in the center of the screen and is returned to the huller through an opening (9b). Figure 8.43 details the hulling process. The circular screen shown in this diagram contains a special feature that, in addition to performing the above hulling process, has an inclined, perforated bottom (9a) that allows for the elimination of beans with a lower screen size through an opening (9c), thus preventing eventual loss.

COFFEE STORAGE, MILLING, AND SORTING 〖 165 〗

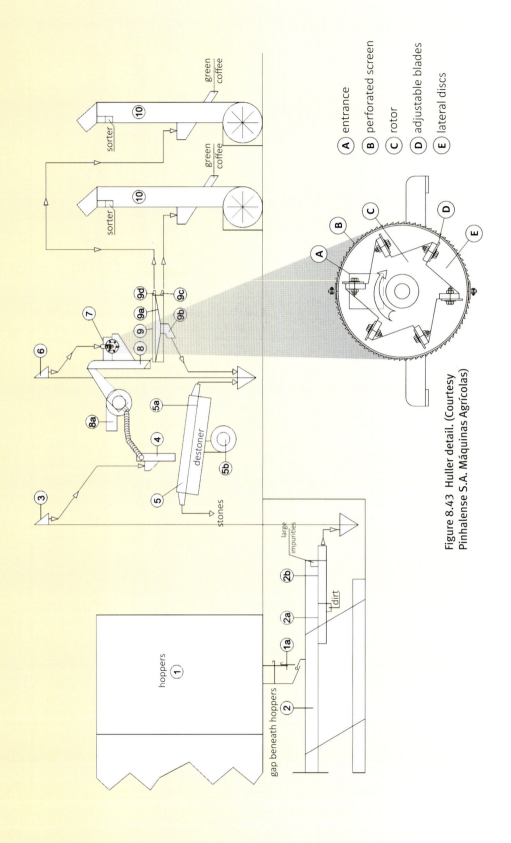

Figure 8.43 Huller detail. (Courtesy Pinhalense S.A. Máquinas Agrícolas)

A entrance
B perforated screen
C rotor
D adjustable blades
E lateral discs

Figure 8.42 Functional schematic of a cross-beater huller with oscillating screen. (Courtesy Pinhalense S.A. Máquinas Agrícolas)

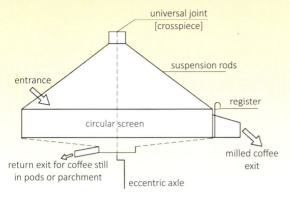

Figure 8.44 Profile of simple circular oscillating screen.

Milled Coffee
↓
Receiving
↓
Cleaning
↓
Size Grading
↓
Sorting
↓
Blending
↓
Shipping

Figure 8.45 Flowchart of processed coffee.

Figure 8.44 shows the profile of a simple circular oscillating screen suspended by a pair of rods that converge to a central hanger that allows for circular movement produced by a rotating eccentric axle.

After hulling, the coffee goes through a preliminary sorting in the pneumatic columns (10, Figure 8.43). The selection is made based on weight.

The final steps of dry milling are the processes of cleaning, size grading, and sorting. These steps are performed to improve the coffee's physical appearance, lot uniformity, and sensorial characteristics, as well as to ensure that a specific coffee meets appropriate national and/or international standards. Figure 8.45 shows a flow chart of these operations

In receiving, the coffee is sampled and weighed, primarily to confirm lot characteristics. The received product arrives in jute bags, big bags, or in bulk.

8.5.5 Size Grading

Before size grading, the coffee passes through another cleaning process, beginning with a pre-cleaner machine (Figure 8.46), where light impurities are removed in pneumatic columns, and impurities that are both larger and smaller than a coffee bean are removed using screens. The coffee then passes through a destoner, where heavier impurities are eliminated by an "air float" system, similar to that described for hulling. Figure 8.47 shows a destoner machine. The upper part is the "air float" system, essentially an air column that removes light impurities and collects dust in a dust collector. A metal detector is sometimes added to the cleaning operation. After cleaning, the green coffee beans are accumulated and put in a series of hoppers to await size grading.

Figure 8.46 Pre-cleaner.*

The coffee is then size-graded by dividing it into granulometrically uniform lots, according to rigorous classification standards. Ideally, there should be no more than 3% variance in the beans in screen size testing. For example, a lot of coffee of screen size 16 is composed of beans that pass through the 17 screen and are retained by the 16 screen. When a 300 g sample of this lot is submitted to a laboratory test, it cannot

Figure 8.47 Destoner.*

contain more than 9 grams retained by the 17 screen, or 9 grams that passed through the 16 screen. Granulometric classification of coffee creates lots with homogeneous screen size, which is necessary mainly to guarantee a uniform roast, as well as to ensure efficient densimetric and electronic sorting.

18	Ø 18
17	Ø 17
PB 1	Ø 11
16	Ø 16
PB 2	Ø 10
15	Ø 15
PB 3	Ø 9
Pan	Pan

Figure 8.48 Schematic of seven screens.

Graders are machines with overlapping flat screens that self-clean by means of rubber spheres. Figure 8.48 is a schematic of seven screens. Figure 8.49 shows a size grading machine with two boxes, each functions as a separate grader that operates based on the above schematic, with an entrance box dividing the coffee between the two systems. A newer type of grader that works with an ascendant product flow is highly efficient, with absolute precision (Figure 8.50). This machine contains two boxes, each with three screens and a bottom pan that work in sequence, producing classification into seven screen sizes and one bottom pan.

8.5.6 Sorting

Sorting is performed on lots that have been graded by size, with the objective of separating these lots into two sub-lots, the non-defective lot and the defective lot. The coffee is sorted in two ways. The first, densimetric, or selection based on density, works on the principle that more dense coffee is of higher quality. The second, electronic selection, is normally performed in addition to densimetric selection, and identifies defective beans by color.

Densimetric selection is done by machines that use air to "float" the coffee over a perforated table. A current of rising air flows from the bottom of the table through the perforations and the product, causing less dense beans to float but leaving denser beans in contact with the table. The table is inclined along the longitudinal axis of the table (Figure 8.51), which moves the product toward the exit at the front of the machine. The table is also inclined along its transverse axis, which, combined with vibration, causes the coffee to separate based on density. The heaviest product, which remains in contact with the table, is pushed toward the upper part of the table. Below these are coffees that are heavier still, but with some defects that can be eliminated through electronic separation. The mid- to low range of the table contains a mix of lighter and heavier beans that must be put through the machine again. The lowest part of the tables contains the lightest coffee, generally the defects.

Figure 8.49 Size grading.*

Figure 8.52 shows a densimetric table with, along with the features described above, an air blowing mechanism at the entrance that blows off dust and a chamber that collects the dust.

Electronic selection, which normally complements densimetric selection, is performed by machines such as

Figure 8.50 Upward flow grader.*

* (Courtesy Pinhalense S.A. Máquinas Agrícolas)

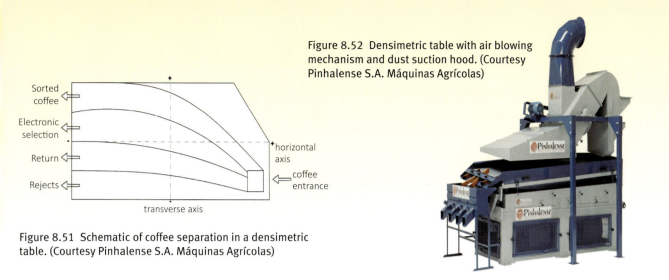

Figure 8.51 Schematic of coffee separation in a densimetric table. (Courtesy Pinhalense S.A. Máquinas Agrícolas)

Figure 8.52 Densimetric table with air blowing mechanism and dust suction hood. (Courtesy Pinhalense S.A. Máquinas Agrícolas)

In Figure 8.51 labels:
Sorted coffee
Electronic selection
Return
Rejects
horizontal axis
coffee entrance
transverse axis

those in Figure 8.53, in this case manufactured by Sammak Industria de Maquinas S/A. Figure 8.54 explains how an optical sorter machine operates.

8.5.7 Preparation for Shipment

After sorting, the clean product, graded by screen size and sorted for defects, is deposited in silos. According to market demand it can then be blended, weighed, and bagged in jute bags or in big bags. Another option is to load the coffee, using a pneumatic loader, directly into containers that are lined internally with a material similar to big bags.

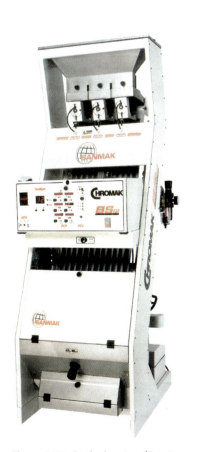

Figure 8.53 Optical sorter. (Courtesy Sanmak Indústria de Máquinas S.A.)

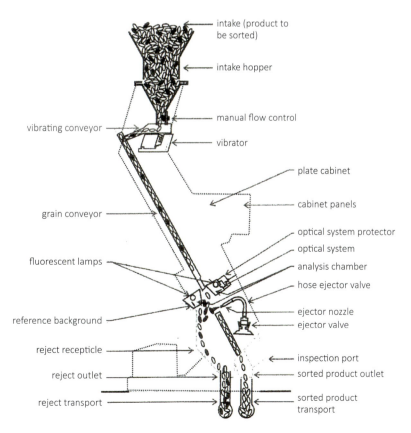

Figure 8.54 Schematic of an optical sorter, also called electronic separator. (Courtesy Sanmak Indústria de Máquinas S.A.)

Labels in Figure 8.54:
intake (product to be sorted)
intake hopper
manual flow control
vibrator
plate cabinet
cabinet panels
optical system protector
optical system
analysis chamber
hose ejector valve
ejector nozzle
ejector valve
inspection port
sorted product outlet
sorted product transport
vibrating conveyor
grain conveyor
fluorescent lamps
reference background
reject recepticle
reject outlet
reject transport

8.6 Changes in Coffee Quality During Storage

The attributes that define the final quality of coffee depend on the species, cultivar, soil, production environment, cultural treatments, and harvesting time and method, as well as the processing and drying method chosen. Nonetheless, preservation of these attributes until arrival at the customer depends primarily on storage conditions.

Depending on these storage conditions, the initial characteristics of the coffee will undergo physical, chemical, biochemical, and sensorial transformations[8]. In order to store coffee for a long period of time while preserving its initial quality, it is necessary to understand the changes that can take place during storage under different conditions, and then identify the necessary controls to ensure its preservation.

8.6.1 Changes in color

The color of coffee beans is important from an economic perspective, since beans that are discolored or different variations of white will obtain a lower price on the market. Changes in color are a strong indication that oxidative processes and enzymatic biochemical transformations have occurred, which will cause quantitative and qualitative changes in the composition of precursors to coffee bean flavor and aroma, resulting in loss of cup quality of the coffee[9].

The origin of coffee color is controversial. Some authors state that the color of the beans is formed post-harvest and depends on the methods of processing and drying employed. The color is a result of a combination of chlorogenic acid and various pigmented compounds[10]. However, other authors state that the green color of coffee comes from the mixture of blue pigment, the product of the reaction between quinone and a grouping of alpha-amino acids, with an excess of yellow-colored quinones present in the bean[11]. Furthermore, studies conducted using coffee bean extracts[12] have described the importance of the magnesium ion and of chlorogenic acids in the formation of color, as well as the presence of chlorophyll in the beans. However, it has not yet been proven that chlorophyll is related to the formation of the green color in coffee[13]. Although the formation of color in coffee beans cannot be completely explained, the processes of deterioration that result from the bean's color alteration, or its complete whitening, should be understood in order to improve the control of storage conditions.

During storage, the color of coffee beans can change from bluish-green to yellowish-green, yellow or completely white[14]. The intensity of the whiteness is a function of environmental conditions. Factors include product damage, light, relative humidity, moisture content, duration of storage, and type of packaging[15].

Whitening can begin in different places on beans that have suffered mechanical damage, and later spread across the entire surface. The time required for these spots to appear and spread depends on storage conditions. Depending on conditions, a damaged bean can become completely white-opaque in three to four days, while beans that remain intact with no damage retain their original color proportionally longer than damaged beans[16].

Light influences the color and flavor of coffee. However, not all wavelengths have this effect. Only beans exposed to white light or light transmitted in wavelengths in the range of blue-violet will exhibit changes in color and cup quality[17]. It is recommended that coffee be stored in locations with low natural light.

In addition to light, relative air humidity plays an important role in the preservation of the original color of coffee. Environments with low relative humidity are preferred[18]. The whitening process is accentuated when relative humidity is higher than 80% and the beans, in equilibrium with the air, have moisture content higher than 13% (wb)[19]. Coffee beans with 11% (wb) moisture content stored in relative air humidity of 52% at a temperature of 10 °C will undergo no color changes[20]. However, beans with moisture content above this limit change color more easily when stored at higher temperatures[21].

The temperature of the storage environment also plays a role in the whitening process; the higher the temperature, the greater the change in color of the coffee beans[22]. The changes will occur more quickly when the temperature of the storage facility is higher than 40 °C. Furthermore, the interaction between higher temperatures and relative humidity causes the beans to begin to lose their color soon after they are stored[23]. On the other hand, in environments where temperatures range from 10 °C to 17 °C, the initial characteristics of the coffee are maintained for a longer period[24].

In Brazil the storage period for coffee begins soon after the first months of harvest. Thus, coffee remains stored both during cold periods and hot, humid periods, with the risk of being affected by conditions that can lead to whitening.

Once the color of coffee changes, it will not regain its desirable characteristics. Therefore, environmental conditions of storage should be rigorously controlled, with an adequate ventilation system and monitoring of the moisture content of the coffee. Whenever possible, the storage facility should maintain less than 70% relative humidity at the lowest temperature possible. Bulk storage in aerated silos can provide good conditions for the preservation of both color and the organoleptic and microbiological characteristics of coffee[25].

Since these ideal storage conditions are not always obtained, an artificial atmosphere can provide a viable alternative for the preservation of coffee quality during storage. In contrast with packaging in jute bags, it is possible to maintain the original color of coffee for periods of up to one year when it is kept in vacuum-sealed or CO_2-enriched packaging[26]. Packaging coffee in hermetically sealed big bags will also inhibit bean discoloration for the storage duration[27].

8.6.2 Changes in dry matter

The preservation of coffee quality is closely related to the metabolism of the beans, the rate of which increases as environmental temperature and relative humidity increase and as the moisture content of the product rises[28].

Stored coffee beans are living, and therefore breathing, organisms. As a result, beans are subject to small but continuous transformations. Breathing under aerobic conditions is a process through which living cells oxidize carbohydrates, fats, and proteins, producing carbon dioxide (CO_2), water, and energy in the form of heat (Figure 8.54).

$$C_6H_{12}O_6 + 6\,O_2 \longrightarrow 6\,CO_2 + 6\,H_2O + Heat$$

180 g   134.4 L   134.4 L   180 g   677.2 cal

Figure 8.54 Equation of aerobic respiration with a glucose molecule as the substrate.

An increase in respiratory activity accelerates the consumption of reserve substances in coffee, causing a reduction in dry matter content. The extent of this reduction in dry matter depends on storage conditions and duration[29]. However, the phases prior to storage can also influence this phenomenon. Processing and drying conditions can contribute to the development of microorganisms that directly influence the intensity of respiration. For instance, it has been observed that coffees dried on dirt patios have greater reduction in dry matter than coffees dried on cement patios[30]. However, the levels recorded in literature are low, not surpassing 0.009% after 80 days in storage, with temperature varying between 20 °C and 25 °C and relative humidity varying between 60% and 80%[31]. These variations are usually not considered by the majority of coffee storage companies. Unlike grains, it is rare for coffee to have reductions related to technical breakdown, that is, reductions in the weight of the product during storage caused by the respiration of beans and consumption of dry material.

8.6.3 Chemical and sensorial changes

The intensity of metabolic activities and the appearance of microorganisms during storage will alter the chemical composition of coffee beans and their sensorial characteristics.

The temperature and relative humidity of the storage environment, the type of coffee being stored, and the form of packaging of the beans all influence the levels of sugars and polyphenols in the coffee, the acidity of the oils, and total titratable acidity, as well as membrane integrity indicators such as electrical conductivity and potassium leaching[32]. Degradation in coffee quality is observed when the following occur during storage: reduction in sugar levels, increase in polyphenols, increase in total titratable acidity and acidity of oils, increase in electrical conductivity and in potassium leaching.

The changes that negatively impact coffee quality are greater when it is packaged in jute bags; these changes intensify with prolonged storage time and with increases in temperature and relative humidity of the storage environment[33]. Furthermore, the changes depend on the initial conditions of the coffee based on how the coffee was processed and dried[34]. After storage, higher sugar levels can be found in dried pod coffee, compared with pulped and green coffees.

The breakdown of chemical components in coffee can create compounds that negatively impact taste[35]. Thus, storage can diminish the quality of the final coffee beverage through the reduction of sensorial attributes such as beverage acidity, flavor, sweetness, and body. This impact is accentuated by higher temperatures and relative humidity[36]. These changes are more noticeable in green coffee stored in jute bags. As these bags are permeable, they allow for the exchange of moisture with the environment, increasing the moisture content of the beans and thus increasing the likelihood of reduced cup quality.

However, initial quality can be preserved by storing the coffee in impermeable packaging that is enriched with CO_2 or vacuum-sealed[37]. Packaging coffee in hermetically-sealed big bags with an artificially controlled atmosphere maintains sensorial characteristics and is a viable alternative for preserving coffee quality for up to 12 months of storage[38].

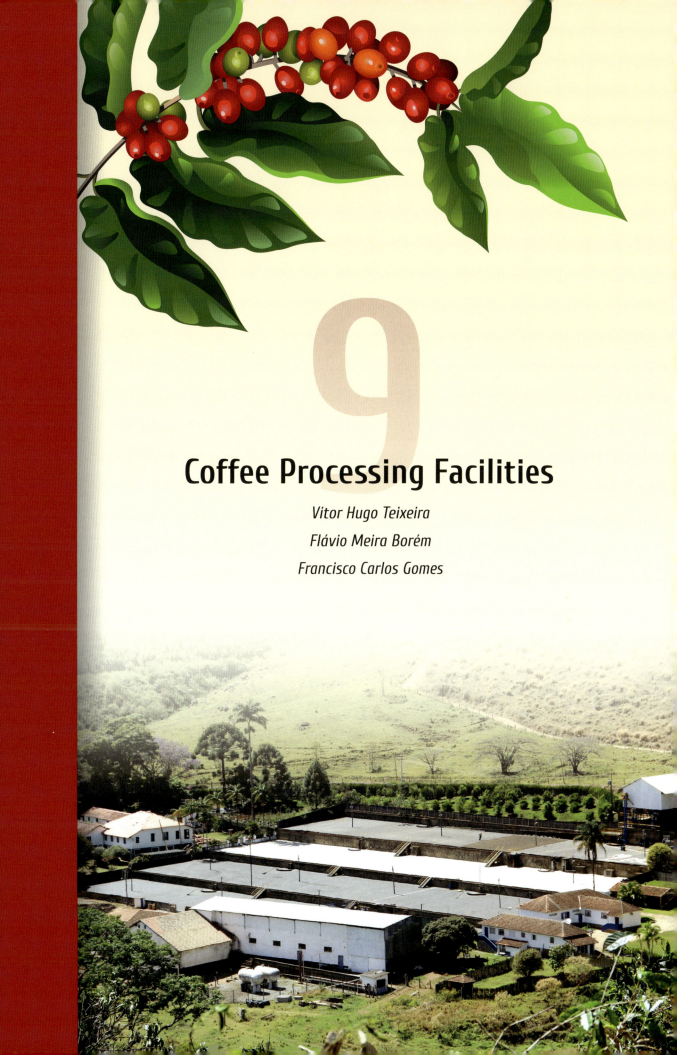

9

Coffee Processing Facilities

Vitor Hugo Teixeira

Flávio Meira Borém

Francisco Carlos Gomes

9.1 Introduction

The coffee processing facility is one of the most important factors in the modern coffee industry. Here, in one location, coffee drying and milling take place in an end-to-end process that enables controlled production of high-quality product that can be more difficult to achieve when these steps are performed in separate locations. For this reason, a well-designed facility will take many factors into consideration, including integrated buildings that support production systemization with the objective of consolidating processing steps, guaranteeing work efficiency, and offering safe, healthy work conditions. An efficient coffee processing facility will maximize equipment and processing capacities and effectively serve market demand.

9.2 Facility Location

It is not always possible to select an optimal area for the construction of a processing facility, and in many cases the producer will prefer a location that is at odds with sound technical advice. Nevertheless, the ideal location should offer conditions that minimize construction costs and should be situated in the right microclimate to preserve coffee quality. In general, to meet the basic needs for coffee processing, requirements in the following categories should be observed: access roads, dimensions and characteristics of the location, water, and energy.

Access roads should facilitate the movement of product and the arrival of raw materials during both wet and dry seasons.

The location should have the appropriate dimensions and characteristics for the construction of the entire infrastructure while allowing for potential future expansion. The ideal site for the facility is a hillside that faces the sun in an area with good drainage, direct sunlight, and good ventilation. Locations at low altitudes that are prone to fog or high relative humidity in the morning should be avoided (Figure 9.1).

To make the most of the location, the land selected should incline between 2% and 5%, allowing for easy flow of rain and processing water. Soil movement during the installation of sheds and patios should be minimal.

Figure 9.1 Patio in an inappropriate location with fog accumulation during coffee drying.

Electrical energy is fundamental in the selection process. Either three-phase or single-phase electrical power should be accessible and it should be correctly configured to minimize costs. Generally, three-phase energy minimizes motor acquisition and maintenance costs.

Water for processing must be of high quality, captured either from a natural well or a waterway. Semi-artesian wells are preferred as they are likely to provide high quality water. Water sourced from a waterway should be treated in appropriate water treatment facilities before use in processing. This can make waterways impractical.

Whichever capture method is used, a central water reservoir should be constructed at the highest point to facilitate flow by gravity to the points of consumption. Reservoir capacity should be determined by the total water consumption needs of the facility.

Figure 9.2 Impurity removal and separation sector completely fed by gravity.

Figure 9.3 Impurity removal and separation facility made of brick with a metal roof, fed by mechanical transporters.

Figure 9.4 Impurity removal and separation facility with metal roof and stone structure.

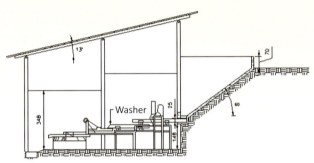

Figure 9.5 Schematic of an impurity removal and coffee separation facility.

9.3 System Components

The coffee processing facility can be divided into the following sectors and components:

- Impurity removal and coffee separation sector: reception tank, bin, or hopper; coffee winnower; hydraulic separator; big screen; pulper; demuciliaging machine; water reservoir; area for wastewater treatment and recycling
- Drying sector: patio; dryer; furnace or boiler; transitory, resting, and unifying hoppers and silos
- Milling and storage sector: storage silos; milling machines; dried coffee hull bin; conventional coffee storage warehouse

9.3.1 Impurity Removal and Coffee Separation Sector

In this sector, for all of the equipment to be fed by gravity the land must have a distinct incline. Otherwise rotary bin dischargers, conveyor belts, or other mechanical transporters will be necessary. (Figure 9.2).

Equipment should be protected from the effects of sun and rain. This can be done by building simple structures of wood, pre-molded concrete, or metal, with a variety of roofing options (Figures 9.3, 9.4 and 9.5).

9.3.1.1 Reception tank, bin, or hopper

Reception depends on the type of hydraulic separator to be used, usually either a mechanical separator or the more traditional masonry separator (in Brazil commonly called the *maravilha*). Both are efficient at removing impurities from and separating coffee, though the traditional hydraulic separator requires a larger volume of water. The mechanical hydraulic separator has higher yield and consumes less water, but it requires a larger investment to acquire and install.

Coffee can be efficiently transported through water channels to a traditional masonry hydraulic separator. However, reception bins or hoppers that do not use water are more commonly employed, particularly when mechanical hydraulic separators are used (Figure 9.6).

Hoppers should have an angled base with a minimum incline of 60° to allow the coffee to move by gravity directly into the winnower or the funnel of the hydraulic separator. Figure 9.7 illustrates width, height, and average height of a hopper.

Figure 9.6 Coffee reception hoppers.

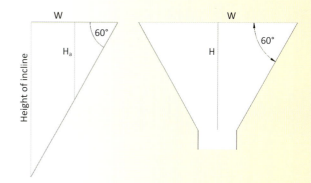

Figure 9.7 Schematic showing the dimensions W, H_a and H.

Hopper volume can be calculated using equation 9.1a for a mechanical hydraulic separator, and 9.1b for a traditional masonry hydraulic separator, or a mechanical hydraulic separator when it is fed by a metallic hopper.

$$Vr = L \times W \times H_a \tag{9.1a}$$

$$Vr = (L \times W \times H) / 2 \tag{9.1b}$$

in which V_r = volume of hopper (m³)
L, W = length and width of bin (m)
H_a = average height of bin (m)
H = height of bin (m)

Whenever possible, the natural slope of the land should be exploited for unloading operations, minimizing manual labor or the installation of mechanical transporters.

A traditional masonry hydraulic separator requires a reception tank. In this case, the water from the separation process is also used to assist the release of coffee into the hopper. A metal grill with 4 mm mesh should be installed to remove heavy impurities from the coffee. This grill can be placed inside the reception hopper (Figure 9.8) and should allow any accumulated dirt to pass

Figure 9.8 Reception hopper with grill for separation of heavy impurities.

through when sprayed at the end of daily operations to avoid clogging.

When a site does not permit the use of gravity to move coffee directly into the hydraulic separator, transporters are required. In some cases, conveyor belts (often cleated) can be used by fitting an adapted device to the unloading chute of the hopper where it contacts the transport equipment (Figures 9.9 and 9.10).

Figure 9.9 Feeding of a conveyor belt during unloading.

To determine the needed volume capacity of a reception hopper given the farm's harvest, equations 9.2 and 9.3 can be used[1].

$$V = A \times Y \times 60 \times R \qquad (9.2)$$

in which: V = total volume of harvested coffee (L)
A = planted area (ha)
Y = average expected yield (bags ha^{-1})
R = relationship between volume of harvested coffee and mass of milled coffee (varies between 7 and 9 kg)

Note: Calculations use bags of 60 kg.

$$V_c = V/P_h \qquad (9.3)$$

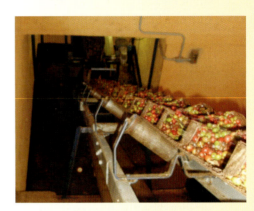

Figure 9.10 Mechanical transport of coffee to feed hydraulic separators.

in which: V_c = volume of harvested coffee (L d^{-1})
P_h = harvest period (days)

Example:

Determine dimensions of both a mechanical and a masonry hydraulic separator for a 100 hectare field of coffee, with average productivity of 40 bags of green coffee per hectare and estimated harvest time of 90 days.

For a mechanical hydraulic separator of 5,000 L per hour:

$V = 100 \times 40 \times 60 \times 8$
$V = 1{,}920{,}000$ L or $1{,}920$ m^3
$V_c = 1{,}920 / 90$
$V_c = 21.333$ m^3 d^{-1} or approx. 21 m^3 d^{-1}

The volume of the hopper would be calculated using equation 9.1a.

Normally one of the bin dimensions is fixed according to the size of coffee transport vehicle. The cart most commonly used for this operation is 2.50 m long, 1.80 m wide, and 0.70 m deep, resulting in a volume of 3.15 m^3. Hoppers usually measure at least 2.50 m in length with a volume 2 to 5 times greater that of the transport vehicle. However, the number of carts used will depend on the transportation logistics for the harvested coffee.

Furthermore, note that when determining the dimensions of a hopper the stated capacity of a hydraulic separator is generally different from its actual yield. The hydraulic separator in the example, with a stated capacity of 5,000 L per hour, actually takes 1 hour 35 minutes to separate the coffee transported by a 3.15 m³ cart, when, according to the stated capacity, this operation should take less than an hour. Thus the calculation can be done using the volume of one or two carts. Note, however, that if only one cart is used, one of the hopper dimensions will be very small. The volume of the hopper with the capacity to receive two cartloads will be:

$$V_r = 3.15 \times 2 = 6.30 \text{ m}^3$$

On the side where the vehicle is unloaded the wall should be 0.2 to 0.5 m above ground level. While this does not limit transport vehicle unloading, it does prevent rainwater from entering the hopper and can also prevent accidents.

Figure 9.7 illustrates that if the total height (H) of the incline is 2.0 m, then the average height (Ha) becomes 1.00 m and the length (L) becomes 1.5 m, resulting in an incline of 60°. Adding 0.5 m off the ground, H_m = 1.5 m. Substituting the values from the previous equation, we get:

$$6.30 = L \times 1.50 \times 1.50$$
$$L = 2.80 \text{ m}$$

Even when using two carts, the dimensions are small.

For determining hopper volume for a traditional masonry hydraulic separator using the previous calculations, the volume would be 6.3 m³. In this case, the location does not need an incline, as it does with the mechanical hydraulic separator, because the hopper for traditional hydraulic separators should not surpass 1.0 m in height.

Given a total height (H) of 1.0 m, width (W) of 1.5 m, 60° angle of base inclination, and a stated yield of 3,000 L per hour, the hopper should have a volume that corresponds to two carts. Substituting these values into equation 9.1b, we get:

$$6.30 = L \times 1.00 \times 1.50 / 2$$
$$L = 8.40 \text{ m}$$

Figure 9.11 Angled base of a masonry hopper that feeds a hydraulic separator using mechanical transporters.

Hoppers can be made of masonry or structural metal. Masonry hoppers should have lateral walls above ground level and be made of cement bricks that are 15 x 20 x 40 cm, fixed in place with mortar made of cement, lime, and sand in the ratio 1:2.5:4. Brick cavities should face up and be filled with a cement, sand, and fine gravel (2.36 to 12.5 mm in diameter) in the ratio 1:3:6, and later be covered with a mortar and sand mixture in the ratio 1:3. The angled base of the hopper should be made of the cement, sand, and fine gravel mixture, this time in the ratio 1:3:4, with a thickness of 5 cm (Figure 9.11). However, modern facility projects normally utilize structural metal in hopper design (Figure 9.12).

Figure 9.12 Hopper made of metal plates and reinforced with transverse bars.

9.3.1.2 Coffee hydraulic separator

There are two kinds of hydraulic separators: masonry and mechanical.

Use of the masonry hydraulic separator is becoming more and more restricted due to its high consumption of water, which can vary between 3 and 10 L of water per liter of coffee. On the other hand, its lower installation cost makes it viable for small coffee producers. In some situations, it is possible to substantially reduce the volume of water required for hydraulic separation by using a decantation tank that recycles used water.

The masonry hydraulic separator includes a tank where the actual separator is situated. This tank is made of masonry bricks with an impermeable coating (Figure 9.13), and the hydraulic separator is made of galvanized plates (Figure 9.14).

Masonry hydraulic separator gutters are 14 to 30 cm wide by 14 cm high and yield 1,500 to 3,500 L h[-1][2]. The yield of the hydraulic separator can be obtained using equation 9.4.

$$Y_{mhs} = V_c / T_{mhs}$$ (9.4)

in which: Y_{mhs} = yield of masonry hydraulic separator (L h[-1])
             V_c = daily volume of coffee (L)
             T_{mhs} = time of operation (h)

Using the daily volume of coffee from the previous example (21 m³) and four hours of operation, the required capacity of the hydraulic separator is determined using equation 9.4 to be 5,250 L h[-1] (5.25 m³ h[-1]). Since the masonry hydraulic separator has a yield of 3,500 L h[-1], under these conditions it is not a viable alternative. Therefore, a mechanical hydraulic separator should be acquired, given its greater capacity.

When the mechanical hydraulic separator is recommended, its dimensions are determined following the specifications of the manufacturer. Unlike the masonry hydraulic separator, mechanical hydraulic separators require electricity.

The calculation of the capacity of the mechanical hydraulic separator is similar to that of the masonry hydraulic separator. Hydraulic separators with varying capacities are available on the market, and the one selected should have capacity just above the calculated processing requirement.

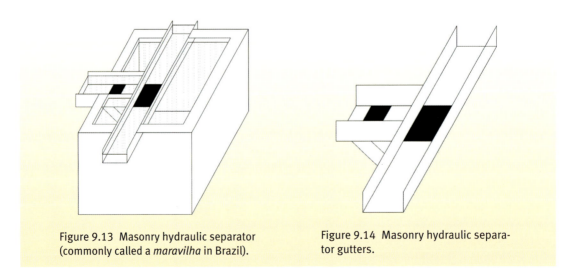

Figure 9.13 Masonry hydraulic separator (commonly called a *maravilha* in Brazil).

Figure 9.14 Masonry hydraulic separator gutters.

9.3.1.3 Water reservoir

The water reservoir can be made of metal, fiberglass, concrete or masonry.

Water reservoir volume dimensions are determined using the quantity of water required for the various post-harvest operations and considering the type of processing (dry or wet method) and form of water capture. Table 9.1 indicates water consumption for different equipment by processing type.

Table 9.1 Average water consumption for different coffee preparation systems.

Coffee preparation system	Equipment used	Average water consumption (L of water L^{-1} of separated coffee)
Dry method	Mechanical hydraulic separator	2.0–3.0
	Masonry hydraulic separator	3.0–10.0
Wet method	Masonry hydraulic separator	5.0–15.0
	Pulper	3.0–4.0
	Mechanical demucilaging machine	0.5–0.7

The facility designer should note the form of water capture. If water comes from an uninterrupted watercourse on the property, it is possible that the volume of water will be sufficient for the specific processing operations, and thus the reservoir can be used to remove debris through sedimentation. Note that most environmental legislation requires a concession for the capture and use of water from watercourses.

Properties without a source of sufficient water volume must have a water reservoir with sufficient capacity for all processing requirements. When water is captured by a pump system, the reservoir must have a volume three times greater than the daily water volume requirement and one and a half times greater when the capture is by gravity, thus maintaining an operational safety margin without system interruptions.

Volume requirement for a water reservoir can be calculated using equation 9.5.

$$V_r = V_c \times \eta \times F \qquad (9.5)$$

in which: V_r = reservoir volume (L)
                  V_c = daily volume of coffee (L)
                  η = average daily water consumption (L L^{-1})
                  F = safety factor depending on type of water capture

In the previous example, using a daily harvest volume of 21,000 L, dry method processing, and a mechanical hydraulic separator with water capture by gravity (Table 9.1), a reservoir with a capacity of 94.50 m^3 would be required. If the masonry hydraulic separator were used with water capture by gravity, the required capacity would be 315.0 m^3.

9.3.2 Coffee drying sector

9.3.2.1 Patio

The ideal location for the patio is on a gentle slope facing the sun, with good ventilation, and far from buildings and dense vegetation. The length of the patio should run east/west, allowing for maximum sun exposure and solar energy. Patios should not be built in low-altitude locations that are humid or where fog is prevalent. Furthermore, the location selected should have minimal soil shifting to reduce the costs of earthmoving.

Earthmoving is done with bulldozers or levellers. These machines help to compact the soil and create a more uniform thickness.

Patios can be made of pressed soil, soil-cement, soil-lime, solid brick, thin layer asphalt, asphalt (hot mix asphalt concrete, or HMAC), simple or reinforced concrete, or raised beds.

a. Dirt patio

The dirt patio, while not recommended, is still commonly used due to its low installation cost; it simply requires clearing the land and levelling the earth.

b. Soil-cement or soil-lime patio

Another low-cost option for uncovered patios is soil-cement or soil-lime. This type of patio can be used when there is a lack of specialized manual labor for construction and in regions with sandy soil, with a clay/silt proportion of less than 40%. The use of soil-lime is restricted to the availability of clay-like soil and in rainy seasons can result in drying problems. The recommended ratio of soil-cement or soil-lime is 1:12 (one part cement or lime to 12 parts soil)[3].

c. Solid brick patio

The solid brick patio was historically very popular due to the availability of construction material and manual labor, its good mechanical resistance, and its ability to provide slow, uniform drying. Solid brick is considered one of the best construction materials for patios due to its thermal inertia: it heats up slowly and retains heat for a long period of time following sundown[4]. However, due to its high permeability and capacity to retain water, this surface can be problematic during rainy seasons (Figure 9.15).

Figure 9.15 Solid brick patio.

d. Concrete patio

Concrete patios absorb water well, minimizing the flooding that can happen with impermeable surfaces. They offer high light reflection and low energy absorption, avoiding the overheating that can occur with asphalt patios. They provide ease-of-use during drying, are easy to clean, and are highly durable, all of which make concrete the best surface for drying coffee. However, the high cost of construction makes the concrete patio a less viable option for the majority of coffee producers.

Concrete patios are normally built in slabs, or sections (Figure 9.16), using either a cement/sand/course gravel ratio of 1:4:6 or a cement/stone ratio of 1:10, with a thickness of around 5 cm.

Figure 9.16 Concrete patio constructed in slabs.

To obtain a better concrete surface (Figures 9.17 and 9.18), construction should follow these guidelines: after levelling the soil, the bed should be compacted, ideally with a gravel sub-base between 0.05 and 0.10 m thickness; the sub-base should then be wetted and covered with a plastic polyethylene tarp until the concrete is poured; the concrete surfaces should be made with a cement / coarse sand / coarse gravel ratio of 1:3:5 for medium resistance concrete and 1:2:4 for high resistance concrete; the surface should be reinforced with metal screens of 0.15 m mesh made of 4.2 mm rods. The area of each slab, defined by the setting of expansion joints, should be no greater than 30 square meters; ideal dimension are 2.0 x 15.0 m. The concrete surface can be made using different types of joints and transfer bars using 12.5 mm iron rods 1 m in length.

The concrete should be cured, maintaining its moisture content at a constant level for a minimum of five days[5]. To facilitate cleaning and minimize damage to the hull or parchment of the coffee, the concrete should have a semi-gloss finish. While it is a higher-quality, more durable product, concrete also has a higher installation cost.

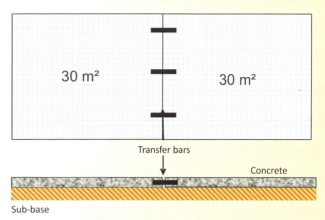

Figure 9.17 Schematic of a concrete patio with metal screens and transfer bars.

Figure 9.18 Concrete patio constructed with a screen, transfer bars and high-resistance concrete.

e. Thin layer asphalt

Thin layer asphalt, also known as cold mix asphalt concrete or slurry seal asphalt, is a low cost surfacing alternative when concrete patios are not viable due to limited resources[6].

When made with a well-compacted base (Figure 9.19), a thin layer asphalt surface supports the pressure of straight-flow vehicle traffic. However, it cannot withstand traction-related maneuvers. The traction resistance of thin layer asphalt can be increased by adding specific aggregates to its composition.

Before compacting, the soil should be analyzed to a depth of about 10 cm in order to verify the need to include aggregate to minimize cracks in the sub-base, which can extend to the pavement and compromise the quality of its construction. In the example illustrated in Figure 9.19, 7% mixed rock powder was added to the soil using a rotary hoe before compacting the sub-base.

As the asphalt layer is thin, it can be perforated by emerging weeds, so it is recommended that weed prevention controls be in place prior to surfacing. Other problems can arise when construction recommendations are not followed (Figure 9.20).

The most commonly used proportions for thin layer asphalt are 55 L gravel powder, 12 L sand, 13 L RL (Asphalt Emulsion Slow Rupture), 7 to 13 L water, and 1 L cement. The ratio of RL to water depends on the humidity of the aggregate. The soil should be made impermeable by adding MC-30 (medium-curing asphalt) 72 hours before applying the thin layer asphalt. It is simple to apply and should be done in 5 mm thick layers. Once completed according to technical recommendations, the patio will have adequate conditions for drying.

f. Asphalt

Asphalt is similar to thin layer asphalt, but with greater resistance. However, asphalt requires machines and specialized labor to construct, which significantly increases its cost.

9.3.2.2 Determining coffee drying patio dimensions

Patios should be constructed in a single group, in an area that is wide enough to minimize soil work and with a slope of 1.0% to 1.5% to facilitate the movement of rainwater into gutters downhill (Figure 9.21). A curb (barrier) 0.2 m in height should be built around the patio to prevent coffee falling off the patio. The barrier should be made of concrete bricks, pre-molded or masonry, in the ratio 1:3:3 (cement, sand, coarse gravel). A grill with holes or 3 mm mesh should be placed every 4 m to allow rainwater to drain from the patio (Figure 9.22).

Equation 9.6 can be used to determine the dimensions of a coffee patio.

$$A = (V_c \times T_d)/t_m \qquad (9.6)$$

in which: A = patio area (m^2)
V_c = volume of daily coffee harvest (m^3 d^{-1})
T_d = average coffee drying time on the patio (day)
t_m = thickness of drying layer (m)

However, to correctly determine the patio area, the planning should also consider different types of coffee, recommended thickness, and particularly the average reduction in volume due to pulping and volumetric contraction during drying (Table 9.2).

When coffee is pulped, a reduction of 40% to 50% of the initial volume of the fruit should be expected. For greater precision, the volume of the ripe fruit or parchment coffee can be calculated in terms of the moisture content of the product, taking into consideration the volumetric contraction that occurs during drying.

Figure 9.19 Phases of compacting the sub-base during construction of thin layer asphalt patios. ⓐ Preparation of the soil with a rotary hoe; ⓑ application of aggregate to reduce cracking of the soil; ⓒ mixing of the aggregate with a rotary hoe; ⓓ compacting with a steamroller.

FIGURE 9.20 Examples of construction problems with emulsified asphalt. **a** Premature surface wear due to inappropriate ratios. **b** RL leaching after precipitation during placement of thin layer asphalt.

Table 9.2 Drying time, reduction of volume, and layer thickness in relation to type of drying and type of coffee.

| Type of drying | Type of coffee | Time (days)[1] | Reduction of volume | | | Thickness[2] (m) |
| | | | Moisture Content (% wb) | | % | |
			Initial	Final		
Pre-drying	Parchment coffee	2–3	40	25	10	0.01–0.02
	Floaters	1–2	30	20	10	0.03–0.08
	Unripe fruit	5–6	70	30	30–32	0.03–0.05
	Ripe + Unripe	5–6	70	30	30–32	0.03–0.05
After half-dry	Parchment coffee	8–12	25	11	‹1	0.02–0.03
	Floaters	5–8	20	11	‹1	0.05–0.10
	Unripe fruit	15–21	30	11	10	0.15–0.20
	Ripe + Unripe	15–21	30	11	10	0.05–0.08

[1] Drying time depends on environmental conditions.
[2] Thickness depends on moisture content of the coffee, with thin layers recommended for humid products, and gradually thicker layers as drying progresses.

Figure 9.21 Patio with curbing and drainage gutter.

Figure 9.22 Barrier and grill for water drainage.

Once area is calculated, the length and width of the patio will depend on the gradient of the given property. A property with a high gradient should be divided into smaller patios. The dimensions should be determined to minimize the cost of installation.

In many cases, the total area required for full drying on patios is prohibitive due to the high cost of installation and lack of suitable flat area. This can necessitate the installation of dryers, which can reduce patio size requirements by 50% through pre-drying.

When coffee is processed using methods that result in two or more different coffee lots, the patio area required for drying will be more precise if the dimensions are calculated separately for each coffee lot, considering the proportion of each coffee lot as a percentage. This is because each lot will require different drying layer thicknesses and drying times.

Following is a calculation of the patio area required to dry coffee harvested from a field of 100 hectares, with average productivity of 40 bags per hectare and an estimated harvest time of 90 days; the fruit received includes an average of 30% ripe, 15% unripe, and 55% floaters.

Calculations:

a. For coffee that is hydraulically separated but not pulped, with a drying time of 15 days for the ripe/unripe lot, 8 days for floaters, and an average drying thickness of 8 cm for ripe plus unripe and 10 cm for floaters:

As seen in the example that determined the dimensions of the hopper, the daily volume of the harvest is $V_c = 21$ m^3 d^{-1}. Thus, the volume of ripe plus unripe (45%) is 9.45 m^3 d^{-1} and the volume of floaters (55%) is 11.55 m^3 d^{-1}.

Area for ripe/unripe lot:

$$A = (9.45 \times 15) / 0.08$$
$$A = 1{,}771.88 \text{ m}^2$$

Area for floaters:

$$A = (11.55 \times 8) / 0.10$$
$$A = 924 \text{ m}^2$$

Total area (ripe/unripe + floaters) = 2,695.88 m²

b. For coffee that is hydraulically separated and pulped, the drying time is 8 days for parchment coffee, 15 days for unpulped unripe coffee fruit, and 5 days for unpulped floaters. The average drying thickness is 2 cm for parchment coffee, 12 cm for unripes, and 8 cm for floaters:

Calculation of volume of parchment coffee, using 30% ripe coffee and a 40% reduction in volume after pulping:

- Volume of parchment coffee = $21 \times 0.3 \times 0.6 = 3.78$ m^3 d^{-1}
- Area for parchment coffee:

$$A = (3.78 \times 8) / 0.02$$
$$A = 1{,}512 \text{ m}^2$$

Calculation of volume of unripe fruits, using 15% fruits after pulping:

- Volume of unripe fruits = 21 × 0.15 = 3.15 m³ d⁻¹
- Calculation of area for unripe fruits

     A = (3.15 × 15) / 0.12
     A = 393.75 m²

Calculation of volume of floaters, using 55% floaters after hydraulic separation:

- Volume of floaters = 21 × 0.55 = 11.55 m³ d⁻¹
- Calculation of area for floaters:

     A = (11.55 × 5) / 0.08
     A = 721.875 m²

Total area (parchment coffee + unripe + floaters) = 2,627.625 m²

9.3.2.3 Dryers

Dryer capacity should be measured as a function of the daily volume of coffee to be dried, the actual available time for the drying operation, and the total drying time.

This is done by determining the volume of the coffee to be processed using equation 9.7. The required volume for the dryer can be calculated using equation 9.8.

$$V_h = V_c / Tmd \tag{9.7}$$

in which V_h = volume of coffee to be dried per hour (L h⁻¹)
             V_c = volume of coffee harvested per day (L d⁻¹)
             T_{md} = operational time of the mechanical dryer (h d⁻¹)

$$V_d = V_h \times T_a \tag{9.8}$$

in which: V_d = volume required for drying (L)
              T_a = actual time of drying (h)

In the previous example (V_c = 21 m³ d⁻¹ or 21,000 L d⁻¹), and considering an actual operation time of 24 h per day, we get:

     V_d = 21,000 / 24
     V_d = 875 L h⁻¹

Given a loss of 10% of the volume of natural coffee when reaching half-dry on the patio, and a total drying time of 32 hours, including the period of rest and the time required to load and unload:

     Volume of coffee after half-dry: 875 × 0.9 = 787.5
     V_d = 787.5 × 32 = 25,200 L or V_d = 25 m³

To select a dryer, consult manufacturer catalogs, which define the drying capacity of each type of dryer. Commercial rotary dryers vary in capacity from 200 to 30,000 L. Thus, as in the previous example, a dryer of 15,000 L capacity and another of 10,000 L capacity could be installed, or three dryers of 7,500 L capacity, to better manage the drying of smaller lots of coffee. Once the type of dryer is determined, the manufacturer's catalog provides the necessary measurements for installation (Figure 9.23).

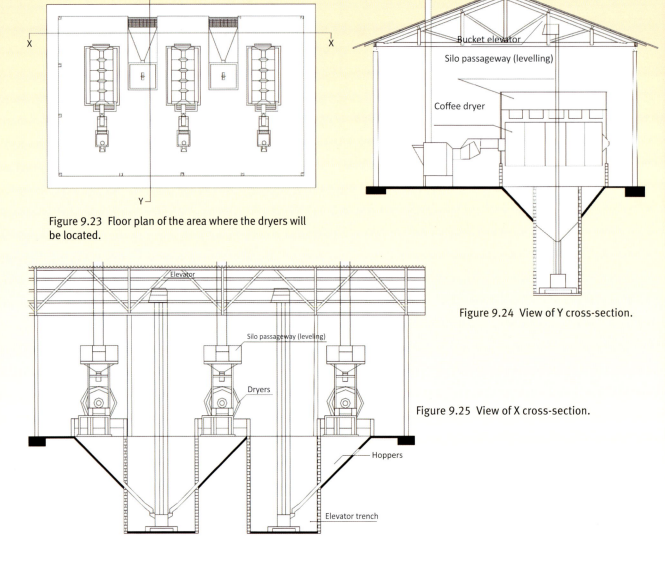

Figure 9.23 Floor plan of the area where the dryers will be located.

Figure 9.24 View of Y cross-section.

Figure 9.25 View of X cross-section.

Figures 9.24 and 9.25 illustrate the set of equipment selected in the recommended layout.

9.3.3 Storage and milling sector

The storage of coffee in wooden coffee silos subdivided into compartments allows for the identification of homogeneous lots according to their classification. The ideal is to have at least two silos, each subdivided into various smaller compartments using wood dividers.

During the drying process coffee can contract (shrink) in volume from 25% to 35%, depending on coffee cultivar. Thus, the post-drying volume after contraction can be estimated from the total harvested volume, and the result can be used to determine the dimensions of the storage unit (Equations 9.9 and 9.10).

$$V_{dc} = V_c \times (1 - s) \qquad (9.9)$$

in which: V_{dc} = volume of dry coffee (L)
s = shrinkage (decimal)

$$V_s = V_{dc} \times P \qquad (9.10)$$

in which: V_s = stored volume (L)
P = percentage retained on the farm

Using the previous example with contraction of 35% and a total harvested volume of 1,920,000 L, we get:

$$V_{dc} = 1,920,000 \times (1 - 0.35)$$
$$V_s = 1,248,000 \text{ L or } 1,248 \text{ m}^3$$

The volume to be stored should be defined by the owner since part of the coffee can be sold immediately after drying. If the producer would like to store 60% of production, we get:

$$V_s = 1,248 \times 0.60$$
$$V_s = 748.8 \text{ m}^3$$

To maintain quality while storing lots of coffee, compartment volumes should be moderate so that lots can be divided according to their various qualities. For large scale production involving both dry and wet processing, half of the calculated volume should be stored in wooden silos with a storage capacity of 150 to 200 bags (most likely for parchment coffee) and the other half in silos with a capacity of 280 to 400 bags (most likely for dried coffee pods); in other words, 43,000 to 58,000 L and 80,000 to 115,000 L, respectively. When all of the coffee is processed using only one processing method, the compartments can be divided equally; however, compartment sizes should still allow for the coffee to be divided by various qualities.

In the above example, two wooden coffee silos could be built and subdivided into five compartments each, the first five with a volume of 58 m³, the second five 115 m³. This subdivision makes it possible to separate lots according to the quality and origin of the coffee. Note that a silo can be built with different dimensions to achieve the same volume. Thus, by knowing the volume, the most cost-efficient dimensions can be determined in order to lower cost of construction.

Most wooden coffee silos are constructed of 2.2 x 1.6 m marine plywood. To maximize the usefulness of the plywood, the silos are generally built to a height and width of 4 m. In the above example, if the volumes of the subdivided compartments are 58 m³ and 115 m³, and the height and width for all are 4 m, then:

For a compartment of 58 m³:

$$58 = L \times 4 \times 4$$
$$L = 3.6 \text{ m}$$

For a compartment of 115 m³:

$$115 = L \times 4 \times 4$$
$$L = 7.2 \text{ m}$$

Note that dimensions will often be determined by standard plywood sizes to diminish material waste when designing the wooden coffee silos. Also, exits should be incorporated for unloading stored product (Figure 9.26).

Unloading normally takes place inside a trench 2 m high by 2 m wide. This trench has a funnel or conveyor belt that moves the coffee to the milling sector.

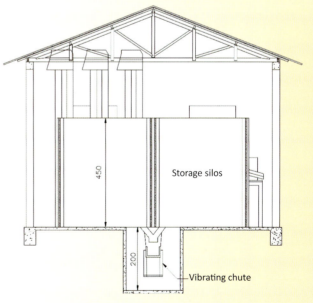

Figure 9.26 Cross-section of a coffee storage facility with a wooden coffee silo and unloading trench.

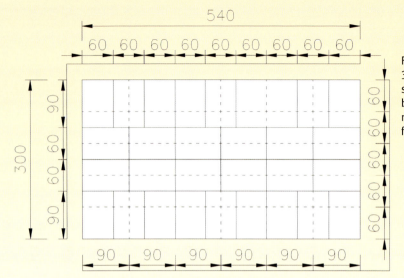

Figure 9.27 Schematic of a base layer of 30 bags and interlaced stacks. The measurements above and on the left are for the bags in the base row, and the measurements at the bottom and to the right are for the bags in the second row.

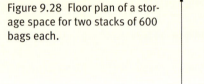

Figure 9.28 Floor plan of a storage space for two stacks of 600 bags each.

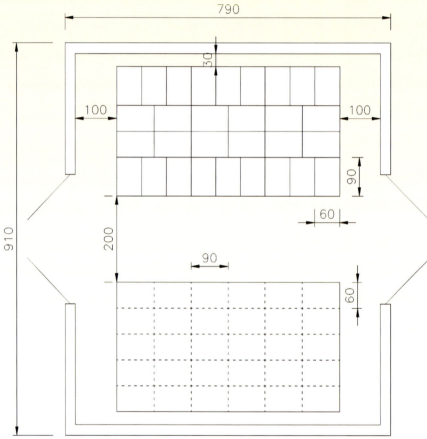

The milled coffee is usually stored in large burlap sacks measuring 0.2 x 0.6 x 0.9 m, which are placed atop base layers (Figure 9.27) in layers that are fixed in place.

Measurements for a storage facility can be obtained using equations that consider the storage capacity needed (number of 60 kg bags), maximum stack height (number of layers of bags in the stack), and the relative number of bags per unit of usable area. Equation 9.11 was developed by the (now defunct) Brazilian Coffee Institute (Instituto Brasileiro de Café).

$$A = SC/(1.6 \times nr) \qquad (9.11)$$

in which: A = total storage area (m²)
              SC = static storage capacity (bags)
              nr = number of rows or layers of piled bags
              1.6 = coefficient (usable bags m⁻²)

Storage facility dimensions should be selected according to available space and the dimensions of the base layers. For example, a base layer of 10 bags occupies an area of 5.4 m²; a base layer of 30 bags occupies an area of 16.2 m² (Figure 9.28). The maximum recommended height for stacking is 25 bags, which means that warehouse ceiling height must be at least 6 m as each bag is 0.2 m high and a minimum of 1 m headroom is needed between the bags and the ceiling.

In addition to an area designated for storage, the producer may also assign an area for milling (Figure 9.29).

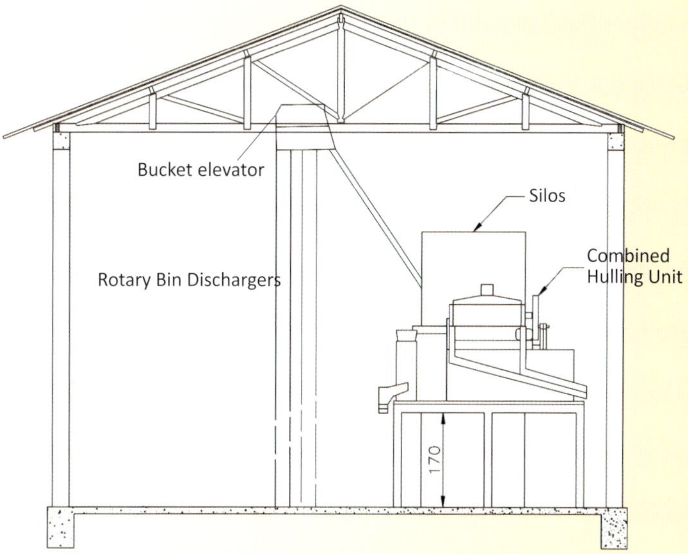

9.3.3.1 Construction materials and techniques

Construction materials and silo linings can interfere with the quality of dried coffee pods, parchment coffee, or green coffee. The recommended material for construction and silo linings is wood, in particular marine plywood[7].

The silos should have an angled base (45°) to facilitate coffee unloading.

Depending on the construction materials and techniques employed, the following parameters should be consid-

Figure 9.29 Cross-section of a milling sector.

ered by the facility designer to preserve the desirable characteristics of the coffee during storage: relative air humidity, moisture content of the beans, temperature, and light.

The moisture content of coffee beans can increase to critical levels during storage due to various forms of infiltration or moisture exchange with the environment, which can compromise coffee quality. Thus, masonry silos should be protected with appropriate mortar and/or weather-resistant paint. Masonry pathologies should be identified and corrected to avoid large losses during storage.

Examples of common masonry pathologies and recommended corrections:

a. Solid brick, ceramic paving stones, or cement blocks, without mortar finishing. In this case, the internal surface exhibits clear stains that are signs of water infiltration. When coffee stored in its dried form directly contacts a wall, it will develop fungi and jeopardize final cup quality.

Correction: Scrub and clean the wall, letting it dry thoroughly. After applying a rough cast of cement and sand, apply sufficient mortar in the ratio 1:6 (cement:sand); the mortar can be 1:2 lime to sand, with cement added at the time of application.

b. Masonry like that in (a), finished with soil-rich mortar in the ratio of 1:8 (cement:sand) with 10% soil, usually with organic material, yields similar results to that of (a).

Correction: Remove the mortar, let the wall dry, and apply mortar according to the description in (a). When the mortar is dry, apply a light-colored, reflective, waterproof paint.

High relative air humidity in a storage facility, when combined with high temperatures, can significantly whiten the coffee beans, depreciating the product. Humidity can also penetrate through the floor on poorly drained properties, primarily when the sub-flooring is made of poor grade gravel that is mostly soil and very little rock. In this case, the following is recommended: construct a passageway around the external stonework of the storage facility; drain the construction perimeter; cover the foundation with soil free of organic material; add a layer of dry sand with a thickness of 8 cm and cover with a plastic tarp; complete the concrete floor in the ratio 1:3:5, using waterproofing.

Once completed, it is still recommended to insulate the two lower layers of the stack with a plastic tarp or pallets.

The effect of construction materials and masonry linings on temperature is related to the thermal inertia indices of the materials, which determine their heat transfer resistance. When of equal thickness, solid brick walls have higher inertia than those of cement blocks and perforated paving stones. For example, solid brick masonry has a thermal inertia of 8 h; that is, heat is absorbed throughout the day, and after 8 hours the masonry begins to release the heat. Higher indices of thermal inertia are desired for the construction of silos, to maintain constant internal temperatures.

> Light is an important factor for coffee preservation during storage. Excess light can compromise the quality of the stored product. The lighting system should be designed so that it can be controlled in individual sectors of the facility.

This is an important consideration since the diffusion of heat in coffee is low, which means that coffee has low thermal conductivity. Therefore, if the construction or lining material has low thermal inertia, transferring heat quickly through the wall of the silo, the temperature of the coffee near the wall will change. If the ambient temperature drops quickly, the air in the spaces between beans will become cooler, which, in some cases where the dew point is reached, will result in humidity condensation and jeopardize the quality of the coffee.

In conventional storage facilities, windows and openings should be designed to allow for natural ventilation and limited temperature control.

When constructing covers for silos designated for coffee storage, the following should be considered: from a thermal perspective, the best tiles are ceramic or metal tiles, or fiber cement tiles that are finished with a layer of thermal insulation. However, use of these tiles is limited by their high cost of installation. Therefore, some owners opt for roofs made of (non-insulated) fiber cement tiles, galvanized plates, or aluminum. It is recommended that roofs made of fiber cement tiles, galvanized plates, or aluminum be designed with an inclination of 20° or 25°, to reduce sun absorption. For roof construction it is also recommended that ventilation gables be designed to release hot air, that ridge vents be avoided due to the difficulty of protecting them from rain and wind, and that the roof be painted bright white.

Light is another important factor for coffee preservation during storage. Certain precautions should be followed when designing and constructing a storage facility, since excess light can compromise the quality of the stored product. Artificial light should allow for work to be done in stages, and the lighting system should be designed so that it can be controlled in individual sectors of the facility. For security reasons, light bulbs should be shatterproof (Figure 9.30).

Transparent tiles should not be used in roof construction, and ventilation gables should be incorporated. It is also recommended that wood or metal plates be placed on ventilation windows to block out light.

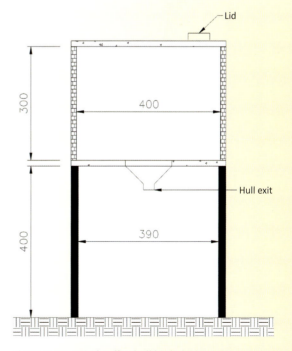

Figure 9.30 Shatterproof light bulbs in a coffee storage facility.

9.3.3.2 Dried coffee hull bins

After hulling, coffee hulls are stored in the dried coffee hull bin. It can be made of masonry bricks, reinforced concrete, or even wood, and should have a minimum height of 3.5 m to facilitate loading of the transport vehicle.

The volume of the dried coffee hull bin is calculated in terms of the time it will be used. In the previous example, for a usage period of 4 days and a volume of coffee of 60 bags milled per day, we get:

$$V_{dh} = PU \times Q \times 65\%$$

in which: V_{dh} = volume of the dried coffee hull bin (m^3)
PU = period of use (days)
65 % = quantity of hulls
Q = quantity of coffee to be cleaned per day (m^3)

However, this value can be calculated using equation 9.9.

$$Q = N_s \times P_s \times Y \times S / 1000$$

in which: N_s = number of bags per day
P_s = weight of the bag (kg)
Y = yield
S = shrinkage after drying (%)

Thus: $V_{dh} = 4 \times (60 \times 60 \times 8 \times 0.60) \times 0.65 / 1000$
$V_{dh} = 44.93$ m^3

For this volume, the dried coffee hull bin could have the following dimensions:

Height = 3.0 m
Area = 44.93 / 3.0
Area = 14.98 m^2
Length = 4.0 m
Width = 3.7 m

Figure 9.31 illustrates a dried coffee hull bin with its respective details and dimensions.

Figures 9.32 and 9.33 on the next page show blueprints of the appropriate configuration for every stage of coffee preparation.

Figure 9.31 Dried coffee hull bin schematic.

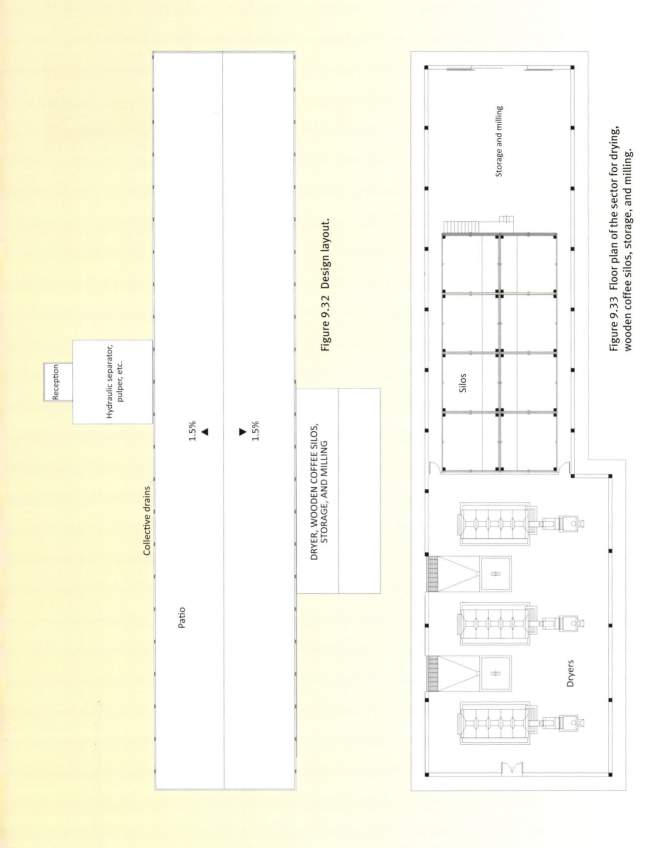

Reception

Hydraulic separator, pulper, etc.

Collective drains

Patio

1.5% ◀ ▶ 1.5%

DRYER, WOODEN COFFEE SILOS, STORAGE, AND MILLING

Figure 9.32 Design layout.

Silos

Storage and milling

Dryers

Figure 9.33 Floor plan of the sector for drying, wooden coffee silos, storage, and milling.

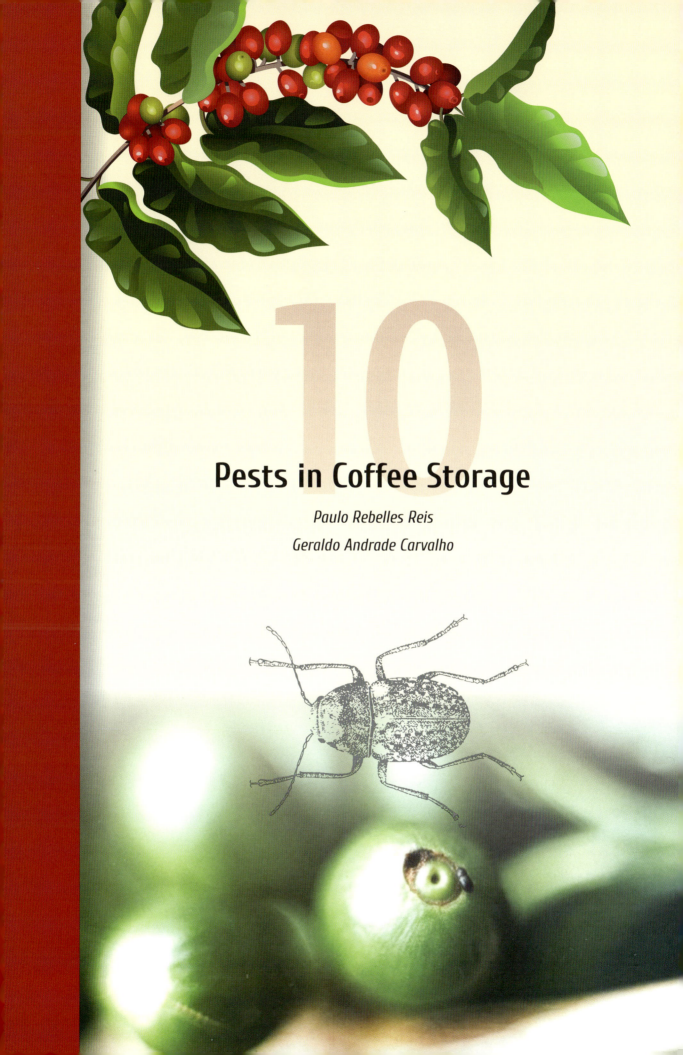

10

Pests in Coffee Storage

Paulo Rebelles Reis

Geraldo Andrade Carvalho

10.1 Introduction

Coffee quality can be initially affected during the lifecycle of the crop through poor crop management practices, then during harvesting and drying procedures, and finally during storage. In addition to addressing insect pests in storage, this chapter will discuss rodents, bats, and birds that can be found in storage facilities and cause damage to the quality of stored product.

10.2 Coffee Berry Borer

Hypothenemus hampei (Ferrari, 1867) | Coleoptera: Curculionidae, Scolytinae

10.2.1 Description and Bionomic Notes

The coffee berry borer is currently considered to be the second most significant pest after the coffee leaf miner for most coffee regions in Brazil. Discovered in Brazil in 1922, the coffee berry borer was the leading coffee pest until 1970, when it dropped to second and then third place in the majority of coffee regions in the country, with the exception of the Matas de Minas region in the state of Minas Gerais, the states of Espírito Santo and Rondonia, and fields close to large bodies of water in regions where there are high temperatures and humidity. The incidence of the coffee berry borer decreased after it was discovered that coffee rust, *Hemileia vastatrix* Berk. et Br., could be combated using more open, ventilated planting systems that create conditions unfavorable to the survival of the beetle. However, with the recent increase in the use of dense coffee planting and shade-grown coffee[1], its incidence may rise again.

The adult insect is a small, shiny, black beetle. Females are approximately 1.7 mm in length and 0.7 mm in width; males are smaller, measuring approximately 1.22 mm in length and 0.5 mm in width. The males do not fly and remain inside the fruits, where they copulate with and fertilize the females. The females perforate the coffee fruit at all levels of fruit maturation, from extremely unripe to mature, generally in the stem region (Figure 10.1), digging a tunnel about 1 mm in diameter to reach the seed. More detailed information about the coffee borer beetle (history, discovery, biology, damage caused, monitoring and control) can be found in Souza & Reis (1997)[2].

Figure 10.1 Adult female coffee berry borer near the hole it made in the crown of an unripe coffee fruit.

In the state of Minas Gerais, Brazil, this infestation rate is usually seen beginning in November in the Matas de Minas region and in January in the Sul de Minas region due to the fact that the fruit still has high water content at these times, making them susceptible to perforation by the adult borer beetles. In the Sul de Minas region, near the Furnas Dam, this level of infestation can occur starting in November because of conditions favorable to coffee borer beetle propagation.

10.2.2 Damage

The attack of the coffee berry borer causes both a quantitative loss, due to the reduction in bean weight and premature falling of the fruit, and a qualitative loss due to the alteration of the fruit around the tip, which can also affect cup quality. The damage is caused by the

Figure 10.2 Coffee fruit with only one seed damaged by the coffee berry borer.

larvae that live inside of the coffee fruit and attack one or both seeds for sustenance. The destruction of the fruit can be either partial or complete (Figure 10.2).

Initially, the damage is caused by the fruit prematurely falling off the tree. For the Arabica plant (*Coffea arabica* L.), it has been reported that the coffee berry borer increases the natural falling rate by approximately 8% to 13%[3]. For Canephora (*Coffee canephora* Pierre and Froehner), the borer beetle can cause an increase in the natural falling rate of about 46%[4].

The bored coffee fruit that remains on the plants has a reduced weight[5] and experiments in Minas Gerais, Brazil, have shown that these weight losses can be up to 21%, or 12.6 kg per 60 kg bag of green coffee. These experiments also concluded that the quality of the coffee was altered, changing its physical classification with a borer infestation. Approximately 20% to 22% of losses occur during milling because of the increased fragility of the beans that have been attacked[6].

Cup quality does not seem to be directly affected by the coffee berry borer, but rather indirectly due to the beetle damage facilitating the penetration of microorganisms, such as fungi of the *Fusarium* genus[7] and the *Penicillium* genus[8], which do affect cup quality. The damage caused by the borer beetle starts when the infestations reach about 3% to 5% in the first blossoming fruits[9].

10.2.3 Control

Normally, after being dried on patios or with mechanical dryers, coffee is stored in its dried pod or parchment state to preserve the physical and chemical characteristics of the beans, such as color, moisture level, etc.

A coffee lot infested with the coffee berry borer should be decontaminated before storage, since the borer beetle can spread within the storage facility and increase losses. Over a period of 30 days—the time necessary for coffee to attain 12% moisture level on a patio—the percentage of coffee fruit attacked by the borer beetle can double compared to the infestations initially brought from the field[10]. This pest does not tend to attack coffee fruit after drying.

Studies of the lethal effect of a temperature of 45 °C on the coffee berry borer in *C. canephora* Pierre verified that the time of lethal exposure for all adult females was four hours for borer beetles in overripe fruit and eight hours for those in ripe fruit.[11] For larvae, the time of lethal exposure was eight hours for both overripe and ripe fruit. For pupae, the exposure time was four hours for both levels of maturation. They also observed that when drying coffees in dryers at 45 °C, specimens in all levels of development were killed on the first day.

The moisture level of coffee seeds can also influence the survival of the borer beetle. Studies have shown that dried Canephora coffee fruit, with 14.7% moisture content, did not provide conditions for the development of the borer beetle, while dried Arabica coffee fruit, with a moisture content of 13.5% to 15.0%, led to a respective decrease of 3.0% to 3.5% in female specimens[12]. However, when coffee seeds are destined for nurseries and planting, their drying time is slower and moisture levels higher—usually around 20%—as this better preserves the seed's germination potential. In these conditions it is common to see a rapid increase in the number of seeds infested with the borer beetle. When this occurs, fumigation or decontamination with aluminum phosphide, a precursor of phosphine (See "Fumigation and Fogging Insecticides for Coffee Bean Weevil Control"), effectively controls borer beetles[13].

To avoid problems with the coffee berry borer during storage, control measures should be applied in the field[14], well before the coffee is harvested and dried on patios or with mechanical dryers. The damage caused by the coffee berry borer starts in the field with the premature falling of the fruit, resulting in a reduction in coffee quality. Bored beans are also eliminated during the milling process, resulting in even greater losses[15].

10.3 Coffee Bean Weevil

Araecerus fasciculatus (De Geer, 1775) | Coleoptera: Anthribidae

10.3.1 Description and Bionomical Notes

The coffee bean weevil is a prevalent pest in stored coffee beans. Initially, it only attacked dried coffee fruit and was only prevalent in farm silos, but it later appeared in green coffee, becoming a significant pest in coffee storage facilities. The attack begins in the field and continues in the storage facility.

The adult coffee bean weevil is a small beetle approximately 5 mm in length and 3 mm in width (Figure 10.3). When it emerges it is a light brown color, which changes to mottled brown. On its elytra, or forewings, there are three visible elongated yellow spots. The entire elytron surface exhibits small, randomly spaced dark and light spots. The tone of the weevil's color varies with its environment. Its face is small and elongated at the base. The pronotum has a transverse carina starting at the base (Figure 10.4). The antennae are long and segmented with three terminal segments being longer than the anterior segments (Figure 10.5). The legs exhibit tarsi with five segments, the third being bilobed, or divided into two lobes, and situated between two bristles (Figure 10.6). In adult males, the pygidium, or posterior body part, faces downward and is not distinctly visible from the dorsal view, while in females, the pygidium faces up and is visible from the dorsal view (Figure 10.7).

The weevil is a polyphagous insect that not only attacks stored coffee but also cocoa, beans, corn, and dried fruits, among others, distinguishing it from the coffee berry borer, which feeds exclusively on coffee.

Under normal conditions, the coffee bean weevil's cycle from egg to adult in coffee beans varies from 46 to 62 days, with egg-bearing to hatching of the larva taking 5 to 8 days, the larva to pupa stage between 35 and 45 days, and the pupa to adult stage

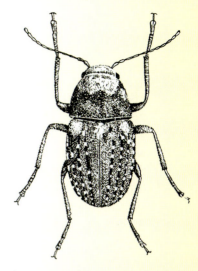

Figure 10.3 Adult *Araecerus fasciculatus* (De Geer), or coffee bean weevil. Source: Autuori (1931), cited by Lima (1956).[16]

Detail of *Araecerus fasciculatus*

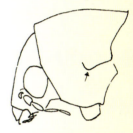

Figure 10.4 Lateral view of head, antenna, and pronotum. Source: Kingsolver (1991), cited by Pacheco & Paula (1995).[17]

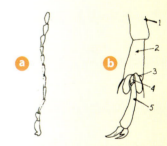

(a) Figure 10.5 Antenna. Source: Halstead (1986), cited by Pacheco & Paula (1995).[18]

(b) Figure 10.6 Tarsus, with five segments. Third segment is bilobed. Source: Halstead (1986), cited by Pacheco & Paula (1995).[18]

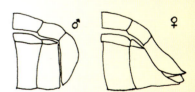

Figure 10.7 Lateral view of pygidium of male and female. Source: Halstead (1963), cited by Pacheco & Paula (1995).[19]

6 to 9 days. Figure 10.8 illustrates the egg, larva, and pupa of the coffee bean weevil.

The larvae initially feed on the mucilage and later penetrate the coffee beans. They measure from 4 to 6 mm in length, are legless, curved, covered with fine hairs, and have several dorsal folds on their abdomen, exhibiting transversal saliences (bulges). They are yellowish-white in color with a light-brown head. The last abdominal segment of the larvae is more pigmented than the rest (Figure 10.8a). The pupae are approximately 4 mm in length and are yellowish-white in color with the exception of darkened areas around the mouth and eyes (Figure 10.8 b, d). The weevil has a life-cycle that varies according to temperature, and in conditions of high relative humidity it completes six generations in 333 days[20].

Figure 10.8 **a** Larva and detail of a salience on the dorsal folds, **b** dorsal view of pupa, **d** ventral view, and **c** egg of Araecerus fasciculatus (De Greer).
Source: Cotton (1921) cited by Lima (1956)[21]; Anderson (1991), cited by Pacheco & Paula (1995).[22]

10.3.2 Damage

The weevil damages dry coffee in every stage: dried fruit, pulped, and green coffee. Since it generally destroys only a portion of the coffee beans and does not alter bean color, aroma, or flavor, damaged beans can still be roasted.

In a period of six months, a coffee bean weevil attack can decrease the weight of stored coffee by 30%[23]. The greatest damage occurs in storage facilities located in hot, humid climates, such as those in Santos (São Paulo), Brazil, and other maritime ports.

10.3.3 Control

In dealing with storage pests, cleanliness is an effective control mechanism, specifically the removal and elimination of leftover stock from the previous harvest before bringing in the new crop. Other control recommendations for prevention and treatment are described below.

10.3.3.1 Misting (or fogging) and Preventive Spraying

As a preventive measure against coffee bean weevil attacks, storage facilities should be misted every 30 days. Also, any coffee arriving at the storage facility that shows signs of coffee bean weevil presence should be fumigated or decontaminated with aluminum phosphide.

10.3.3.2 Preventive Spraying with Liquid Insecticides

Insecticides complement good hygienic practices, focusing on the elimination of existing insects and creating a barrier to prevent future infestations. After thoroughly cleaning the storage facility, the coffee should be sprayed with residual insecticides that are properly registered for their intended use, such as malathion, fenitrothion, or deltamethrin (Table 10.1).

In cases where there is remaining stock from a previous harvest that cannot be removed, it is necessary to test for the presence of infestations. If the tests are positive, at least two decontaminations using aluminum phosphide must be carried out, allowing an interval of 10 to 15 days between each application, as well as at least two surface applications 15 days apart to avoid an infestation in the newly received coffee.

10.3.3.3 Equipment Used in Preventive Spraying

For small storage facilities, generally manual, back-mounted sprayers are used to spray insecticide onto the walls, floors, cleaning equipment, bins, and other items used in the facility.

Motorized, back-mounted, turbo atomizer sprayers, when used at high acceleration levels, generally produce very fine droplets that are easily carried by air currents, or even vaporized by higher ambient temperature in some regions. Therefore, this type of sprayer is not recommended for applying insecticide since the areas to be treated would not receive the required treatment. For medium- and large-size storage facilities, motorized sprayers mounted on carts are recommended. They should be powered by electric motors and have a large tank capacity, and electric cable and hose length adequate for the dimensions of the storage facility to be treated, as well as a spray gun (Figure 10.9). Motorized sprayers provide better insecticide coverage, faster application, and greater ease of use.

Figure 10.9 Sprayer mounted on a cart. Detail of a spray gun. Source: Stresser (2005)[24].

For optimal results in preventive spraying, it is important to calibrate the equipment using water with a preliminary test, called a white test, in a 10 × 10 meter area, or 100 m². This test will determine how many liters of the insecticide mixture are required to cover the 100 m² area, given the flow rate of the nozzle(s) and the work rate of the operator, thus ensuring that the insecticide mixture is applied correctly and in the recommended doses. This can be done by marking a 10 × 10 meter area, filling the sprayer with water (manual back-mounted or motorized), and calibrating it so that the sprayer releases medium size droplets in order to make the selected area completely wet. Once completed, the water left in the sprayer is measured in order to ascertain the quantity necessary to cover 100 m². This is a simple method that ensures ideal hygienic conditions in storage facilities without product loss, through adequate deposits of insecticide residues on the sprayed surfaces.

10.3.3.4 Spraying the Surfaces of Coffee Bags in Storage Facilities

This is an indispensable complementary measure to prevent insect reinfestations in stored products; it is done after completing the preventive treatment, or even after decontaminating with aluminum phosphide. It should be done periodically to restore the surface layer of insecticides that are in direct contact with the environment, since they tend to degrade through oxidation. Likewise, constant movement over the bags by workers conducting inspection and sample collecting, as well as the unavoidable dust and coffee beans that fall on the bags, justify the adoption of this treatment in order to create a permanent protective barrier against the penetration of invasive insects.

Surface sprayings should be repeated every 30 or 45 days, especially in hotter regions and during warmer periods throughout the year when insect activity is more intense.

The surface application onto coffee bag stacks and blocks achieves excellent results. The equipment used and the dosage and type of insecticide applied are the same as in the preventive treatment, namely malathion, fenitrothion, and deltamethrin (Table 10.1).

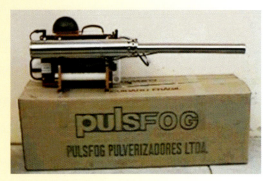

Figure 10.10 Pulsfog Thermal Fogger.
Source: Stresser (2005)[24].

10.3.3.5 Fogging

Fogging is used to eliminate insects in locations that are inaccessible to sprayers, such as building frameworks, roof structures, catwalks, etc. It is especially recommended to combat adult moths (see item 10.6 "Moths").

Since fogging leaves almost no insecticide on the surfaces being sprayed, and the eggs, larvae, and pupae are not eliminated by this treatment, its use is only recommended in hotter regions and during warmer months when insect activity is more intense. In these cases, fogging should be repeated every 15 days.

The operation is performed by a fogger or thermal fogger (Figure 10.10) using a dosage of 80 mL of deltamethrin 25 EC plus 80 mL of fenitrothion 500 EC or malathion 500 EC per liter of emulsifiable mineral oil. Due to the risk of explosion, diesel oil is not recommended. Each liter of mixture is sufficient to fog 4,000 m^3.

A Brazilian-made Pulsfog Thermal Fogger uses approximately 0.5 L of gasoline to apply 5 L of insecticide mixture on an area of 20,000 m^3. The volume of the area (m^3) to be fogged should be measured in advance to calculate the quantity of insecticide mixture to be prepared according to the aforementioned information and the instruction manual provided by the manufacturer. Fogging should begin at the back of the storage facility, moving toward the entrance. Doors and windows should be closed and the facility made as airtight as possible. Three to four hours after fogging the treated space should be aired out.

10.4 Insecticides to Be Used in Coffee Storage Facilities

All insecticides to be used in spraying and fogging should be registered with the relevant government authorities.

10.4.1 Insecticides and Doses for Spraying in Storage Facilities

Clean, preferably filtered, water should be used to prepare the insecticide mixture. The water should have a neutral pH since elevated alkalinity can destabilize the insecticide.

   a. Deltamethrin 25 EC (Table 10.1): 1.25 to 2 g I.A. 100 m^{-2}, or 50 to 80 mL 100 m^{-2}. Storage facilities with constant infestations (known as chronic infestations of resistant insects to normal dosages) should apply between 3 to 5 g I.A. 100 m^{-2}, or 120 to 200 mL 100 m^{-2}.

   b. Fenitrothion 500 EC (Table 10.1): 50 to 100 g I.A. 100 m^{-2}, or 100 to 200 mL 100 m^{-2}. A treatment works on a surface for about 12 weeks. In dusty areas, the effect of the treatment will be compromised. For chronic infestations, after a thorough cleaning apply 100 to 200 g I.A. 100 m^{-2}, or 200 to 400 mL 100 m^{-2} in the first application, and 100 to 200 mL 100 m^{-2} as a maintenance dose when needed.

   c. Malathion 500 EC (Table 10.1): 50 to 100 g I.A./100 m^{-2}, or 100 to 200 mL 100 m^{-2}. Dosages and usage information for malathion are the same as fenitrothion, listed above.

10.4.2 Insecticides and Doses for Fogging in Storage Facilities

   a. Deltamethrin 25 EC (80 ml) + mineral oil (920 mL) (Table 10.1) for each 4,000 m^3.

   b. Fenitrothion 500 EC (80 to 100 mL) or malathion 500 EC (80 to 100 mL) + mineral oil (920 or 900 mL) (Table 10.1) for each 4,000 m^3.

Good results can be obtained with a mixture of 80 mL deltamethrin 25 EC and 80 to 100 mL fenitrothion 500 EC or malathion 500 EC, plus 840 or 820 mL mineral oil. This quantity is sufficient to fog 4,000 m^3.

10.5 Fumigation (Decontamination) and Preventive Misting

With the presence of any pest in any concentration, fumigation or decontamination using aluminum phosphide should be conducted, placing plastic tents over the coffee bags. In the past, methyl bromide was used for fumigation. While it acted more quickly, it left residual bromides on the beans, and today it is prohibited worldwide for the decontamination of coffee beans and other grains. For decontamination, only aluminum phosphide, a powerful and effective fumigant for stored grains, is recommended[25]. The dosages are: one 3 g tablet of aluminum phosphide (Table 10.1) for each 15 to 20 bags or for each 2 to 2.5 m^3 to be fumigated. Decontamination is conducted in a hermetically sealed environment for a minimum of 72 hours.

Fogging complements fumigation, making the entire space unviable for insects. This should be done on a monthly basis. Using the appropriate number of foggers, the space should be fogged in the least amount of time possible. The storage facility should remain closed for 24 hours.

10.6 Moths

Diverse species of insects of the order Lepidoptera (a large order of insects that includes moths and butterflies) can appear in coffee storage facilities. The two most common are the coffee moth and the rice moth.

10.6.1 Coffee Moth

Auximobasis coffeaella Busck, 1925 | Lepidoptera: Blastobasidae

10.6.1.1 Description and Bionomical Notes

These are small moths with wingspans ranging from about 11 to 13 mm. They have a whitish head and thorax, and whitish forewings with dark scales, most notably along the outer edge (Figure 10.11). The caterpillar, when completely developed, measures close to 10 mm in length and is mottled or light brown. It transforms into a chrysalis inside a silk cocoon. The exterior of the cocoon often contains debris and insect excrement. In fact, one sign of a coffee moth attack is the presence of moth excrement in insect webs inside warehouse.

Figure 10.11 Adult *Auximobasis coffeaella* Busck. Source: Busck and Oliveira Filho (1925), cited by Graner & Godoy Júnior (1967)[26].

The cycle from egg to adult stage can take 35 to 50 days, depending on temperature. The caterpillar stage, when attacks on the coffee bean occur, lasts from 25 to 35 days.

10.6.1.2 Damage

The caterpillar feeds almost exclusively on the hull of coffee fruits. It does not attack perfect beans; however, it completely destroys those that are broken.

Fruits perforated by the coffee berry borer or the coffee bean weevil can be completely destroyed by the moth since it utilizes the opening created by these pests to penetrate into the bean. It is only in these situations that the moth is able to feed on the endosperm of the coffee bean, competing with and successfully expelling the coffee berry borer[27].

The presence of cocoons and insect excrement on coffee bags and beans jeopardizes exportation. In addition to damaging the coffee, it also destroys the coffee bags, which then require replacement, thus increasing losses.

10.6.1.3 Control

Control measures used for the coffee bean weevil are also effective for moth control. The elimination of broken beans through milling and coffee classification reduces moth outbreaks; however, storage facilities should also be thoroughly cleaned before the arrival of a new harvest.

10.6.2 Rice Moth

Corcyra cephalonica (Stainton, 1865) | Lepidoptera: Pyralidae

10.6.2.1 Description and Bionomical Notes

The rice moth is a species of moth that is common to coffee in Santos and São Paulo[28], in Brazil. The adult insects are small moths with a wingspan of about 19 mm and a length of about 9 mm. The body and forewings are grey in color. It is nocturnal and does not fly well.

The eggs, which are elliptical and pearly white in color, are either divided into small groups or dispersed onto the walls and product stock of the storage facility. A female lays 180 eggs, on average. The embryonic period lasts from five to nine days.

The caterpillars are cylindrical and a dirty white color, with the head, thoracic shield, and last abdominal segment darker in color. They reach 12 mm in length when fully developed. The complete cycle lasts from 45 to 50 days.

This species can be identified by examining wing veins or genitalia (Figures 10.12, 10.13 and 10.14).

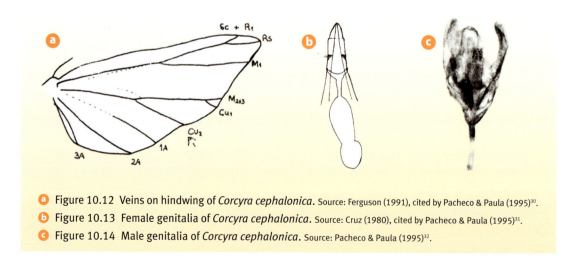

- ⓐ Figure 10.12 Veins on hindwing of *Corcyra cephalonica*. Source: Ferguson (1991), cited by Pacheco & Paula (1995)[30].
- ⓑ Figure 10.13 Female genitalia of *Corcyra cephalonica*. Source: Cruz (1980), cited by Pacheco & Paula (1995)[31].
- ⓒ Figure 10.14 Male genitalia of *Corcyra cephalonica*. Source: Pacheco & Paula (1995)[32].

10.6.2.2 Damage

The damage caused by the rice moth is similar to that of the coffee moth. Weight loss caused by a rice moth infestation is close to 2%[29]. Other forms of damage are similar to those reported in the previous section regarding coffee moth damage.

10.6.2.3 Control

Control measures for the rice moth are the same as those recommended for the coffee moth.

10.7 Registered Products for Coffee Storage Pest Control

Table 10.1 shows some insecticides used for pest control in coffee storage and their principle characteristics.

Table 10.1 Some products that can be used to control pests in coffee storage and some of their characteristics.

Name		Formula Concentration[1,2]	Class		Chemical Group
Technical	Commercial		Toxic[3]	Environ[4]	
Deltamethrin	Decis	25 EC	III	I	Pyrethroid
	Keshet	25 EC	I	II	
	K-Obiol	25 EC	III	I	
Fenitrothion	Sumibase	500 EC	II	II	Organophosphate
	Sumigran	500 EC	II	II	
	Sumithion	500 EC	II	II	
Aluminum Phosphide	Degesh Aluphos	560 FF	I	I	Inorganic precursor of phosphine
	Fertox	328 FF	I	III	
	Gastoxin	570 FF	I	I	
	Gastoxin-B 57	570 FF	I	I	
	Phostek	570 FF	I	III	
Malathion	Malathion Chab	500 EC	III	III	Organophosphate
	Malathion 500 Cheminova	500 EC	II	II	
	Malathion Sultox	500 EC	III	III	
Mineral Oil	Assist	750 EC	IV	IV	Aliphatic hydrocarbons
	Dytrol	756 EC	IV	III	
	Iharol	760 EW	IV	III	
	Miner Oil	800 EC	IV	III	
	Fersol Mineral Oil	800 EW	IV	IV	
	OPPA-BR-EC	800 EC	IV	III	
	Spinner	672 EW	III	III	
	Spraytex S	756 EC	IV	III	
	Triona	800 EW	IV	III	

1 Concentration: I.A. in g L-1 or g kg-1. | 2 Formulation: EC Emulsifiable concentrate; FF – Fumigant; EW – Emulsions in water; SO – Spraying Oil | 3 Toxicological Classification: I – Extremely Toxic; II – Highly Toxic; III – Moderately Toxic; IV Slightly Toxic | 4 Potential of Environmental Danger (PAD) Class I –Highly Dangerous; Class II – Very Dangerous; Class III – Dangerous; Class IV – Slightly Dangerous. | Source: Agrofit (2013), Compêndio... (2009) and SIA (2005).[33]

10.8 Rodents (Rats)

10.8.1 Description and Bionomical Notes

There are three species of rodents that can damage coffee during storage: the brown rat (*Rattus norvegicus*), also known as the Norwegian rat or sewer rat; the black rat (*Rattus rattus*), also known as the roof rat, common rat, or house rat; and the house mouse (*Mus musculus*) (Figure 10.15).

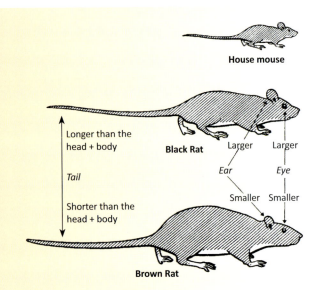

Figure 10.15 Morphological characteristics of the principle rodent species found in storage facilities. Source: Puzzi (1977).[34]

The brown rat is the largest and heaviest of these rodents. It has smooth fur, robust thighs, small ears, and a tail that is smaller than its body; in its adult phase it weighs about 350 grams and measures about 25 cm in length.

The common rat belongs to a smaller species of rats with thin thighs, coarse fur, large ears, and a tail that is longer than its body; in its adult phase it weighs about 220 grams and measures about 20 cm in length.

The house mouse measures 10 to 15 cm in length, is dark brown, and has a characteristic unpleasant odor that is transmitted to the beans and, generally, to the places it inhabits. The house mouse is a rodent with smooth fur, large ears in relation to its size, and a tail that is similar in length to its body.

The brown rat prefers to live near water and makes its burrows and nests in the earth, around 30 cm deep, next to walls and fences as well as in sewers and storm drains. It scares easily and is active at night.

The black rat, or roof rat, is smaller than the brown rat. It makes its nest above ground, preferring attics, ceilings, trees, and upper regions of buildings. It can jump up to 1.5 m high and about 2 m forward.

House mice live inside houses, making their nests in cupboards, drawers, upholstered chairs, pantries, and seldom-used compartments.

Rats are almost wholly blind, which is why they move along walls, fences, and other structures. While they can see forms and movement up to 10 m away, they cannot distinguish colors.

According to the World Health Organization (WHO), a pair of rats and their descendants can produce approximately 10,000 offspring in a year. The gestation period of the brown rat and the black rat is 22 days; for the house mouse it is 19 days. Litter sizes vary from 5 to 12 for brown rats, from 6 to 10 for black rats, and from 5 to 6 for house mice.

10.8.2 Damage

In addition to losses caused by rodent feeding, the risk to human health is also a very important factor that should be considered when developing an undertaking such as coffee production.

The subject of worker safety has recently received significant attention from producers and workers on rural properties as a result of ongoing concern regarding activities that cause, or can potentially cause, health (e.g., diseases transmitted by rodents) and safety risks for workers.

As a result of these concerns, coffee producers have developed systems for food safety and traceability designed to meet the demands of foreign consumer markets. These systems seek to eliminate existing consumer concern about the origin and type of coffee they consume as well as the conditions under which it was produced.

In this context, rats are inarguably one of man's biggest enemies. On a daily basis, they eat 10% of their weight. They consume and destroy millions of tons of food on a monthly basis. They transmit innumerable diseases to humans, some of which are fatal, including leptospirosis, bubonic plague, typhoid, rabies, salmonellosis, foot and mouth disease, and various ectoparasites such as fleas, lice, and ticks. In the Middle Ages, for instance, mortality from the bubonic plague, a disease transmitted by fleas carried by rats, reached an estimated 30% to 60%.

The economic losses resulting from contamination by rat fur, feces, and urine are ten times greater than those caused by what they consume. They also damage doors, walls, wooden windows, electric and telephone wiring, and can even cause fires by eating away the insulation around electrical wires.

10.8.3 Control

As with insects, preventive measures are critical for combating rats. Some preventive measures that should be adopted include removing leftover food, debris, wood and other materials; eliminating vegetation that can provide shelter; cementing fissures and holes in cement; and closing gaps and openings in storage walls.

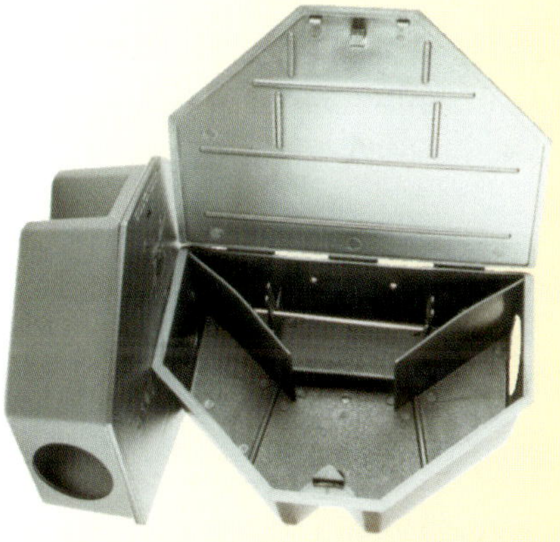

Rats are complacent, preferring not to venture far from where they live. Therefore, it is necessary to perform a thorough, detailed inspection along the periphery of storage facilities to locate where they reproduce, where they travel, their trails, etc. In addition to preventive measures, efforts should made to kill the rats. Bait stations or rat traps, preferably made of plastic (Figure 10.16) and with

Figure 10.16 Bait trap or bait station, closed and open. Source: Available at: http://www.albrasil.com.br/produtos_descricao.asp?lang=pt_BR&codigo_produto=181 (Accessed: July 23, 2013.)

a single dose of raticide in a paraffin block, should be placed as close as possible to their burrow. Inside these boxes, which are locked with a special key, the rats feel secure and protected.

Paraffin or resin blocks are ideal for combating rats in external areas and in areas of high humidity or potential for rainfall. They are especially recommended to form a barrier against rodents entering a particular area.

Bait stations should be positioned in external and internal perimeter areas, near locations with evidence of rat presence, such as trails, feces, urine, etc. They should also be placed to form a protective barrier to inhibit the invasion of rats from nearby areas that have not been treated.

Innumerable raticides, as well as their active ingredients, are available on the market. Acute raticides are currently prohibited, and the use of multi-dose rat baits is discouraged. Thus, single dose anticoagulant raticides are the preferred option. They are safe, effective, and have low toxicity for humans and domestic animals. Preference should be given to raticides that have a lower LD50, a lethal dose that, when applied to a population, kills 50% of its members, expressed in mg or g kg^{-1} of living mass (Figure 10.17).

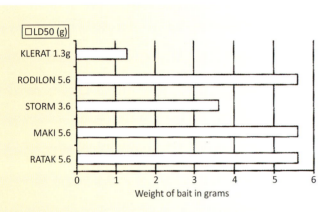

Figure 10.17 LD50 in grams of different commercial raticide baits, for a 250 g rat. Source: Matias (2003).

A highly palatable and effective raticide is made with the anticoagulant bromadiolone. Due to its slow-acting nature, rats do not associate its ingestion with the death of their companions. Since the effect is not immediate, many rats die inside their burrows and shelters[35].

Rats are neophobic, or fearful of new things or experiences, so they are likely, at first, to resist entering bait stations. For this reason, stations should be installed before adding the toxic bait, so that the rats can become accustomed to them. After five days, the rats lose their fear and will begin eating the raticide. After the fifth day of feeding on the bait, the number of rats eating the bait decreases as sick rats begin to appear. After the sixth and seventh days, the effects are even more pronounced. Dead rats should be removed daily and buried.

After seven to ten days, bait that remains untouched should be removed and placed in another location. Numbering the bait stations facilitates collection and greater control.

After collecting the stations, they should be placed again and the process repeated every eight days, until evidence of rodents is eliminated.

To combat roof rats, resin or paraffin bait blocks should be secured to the roof structure, as well as along the rats' travel routes to and from the roof. These routes can be found by looking for grease marks left by the rats.

When treating for mice, the bait, usually pellets, should be placed 30 to 40 cm apart, since mice are curious creatures and will only nibble on the poison.

To control rodent populations, control measures should be undertaken periodically. During times of food scarcity in the field, such as non-harvest periods, it is common for the number of rats to increase.

It should be noted that all raticides are toxic, and therefore special care should be taken when using them, following all manufacturer usage recommendations. They should always be kept out of the reach of children and domestic animals. Also, raticide should never be directly applied to coffee or any other stored product.

10.9 Bats

10.9.1 Description and Bionomical Notes

Bats belong to the class *Mammalia*, vertebrates with hair or fur covering the body and that feed their young with milk secreted by female mammary glands. This class is subdivided into 19 orders, including Chiroptera, which contains bats[36].

Bats are the only mammals that can fly, which they do using a membrane that connects four of the five fingers on their forearm[37], forming a wing (Figure 10.18). They are generally

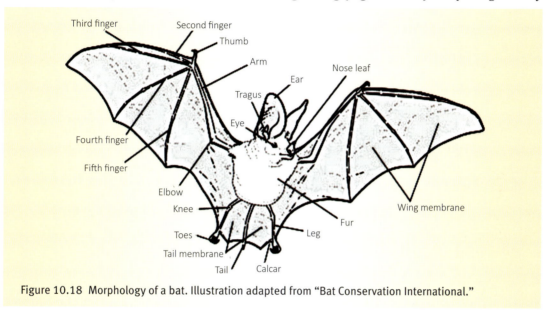

Figure 10.18 Morphology of a bat. Illustration adapted from "Bat Conservation International."

small, not exceeding a weight of 100 g.

The gestation periods of bats vary from two to seven months, depending on the species. Insectivore bats have a shorter gestation period, about two to three months, while hematophagous bats are known for their long gestation period of about seven months. The gestation period of phytophagous species (frugivores and pollinivores) varies from three to five months[38].

Offspring, or pups (generally one per gestation), are born hairless or with a sparse coat. Nursing lasts two to four months. In hematophagous bats, the transition from a milk diet to blood occurs slowly and gradually. The pup is nursed for 10 months; however, in the second month, the mother begins to regurgitate blood directly into its mouth, and beginning in the fourth month, the pup begins to accompany her on her nocturnal search for prey.

In many species of bats, social organization is based on the presence of a dominant male in relation to a group of females. Monogamy occurs in few species. Hematophagous bats tend to live for 20 to 30 years; the average for other species is 20 years.

To date, approximately 1,100 species of bats have been identified worldwide, with presence on every continent except Antarctica. In Brazil, close to 138 species of bats have been identified, of which three are hematophagous—the common vampire bat, *Desmodus rotundus* (E. Geoffroy, 1810); the white-winged vampire bat, *Diaemus youngi* (Jentink, 1893); and the hairy-legged vampire bat, *Diphylla ecaudata* (Epix, 1823)—with only one of these species feeding on the blood of other mammals. The remaining species of bats, in terms of their feeding habits, can be classified as omnivores, frugivores, insectivores, piscivores, ranivores (frog-eaters), carnivores, etc.

Contrary to common belief, bats are not blind. In addition to having excellent vision, as with carnivorous bats, many species are able to fly and capture their prey in complete darkness due to their ability to emit and detect ultrasonic pulses (echolocation).

While bats are important for pollination, seed dissemination, and insect control, they also transmit various diseases, including rabies, histoplasmosis and salmonellosis.

Rabies is a disease where the etiological agent is a virus that strikes the central nervous system, resulting in encephalitis, and on occasion, inflammation of the spinal cord. Once the incubation period is over (usually a few months in humans) and symptoms appear, the disease is almost always fatal.

Cases of rabies transmission by bats to humans are rare, given the low rate of rabies infection among bats. Much more common is transmission by rabid dogs and cats. For an infection to occur, the rabies virus must enter the blood stream of the host through a wound, cut, bite, scratch, or contact with contaminated body fluids, such as saliva. There is no evidence of rabies transmission from contact with animal urine or feces.

However, contact with animals that appear to be sick or injured should be avoided since they are likely to bite in self-defense. It should be noted that any species of bat has the potential to transmit the disease, not only hematophagous bats.

Histoplasmosis is a respiratory infection caused by the fungus *Histoplasma capsulatum*, which occurs naturally in the soil and proliferates with the presence of bat and bird feces. Transmission occurs through the inhalation of particles containing the fungus spores in the dried feces of these animals. When the feces are stirred up, dust and other particles suspended in air attract the spores. The gravity of the disease is directly related to the quantity of spores inhaled. Elevated humidity and temperature can significantly increase the proliferation of the fungus.

Inhalation of feces particles containing fungus spores can be avoided by using respiratory masks capable of filtering microscopic particles as small as two micrometers in diameter.

Another disease transmitted by bats to humans is salmonellosis, an infectious disease caused by microorganisms of the *Salmonella* genus, enterobacteria found in both humans and animals, and which provoke fever and intestinal disturbances.

Infection occurs through the ingestion of contaminated water or food, or through the handling of contaminated objects. Therefore, thorough hand washing is always recommended after handling tools used for coffee production since they could be contaminated with these bacteria due to the presence of bat feces. Potable water quality should also be strictly monitored, and food must always be hygienically handled.

The presence of bird and bat feces and urine favor the development of fungi that are already present on coffee fruit and beans, such as *Aspergillus* and *Penicillium*. Most of these fungi appear because of the way the coffee is harvested. Therefore, coffee harvested using a picking mat, either manually or with machines, should not be mixed with coffee harvested off the ground.

Studies of 19 species of *Aspergillus* associated with coffee confirmed that six of the species (*A. elegans*, *A. ochraceus*, *A. sclerotiorum*, *A. sulphureus*, *A. auricomus*, and *A. ostianus*) produce ochratoxin A, while two species (*A. flavus* and *A. parasiticus*) produce aflatoxins[39].

Some countries in the European Union, such as Spain, Italy, and Holland, have been evaluating the quality of Brazilian coffee lots exported to Europe. These analyses revealed that some lots of green coffee exhibited unacceptable levels of ochratoxin A (>10 mg kg^{-1}), which resulted in two High Alerts being sent by the European Union to Itamarati—the Ministry of Agriculture, Farming and Logistics—and the lots being returned to their respective exporters[40].

Immediate measures should be taken to minimize food product contamination and health risks to consumers and rural workers resulting from the presence of bats and other pests throughout the processes of coffee production, storage, and milling. Failure to do so could result in coffee not being accepted on the international market.

10.9.2 Damage

Just like rodents, bats and birds (primarily pigeons) can also be characterized as organisms that can cause economic losses for the coffee industry. They can cause the value of stored coffee to depreciate due to the presence of their feces and urine, and by creating conditions for fungi to develop, such as the genera Aspergillus and Penicillium, which affect the ultimate quality of the coffee beverage. Bats and birds are also responsible for the transmission of some human diseases.

10.9.3 Control

Bats should be controlled without harming them since they also play a significant role in the agroecosystem. In order to remove them from coffee storage locations, the first step is to identify where they are entering and leaving the facility, usually at sundown. Once the bats' entry and exit points are determined, a screen or plastic net should be placed over the openings or cracks, and every side of the screen or net should be sealed off except the bottom (Figure 10.19), so that the bats can leave but cannot enter.

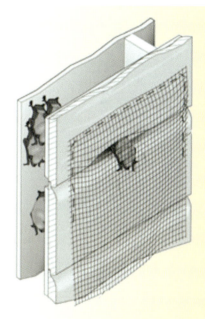

Figure 10.19 Schematic of how to remove bat colonies from their lodgings.

The screen should remain over the openings or cracks for at least one week, and until such time as all of the bats have left. Then all points of entry should be sealed off so that other animals cannot enter.

Before beginning the process of expulsion, it must first be confirmed that there are no young living in the space in question. Since they cannot yet fly, they would be unable to leave and would die of starvation, resulting in strong odors and contamination of the locale.

Figure 10.20 Electronic equipment that emits sound waves to drive away bats. Source: Available at: http://www.casados-repelentes.com.br/morcegos/afasta-ratos-por-ultrasom-2.html#. Accessed on: 3 March 2014.

Another technique used to control bats involves the use of electronic equipment that emits sound waves above 20 kHz (ultrasound) (Figure 10.20), which interfere with the bats' echolocation system, leaving them disoriented and unable to communicate with each other. This equipment is recommended for both internal and external use and can cover a territory of about 800 m². Bats can also be controlled with repellents, which have an unpleasant odor for bats, forcing them to leave their shelters. This is also a recommended technique for controlling pigeons and sparrows.

There have also been reports of wind chimes, illustrated in Figure 10.21, having contributed to the control of bats when distributed throughout the external periphery of storage sheds. This technique keeps bats away without harming them.

10.10 Pigeons

10.10.1 Description and Bionomical Notes

Figure 10.21 Wind chime.

Pigeons, *Columba livia domestica*, which originated in Europe, belong to the class Aves and the order Columbiformes. Today, they are found in almost every city in the world. They feed on grains, seeds, and garbage and can live up to 35 years. They have very acute vision, which allows them to locate a grain of corn up to 200 meters away. They are monogamous and each female can have from four to six young per year.

These birds can hear ultrasound frequencies and also have the capacity to see infrared and ultraviolet light. They can fly at a speed of approximately 80 km per hour.

10.10.2 Damage

As with bats, birds—primarily pigeons—can jeopardize the quality of stored coffee and the health of workers. They damage stored coffee by leaving feces and feathers, which attract high numbers of roof rats that feed on the eggs and young of adult pigeons.

For some researchers, pigeons are classified as synanthropic, as they live close to human beings. As such, they can cause health problems and transmit diseases to humans, including cryptococcal disease or cryptococcosis, histoplasmosis, salmonellosis, allergies, and meningitis.

Diseases such as cryptococcosis and histoplasmosis are transmitted through the inhalation of dust that contains dry feces contaminated by fungi. These diseases compromise respiratory function and the central nervous system.

Cryptococcosis is caused by the fungus *Cryptococcus neoformans*, in the forms neoformans (serotypes A and D) and gatti (serotypes B and C). This type of fungal infection has two forms: cutaneous and systemic. The cutaneous form appears in 10% to 15% of cases and is characterized by acne-like lesions, ulcers or subcutaneous lumps that resemble tumors. The systemic form frequently resembles subacute or chronic meningitis and is characterized by fever, weakness, chest pain, neck stiffness, headache, nausea and vomiting, night sweats, mental confusion, and changes in vision. It can also compromise vision, respiration, and bone strength.

The incubation period of cryptococcosis is unknown, and respiratory dysfunction can precede a cerebral attack by years. Meningitis caused by Cryptococcus can lead to death if not treated quickly enough.

10.10.3 Control

Control of these animals is necessary in order to hinder their proliferation. Thus, in areas with large pigeon populations, steps should be taken to reduce their numbers—preventing access (protection screens), removing sources of water, and primarily, hindering access to shelter and food. Another control mechanism is the use of electronic equipment that emits sound waves (ultrasound), which prevent them from sleeping and communicating. Each machine can cover an area of 900 m^2.

Liquid or gel repellents can also be used to control pigeons. The odor of these repellents is unpleasant to the pigeons and drives them out of their habitats. After abandoning the area sprayed with repellent, the affected birds warn other birds to not visit the location protected by the product.

Landing destabilizers are also considered an effective control mechanism. Examples include spikes (Figure 10.22), springs, or nylon wire. These devices, when installed on landing surfaces (e.g. roofs), cause pigeons to feel unstable and drive them away.

The plant known as "crown-of-thorns" or "crown-of-Christ" can be used in place of spikes with similar results. When using nylon wire, fishing line is a good choice. It is secured 10 cm from the edge of the landing surface with nails placed 3 cm apart.

Figure 10.22 Spikes used to repel pigeons.

None of the above-mentioned control mechanisms, however, has the ability to prevent inhalation of feces that contain disease-causing agents in areas already infested with pigeons. In order to minimize the risks of disease transfer in locations where these animals exist, along with accumulation of their feces, efforts should be made to humidify the air so that fungus spores can be safely removed, preventing their dispersion by dust.

10.11 Final Considerations

The external consumer market has relatively high standards for food safety, as well as concern for the environmental impact of agricultural activities and the conditions in which the products they consume were produced and stored. It is a market that is concerned with physical, mental, and social well-being as well as the conditions and organization of work processes.

These demands, along with growing consumer concern, motivated Brazilian producers, along with public and private institutions associated with the coffee industry, to enforce the Rural Regulatory Norms (RRN) and Best Practices in Cultivation and Preparation (BPCP) through the development and implementation of various initiatives for quality, safety, and health in the workplace. Producers have also been motivated to seek certification for their products from certifying companies.

In Brazil, certification is undertaken by companies such as BCS Oeko Garantie, with headquarters in Piracicaba, Sao Paulo; OIA, Organización Internacional Agropecuária, headquartered in São Paulo; IMO, Instituto de Mercado Ecológico, also in São Paulo; and Imaflora with headquarters in Piracicaba, Sao Paulo. International certification companies include Utz Kapeh, Fair Trade, Transfair, and Eurep Gap, among others.

In this respect, the entire coffee production chain, from planting to sales, should be built using production systems that are fair from social, economic, and environmental perspectives. Thus, controlling the organisms mentioned herein must be done in a very calculated manner, while making food safety and worker health a top priority.

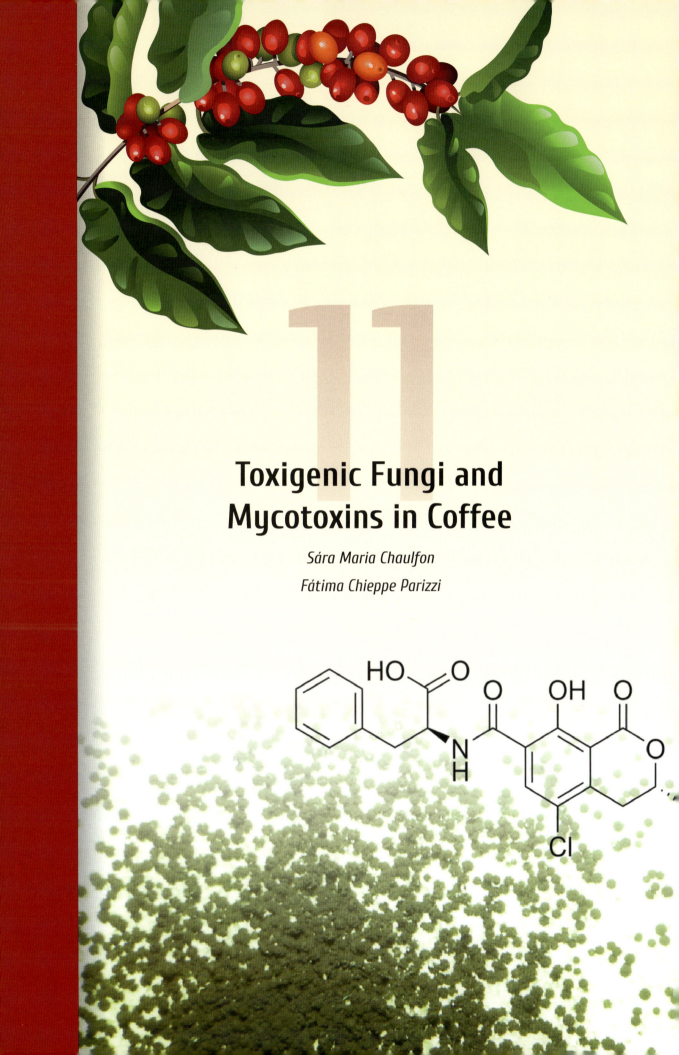

11

Toxigenic Fungi and Mycotoxins in Coffee

Sára Maria Chaulfon

Fátima Chieppe Parizzi

11.1 Introduction

Several fungi are found in coffee throughout the production cycle; these fungi can, under specific conditions, cause quality loss by producing unpleasant odors and flavors. In some cases these fungi can produce toxic metabolites called mycotoxins, which compromise the safety of the final product. The mycotoxin most commonly found in coffee is ochratoxin A (OTA). Aflatoxins and sterigmatocystin may also be found in coffee, though they are far less common.

Ochratoxin A is a highly toxic metabolite produced mainly by fungi of the genera *Aspergillus* and *Penicillium*, which are opportunistic biodeterioration agents of agricultural commodities rich in carbohydrates. They are found throughout the world, in latitudes that vary in temperature from cold to tropical.

Since fungi that produce aflatoxins, among them *A. flavus* and *A. parasiticus*, are microbials associated with coffee, why is there not a problem of aflatoxin being found in coffee? Recent results have confirmed previous research indicating that several of the chemical components of coffee, among them caffeine (1,3,7 tri-methylxanthina), exercise an elevated inhibitive effect on the development of aflatoxigenic fungi and on the synthesis of aflatoxins[1]. The same is not true for OTA. The effect of caffeine on fungi that may potentially produce OTA was partial and was dependent on the dose tested within normal levels of this alkaloid in species of Arabica and Canephora plants (varying from 1% to 2%).

> Globally, around 145 million bags of coffee are produced every year. The market currently faces the challenge of providing a product free of contaminants, especially ochratoxin A.

Globally, around 145 million bags of coffee are produced every year[2]. The market currently faces the challenge of providing a product free of contaminants, especially ochratoxin A. Besides representing a potential health risk to the consumer, the recall of a shipment of coffee contaminated by ochratoxin A can incur a substantial economic loss.

11.2 Occurrence and Significance of Ochratoxin A

The mycotoxin ochratoxin A was first identified in *Aspergillus ochraceus*, the first fungus from which it was isolated. It is also produced by other fungi of the genera *Penicillium* and *Aspergillus* that are common and largely disseminated. OTA has been found naturally occurring in cereals such as corn, wheat, barley, rice, and in animal rations, as well as in a large variety of grains, coffee, cacao, and dried fruits. It has also been found in processed food products such as flour and bread, grape juice, wine, and beer.

In Europe, OTA enters the human diet through grain and cereal products, meat and meat products, wine, beer, dried fruits, and coffee. The estimated average ingestion of OTA in Europe is 0.2–4.6 ng kg kg^{-1} per person every day based on an analysis of diet and blood plasmas, with notable regional differences[3].

The most important source of mycotoxins in the Temperate Zone is *Penicillium verrucosum*, and in tropical and subtropical regions is *Aspergillus ochraceus*. When present, however, the quantities of mycotoxins linked to food products vary considerably from a few dozen µg kg^{-1} to the detectable limit, which is currently less than 1 µg kg^{-1}. Occasionally, extreme quantities are found, and in the laboratory OTA levels of hundreds of milligrams can be produced. There is potential for high exposure of a food product to contamination when it is produced and stored in unsuitable conditions.

It is significant that mycotoxins rarely appear in uniform distribution, rather occurring unequally and irregularly. As such it is important to adopt an appropriate sampling plan that will yield a sample representative of the entire lot.

Ochratoxin A is a nephrotoxin for all animal species studied, but is more toxic to human kidneys than any of the other studied species. When ingested through contaminated food products, OTA does not accumulate in the human organism, but can have a half-life in the blood of up to 35 days. The duration of OTA in animal systems varies considerably, with a half-life of 3 days in rats, 3 weeks in monkeys, and 7 weeks in humans. The presence of ochratoxin A in humans is undoubtedly worrisome, in spite of the fact that for most of its time in the body it is connected to proteins, resulting in less serious consequences than if it were moving freely in the blood.

Ochratoxin A is on the International Agency for Research on Cancer (1993) list of Group 2B carcinogens, defined as a substance that is "possibly carcinogenic to humans." The agency also recognizes sufficient evidence that Ochratoxin A can cause renal toxicity, nephropathy and immunosuppression in various animals.

The effects on human health were observed by the detection of OTA in biological fluids such as human milk and plasma. This was traced back to the direct ingestion of contaminated products and by the consumption of meat and meat derivatives of animals fed with rations that contained the toxin.

Due to the physiological and functional similarities between the lesions of nephropathy in swine cause by ochratoxin A and the nephropathies endemic to humans, it is thought that this mycotoxin could, in fact, be a causal agent. Mortal nephropathy has greatly affected the populations of Serbia, Bosnia-Herzegovina, Croatia, Bulgaria, and Romania, where it is estimated that around 20,000 people suffer from the sickness known as Balkan Endemic Nephropathy (BEN), and the rate of malignant tumors in the urinary tract is 100 times greater than in non-endemic regions[4].

In 1982, ochratoxin A was first detected in the blood of humans residing outside the Balkans. Since then, various studies in non-endemic countries have shown that while the average ochratoxin A levels in Croatia were 0.39 ng mL^{-1} and 0.53 µg kg^{-1} of body weight per day, the average in other European countries was 0.9 µg kg^{-1} of body weight per day. These results show the necessity to reduce OTA contamination in grain and grain-based products as well as human and animal exposure, not only in Balkan Endemic Nephropathy (BEN) areas but also in other European countries where the disease is less prevalent.

Experiments have shown that ochratoxin A is teratogenic, causing craniofacial defects in rodents and birds. It is also a renal carcinogen. OTA is immunosuppressive due to its inhibiting effect on the synthesis of proteins containing phenylalanine and by acting as an enzymatic inhibitor, particularly with respect to carboxypeptidase and several enzymes found in the kidneys, most notably phosphoenolpyruvate carboxykinase (PEPCK). Through its effect on lipid peroxidation, OTA can rupture the endoplasmic reticulum.

Many of the different toxic effects of Ochratoxin A can be partially controlled by the addition of antioxidant compounds in the diet, particularly vitamin E and possibly other carotenoids and flavonoids. Of particular importance is the ability of these compounds to neutralize the immunotoxic and carcinogenic effects of mycotoxins. The discovery that vitamin E reduces toxicity, lipid peroxidation, and the creation of free radicals *in vitro*, while neutralizing free radicals *in vivo*, raises some interesting questions about the role that free radicals play in the etiology of ochratoxicosis.

11.3 Fungi that Produce Ochratoxin A

The toxigenic potential of a fungus is the ability of a lineage to produce toxic metabolites. Mycotoxins are characterized as toxic secondary metabolites and are produced between the end of the exponential phase and the beginning of the stationary phase of fungi; however they are not related to the biological role of the microorganisms.

As shown in Table 11.1, the conditions that permit the production of mycotoxins by a potentially toxigenic lineage are more limiting than those that permit the growth of fungi. Within a species that is considered toxigenic, many lineages do not have this property. In potentially toxigenic species, the production of toxins can vary by up to 1,000 times.

Table 11.1 Minimum water activity (a_w) and moisture content levels for fungi and toxigenesis development.

Event	Water Activity (a_w)	Moisture Content (% wb)
Development of the fungus *A.ochraceus*	0.76	14.2
Development of the fungus *P. verrucosum*	0.81	15.5
Development of ochratoxins	0.85	20.0

Source: Moss (1996).[5]

Different species have been responsible for the presence of ochratoxin A in different products. For example, OTA in figs is attributed to *Aspergillus alliaceus* Thom & Church, in grapes and wine to *Aspergillus carbonarius* (Bainier) Thom and *A. niger*, in European grains to *Penicillium verrucosum* Dierckx, and in cheese and meat products to *Penicillium nordicum* Dragoni & Cantoni ex. C. Ramírez. The OTA found in coffee has been mostly attributed to molds in the genus *Aspergillus*, mainly those in the *Circumdati* and *Nigri* sections.

11.3.1 *Aspergillus ochraceus* and species of the genus *Aspergillus* section *Circumdati*

Aspergillus section *Circumdati* includes species taxonomically related to *A. ochraceus* (Figure 11.1). Fungi of this species have conidia of varying sizes, do not narrow near the vesicle, have walls that range in texture from smooth to noticeably wrinkled, and have pigmentation that can be yellow, orange, brownish, or even transparent. The conidial head is usually yellowish to ochre in color. Vesicles are round and rarely elongated, biseriate, with long metulae in the shape of a wedge, occasionally septate. The conidia with thin walls are smooth to finely wrinkled,

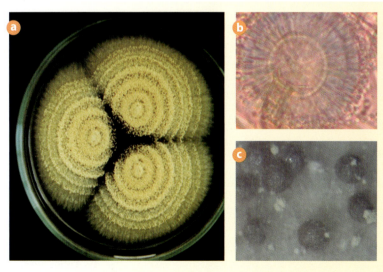

FIGURE 11.1 *Aspergillus ochraceus*: **a** ochre-colored colonies on CYA; **b** biseriate conidial head; **c** burgundy to purple-colored sclerotia. (Chalfoun & Batista, 2003)[6]

rounded to ellipsoidal in shape, and never exceed 4–5 µm in diameter. Scleroid nuggets are common, with a cream, yellow, reddish, burgundy, or black coloration with varied maturation. This species does not have Hulle cells[7].

Desert soils, cultivated fields, and forests, principally in latitudes from 24° to 45°, are the main habitat of these species. That they are found in decomposing material and stored grains is due to the fact that they are biodeteriorating fungi, not phytopathogenic fungi.

Ochratoxin A was originally isolated from *A. ochraceus* in a lab in 1965[8]. Since then, another six species belonging to the section *Circumdati* (*A. alliaceus, A. melleus, A. ostianus, A. petrakii, A. sclerotiorum*, and *A. sulphureus*) have been reported to cause mycotoxin.

Table 11.2 shows some species of the genus *Aspergillus* section *Circumdati* identified as producers of ochratoxin A. It also shows that these species are frequently isolated from coffee fruits and grains. The percentage of isolated ochratoxin A producers varies greatly, depending on the number of samples tested. The percentage of *Aspergillus ochraceus* that produces ochratoxin A has varied between 25% and 75%.

Table 11.2 Species of the genus *Aspergillus* section *Circumdati* cited as producers of ochratoxin A.

Species	Number of fungi tested / producers of ochratoxin A	Origin/product	Reference
A. alliaceus	1/1	Fungus collections	Abarca et al. (1997)[9]
	1/1	Soil of Australia	Bayman et al. (2002)[10] Varga et al. (1996)[11]
	6/6	Figs	
A. auricomus	1/1	Coffee beans	Batista et al. (2003)[12]
	3/1	Peanuts	Varga et al. (1996)[11]
A. elegans	2/2	Coffee beans	Batista et al. (2003)[12]
A. ostianus	1/1	Coffee beans	Batista et al. (2003)[12]
A. ochraceus	41/27	Coffee beans	Abarca et al. (1997)[9]
	42/37	Coffee fruit/beans	Accensi et al. (2004)[13]
	15/2	Fungus collections	Batista et al. (2003)[12]
	19/3	Coffee fruit/beans	
	18/14	Coffee beans	Nasser (2001)[14]
	12/9	Coffee fruits	
	20/1	Animal feed	Prado et al. (2004)[15] Silva (2004)[16] Urbano et al. (2001)[17]
A. sulphureus	23/22	Coffee beans	Batista et al. (2003)[12]
	2/1	Soil of India	Nasser (2001)[14]
	34/12	Coffee beans	Varga et al. (1996)[11]
A. sclerotiorum	3/2	Coffee beans	Batista et al. (2003)[12]
	3/1	Soil of Thailand	Nasser (2001)[14]
	3/3	Coffee beans	Varga et al. (1996)[11]
A. petrakii	1/1	Coffee beans	Batista et al. (2003)[12]

11.3.2 *Aspergillus carbonarius* and Species of the Genus *Aspergillus* section *Nigri*

The species belonging to the genus *Aspergillus* section *Nigri* have the principle characteristic of producing black, brown, and dark-brown conidia. This group of fungi is thus known as "Black Aspergillus." Morphologically, some species in this section possess noteworthy structure characteristics that easily differentiate them from other species. *A. carbonarius* (Figure 11.2) is noticeable because of its conidia, which are black, large, distinctly wrinkled, and attached to long conidiophores. *Aspergillus japonicus* Saito, on the other hand, is the only species that has a uniserial conidial head. In the intermediate species, *Aspergillus niger* aggregate stands out since it has a biserial conidial head that is relatively small. Also, the scleroid nuggets are not present in most isolated strains.

Black Aspergillus (*Aspergillus* section *Nigri*) has been isolated throughout the world, with *A. niger* being the most commonly found. The species of this section are considered opportunistic pathogens, causing deterioration of food. In nature they are found in soil, cultivated areas, and on organic compounds that are in a state of decomposition.

A list of species of the section *Nigri* accepted by the International Commission on Penicillium and Aspergillus (ICPA) includes *A. awamori, A. carbonarius, A. ellipticus, A. foetidus, A. helicothrix, A. heteromorphus, A. japonicus, A. niger, Aspergillus pallidus* Kamyschko, *A. phoenicis* and *A. pulverulentus* (McAlpine) Wehmer[18].

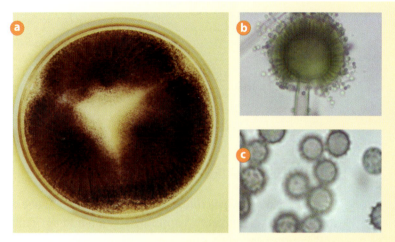

FIGURE 11.2 *Aspergillus niger* van *Tieghem* var. *niger*: Dark brown to black-colored colonies on CYA **a**; biseriate conidial head **b**; subgloboid and wrinkled conidia **c**. (Chalfoun & Batista, 2003)

The ability of *A. niger* to produce ochratoxin A presents a risk to human and animal health, especially since this species is amply used in the food industry and is classified by the FDA (Food and Drug Administration) as GRAS (Generally Recognized as Safe).

A. niger develops in temperatures between 6 °C and 47 °C, with optimal temperatures for its development between 35 °C and 37 °C. The minimum free water necessary for its growth is 0.88, which is relatively high compared to other species of *Aspergillus*[19]. *A. carbonarius* can develop in temperatures between 10 °C and 40 °C, with optimal growth temperatures between 30 °C and 35 °C and optimal water activity for its growth varying between 0.93 and 0.987. Ideal temperatures for the production of ochratoxin A are between 15 °C and 20 °C, with water activity between 0.95 and 0.98.

Black spores of this group of fungi provide protection against solar radiation and UV rays, providing a comparative advantage in these habitats[20] for development, for example, during the drying phase of coffee.

Studies have shown that occasionally isolated strains of *A. niger* can produce OTA. *A. carbonarius*, a species that is very close, morphologically, to *A. niger*, is more commonly a source of ochratoxin A, since it is able to synthesize high concentrations of the toxin, but this species is far less common than *A. niger*. Table 11.3 shows some species of the genus *Aspergillus* section *Nigri* identified as producers of ochratoxin A.

Table 11.3 Species of the genus *Aspergillus* section *Nigri* that produce ochratoxin A.

Species	Number of fungi tested / producers of ochratoxin A	Origin/Product	Reference
A. carbonarius	18/18	Grapes	Cabañes et al. (2002)[21]
	6/6	Coffee fruit and beans	Heenan et al. (1998)[22]
	33/30	Growth medium	Joosten et al. (2001)[23]
	12/5	Growth medium	Téren et al. (1996)[24]
A. niger	87/10	Coffee fruit and beans	Abarca et al. (1994)[25]
	19/2	Grains, soybeans, and peas	Heenan et al. (1998)[22]
	115/2	Growth medium	Urbano et al. (2001)[26]
A. foetidus		Rice grains	Nakajima et al. (1997)[27]
A. japonicus	45/9	Growth medium	Téren et al. (1996)[24]
A. niger aggregate	304/21	Coffee fruits and beans	Prado et al. (2004)[28]
	100/3	Growth medium	Téren et al. (1996)[24]

11.3.3 Species of the genus *Penicillium*

Within the genus *Penicillium*, the only species that are considered to be producers of Ochratoxin A are *P. verrucosum*, common in grain and agricultural products in temperate climates, and *P. nordicum*, more commonly found in meat and cheese products[29].

11.4 Coffee as a Source of Ochratoxin and Its Implications on Human Health

After the first account of the natural occurrence of ochratoxin A in coffee in 1974[30], data has been obtained indicating product contamination concentrations varying from 0.2 to 360 µg kg^{-1} [31]. The fear of OTA contamination in coffee has become more predominant in the last few years since analytical methods have become more sensitive.

A study analyzed 633 samples of products acquired from retail locations in various European countries[32]. These samples were analyzed in nine different laboratories that freely chose the analytical methodology that they would use. The detection limits varied from 0.2 to 1 ng g^{-1} of OTA for soluble and roasted coffee. Of these samples, 299 were contaminated, with the highest level of OTA, 27.2 ng g^{-1}, found in soluble coffee.

In 1996 the Ministry of Agriculture, Fisheries and Food (MAFF), in the United Kingdom, presented an analysis of 291 Arabica and Canephora green coffee samples from 27 different producing coun-

FIGURE 11.3 ⓐ Coffee bean colonized by fungi of the genus *Aspergillus* section *Circumdati*; ⓑ structural details.

tries. A total of 110 samples were contaminated. The maximum level of OTA, found in a sample of Canephora, was 27.3 ng g^{-1}. The highest level found in Arabica samples was 9 ng g^{-1}.

In 2004, a study, conducted by the *Consórcio Nacional de Pesquisa e Desenvolvimento do Café* (CNP&D–Café) in Brazil, analyzed 289 samples from the State of Minas Gerais, the largest coffee producing state in Brazil, and found that 75.09% of the samples had values below 5 ppb (the proposed European Union limit). The highest OTA levels were found principally in samples that had been in contact with the soil, a practice that is not recommended for the production of coffee.

The ingestion of OTA is evaluated by looking at the average consumption levels of agricultural products by a population and then determining the average amount of OTA found in these products. In Europe, according to data from the Food and Agriculture Organization of the United Nations (FAO) and the World Health Organization (WHO), the average level of OTA ingestion was estimated to be 45 μg kg^{-1} of body weight per week, assuming an average body weight of 60 kg. Grains and wine contributed 25 and 10 μg kg^{-1} of body weight per week, respectively, for average consumption, while grape juice and coffee each had levels of 2 to 3 μg kg^{-1} of body weight per week. Other food products, such as dried fruit, beer, tea, milk, and poultry, contributed 1.0 μg kg^{-1} of body weight per week[33].

An evaluation of the incidence of OTA in coffee-derived products sold in Europe verified that roasted and ground coffee presented an average contamination of 0.8 ng g^{-1}, while soluble coffee was 1.3 ng g^{-1} [34]. Based on this information, the authors inferred that the consumption of four cups of coffee per day, considered average per capita consumption in most European countries, corresponded to 24 g of roasted and ground coffee and 8 g of soluble coffee, contributing to the ingestion of 19 and 10 ng d^{-1} of OTA, respectively. These values correspond to approximately 2% of the weekly allowable limit of OTA ingestion as established by the Joint FAO/WHO Expert Committee on Food Additives. The defined limit is up to 100 ng kg^{-1} of body weight per week[35].

The relative quantity of the contribution of each food product or group of products can vary each year, depending on climatic conditions during the harvest.

Brazil does not yet have a representative evaluation of the contribution of coffee to the daily ingestion of OTA, since relative data on the incidence of this mycotoxin in green and roasted coffee is not considered significant in the overall national agricultural production chain.

In general, the selection process used to choose export-level coffees in Brazil includes a sensorial dimension that eliminates any material with fungi, and therefore OTA is quite rare. Furthermore, highly contaminated coffee shipments are unlikely to be accepted since undesirable odors and flavors will make these coffees unacceptable. In other coffees that do not pass through the same rigor in product selection, significant levels of OTA can be detected. These coffees are more likely to be consumed internally in the producing country.

Adding coffee byproduct—hulls and parchment—to roasted coffee is a common practice at some roasting facilites. This constitutes consumer fraud and also exposes the consumer to a higher contamination risk, since in conditions favorable to OTA development, the toxins tend to concentrate in the hull.

Coffee is not a major source of OTA in a normal diet[36]. However, the presence of ochratoxin A in coffee is undesirable because it can be used as a non-tariff barrier. Countries that are not producers of products that are susceptible to mycotoxin contamination for certain products can place limits for minimum tolerances on them. In Denmark, where Porcine Nephropathy caused by the consumption of rations contaminated with OTA is endemic, the OTA limit in pig kidneys is 25 µg kg[-1], while for coffee, an imported product, it is 5.0 µg kg[-1].

The concept of a free market, aside from involving the question of competition between markets, is based on the restraints of trade barriers and control systems that guarantee the safety and sanitation of products. The principle tools of these systems are non-tariff trade barriers, especially those related to phytosanitation and food safety.

11.5 Levels of Ochratoxin A in Coffee Beans and Proposed Limits

Since microbiological studies have demonstrated that the species of fungi that can potentially produce ochratoxin A can be found in coffee fruit and beans as well as other agricultural products, these products can be subject to contamination in all phases of agricultural production.

The risk of OTA exposure correlates with how the crop is treated in the field and in the post-harvest process. This is clear from the great variability of ochratoxin A levels observed in coffee beans, depending on the type of sample analyzed. In the first study on ochratoxin A[37], the coffee beans were visibly contaminated by the fungus *Aspergillus ochraceus*. The study found low levels of contamination despite the presence of the fungus, leading the authors to conclude that coffee did not appear to be a good substrate for OTA production compared to other products.

Subsequent work, using high performance liquid chromatography, a method more sensitive and precise, corroborated these conclusions, demonstrating a large variation in the results obtained, as shown in Table 11.4.

Table 11.4 Levels of ochratoxin A in green coffee samples from various origins.

Origin	Total number of samples	No. of contaminated samples	Level of contamination	References
Various	40	9	0.50 – 23.00	Cantáfora et al. (1983)[38]
Various	22	4	9.90 – 46.00	Tsubouchi et al. (1984)[39]
Various	29	17	0.20 – 15.00	Micco et al. (1989)[40]
Various	25	13	1.20 – 56.00	Studer & Rohr et al. (1995)[41]
Various	25	13	0.90 – 50.00	Studer & Rohr et al. (1995)[41]
Various	47	14	0.10 – 17.40	Nakajima et al. (1997)[42]
South America	19	9	0.10 – 4.60	Trucksess et al. (1999)[43]
Africa	84	76	0.50 – 48.00	
The Americas	60	19	3.00*	Romani et al. (2000)[44]
Asia	18	7	1.06*	
Brazil (South Minas Gerais)	40	5	0.64 – 4.14	Batista et al. (2003)[45]
Brazil (São Paulo and Minas Gerais)	135	9	> 5.00	Taniwaki et al. (2003)[46]
	90	52	0.20 – 165.00	Iamanaka & Taniwaki (2003)[47]
	54	22	0.30 – 160.00	Moraes & Luchese (2003)[48]
The Americas	41 (Arabica)	22	> 5.00 (6.90*)	Prado et al. (2004)[49]
Africa and Asia	16 (Robusta)	8	> 5.00 (6.10*)	
Brazil (Minas Gerais)	289	89	0.10 – 5.00	Batista (2005)[50]

*Average values

This information is of more importance when it can be associated with the particular processing stages in which the contaminations, fungus development, and ochratoxigenesis could have occurred, something that most studies do not verify. It is clear, nonetheless, that the majority of samples registered zero or low levels of OTA contamination, confirming that coffee is a less favorable substrate for the biosynthesis of ochratoxin A. On the other hand, high levels of OTA registered in these studies have been associated with poor climatic conditions, such as high levels of relative humidity during fruit maturation and inadequate processing methods.

The large variation in the levels of OTA detected in green, roasted, and soluble coffee originating in different countries reflects a problem in the heterogeneous distribution of contaminated beans that leads to sample variance, as well as the detection sensitivity of the various analytical methods used. Organizations responsible for the regulation and quality control of food products, above all coffee, have in recent years focused on this subject so as to guarantee the reliability of the adopted measures. This care is necessary since it involves decisions in evaluating the risk of the population exposed to danger, monitoring of food quality, and meeting evermore demanding trade standards.

The legislation for food products serves to both assure consumer health and the economic interests of producers and purchasers of food products. The establishment of a regulatory norm for contaminants, especially the mycotoxins in food and agricultural products, aims to facilitate international commerce while at the same time protecting populations from the risk of ingesting contaminated products. Various factors, both scientific in nature and not, can influence the establishment of regulations and limits on mycotoxins: availability of toxicological data, availability of data on the presence of mycotoxins in basic food products, knowledge of the distribution of mycotoxin concentrations in product lots, availability of analytical methods, and legislation in countries in which there is a commercial market and a need to supply certain food products. The size of a country, its geopolitical characteristics, state of industrialization, and economic situation all affect the way in which food safety regulations are implemented.

The scarcity of reports of acute or chronic toxicity of mycotoxins in humans is, in general, a consequence of the lack of opportunities for health professionals and toxicologists to work together in order to associate the exposure of mycotoxins to disease and human fatalities. Mycotoxin occurrences are not restricted to developing countries, as studies conducted on an international level have shown that the occurrence of mycotoxins in food products is common, though it varies by region and time of year. However, occurrences of mycotoxins in developed countries are lower than in developing countries due largely to broader ranging and more rigorous legislation as well as better alimentary habits.

When a contaminant, such as OTA, is a recognized carcinogen, it is not possible to know what a safe level of exposure would be. The advice of toxicologists who control it must be very rigid, and the admissible index should be as low as possible. This value is normally interpreted as the lowest level that can be analytically detected[51].

Vigorous regulation in countries with a significant amount of international trade should be established using well-defined priorities and regulatory criteria. The determination of maximum tolerable limits is generally based on the toxicology evaluation, available analytical methods, supporting data, distribution of mycotoxins in the products, and composition of the basic diet of the population. For any given country, it is easier to establish tolerable mycotoxin limits for a product that is exclusively for exportation and that has insignificant levels of internal consumption, since these items are usually "luxury items" that have a high aggregated value that can in some ways compensate for the investment in control measures applied throughout the supply chain. In situations such as these, the tolerable limits are normally lower and the decisions depend almost exclusively on the demands put into place by the importing countries.

For those products that are both largely consumed internally and exported, e.g., coffee in Brazil, the establishment of tolerable limits is a complex decision that involves diverse interests, including commitments that fall under the purview of the World Trade Organization (WTO). Specifically, WTO members cannot hold a higher standard for imported products than they do for that same product internally.

Due to this, in establishing a maximum acceptable limit for OTA in coffee and other grain products, it is not recommended that a country establish stricter regulatory levels for internal markets than for the same products for export. The simple reason for this is that if it did not do this, the country would limit its ability to put OTA limit restrictions on imports.

At the same time, producing countries have the obligation to meet the demands of their customers, the importing countries that rely on coffee as a raw material in their food industries to be used for soluble coffee, instant coffee, soft drinks, etc.

The first guidelines aimed at preventing the development of fungi that produce ochratoxin A in coffee were established during the International Conference on Coffee in 1997, in Nairobi, Kenya[52]. According the Joint FAO/WHO Expert Committee on Food Additives (JECFA), tolerable ingestion levels have been estimated at 100 ng k^{-1} of body weight per week and 1.5–5.7 ng kg^{-1} of body weight per day by the European Union [53].

In 2004, the Committee of Permanent Representatives of the European Union approved the following maximum limits of ochratoxin A:

- Roasted Coffee: 5.0 μg kg^{-1}
- Soluble Coffee: 10 μg kg^{-1}
- Wine and other grape must–based beverages: 2.0 μg kg^{-1}
- Grape juice and grape juice ingredients in other beverages: 2.0 μg kg^{-1}

Once approved, these limits have a tendency to be adopted by the EU countries. The limits are so low, however, that they are seen by some producing countries as being trade barriers by those that adopt them.

Brazilian legisilation establishes a maximum tolerable limit of 10 μg kg^{-1} of mycotoxins for both roasted coffee (ground and whole bean) and soluble coffee as approved by resolution RDC No. 7 of the National Health Inspection Agency (Agência Nacional de Vigilância Sanitária or ANVISA) on February 18, 2011.

This limit is based on results obtained using methodologies that met the criteria established by the Codex Alimentarius, and applies to all products, including all food products and raw materials that are imported, produced, distributed, and sold in Brazil.

In order to identify the presence of residues of chemical substances that are potentially harmful to consumer health, complementary monitoring actions and investigations have been adopted by the Ministry of Agriculture, Livestock, and Supply through its National Plan for the Control of Residues and Contaminants of Plant origin (PNCRC/Vegetal), a federal program for the inspection and regulation of food products, based on risk analysis. These residues include pesticides and other agricultural toxins, as well as environmental contaminants such as aflatoxins and inorganic contaminants (heavy metals) in various production chains, including coffee.

The program allows for the evaluation of Good Agricultural Practices (GAP), Good Manufacturing Practices (GMP), and other self-regulating measures along the food chain. Its objective is to identify probable causes of contaminants present in levels over the maximum permitted by legislation.

Annually, based on the violations detected, recommendations can be adopted by the production sector for health safety education measures in order to better meet good agricultural practices, as well as to fortify self-regulating and preventive measures. Furthermore, an analysis of these violations can help improve official education and corrective policies. These measures attempt to supply system-wide food safety guarantees to the consumer that meet the international demands established by MERCOSUL, CODEX, OMC, and auxillary bodies (FAO, OIE, WHO).

The establishment of regulatory measures is a difficult decision that involves competing interests from diverse segments. For this reason the FAO and the WHO created the Codex Alimentarius Commission, in 1963, to standardize international food standards. The CAC began with more than 30 participating countries, a number that has grown to over 150 in its 50-year history. The objective of the Codex in terms of food contamination is to inform participating countries about their available options and to promote the harmonization of international standards.

While the introduction of maximum limits for cereals can be justified, given that "commodities" are the major sources of exposure, the protection of human health is better assured through prevention, which is best done by minimizing both exposure to fungi that produce ochratoxin A and conditions favorable to the production of the toxin. The FAO created the "Enhancement of Quality in Coffee by Prevention of Mould Formation" program to identify and control the causes of coffee contamination. The program is based on Good Agricultural and Processing Practices and is implemented through management practices such as Hazard Analysis Critical Control Point (HACCP).

Despite the diversity and technological limits observed in producing countries, adoption of good management practices is still the best alternative for preventing mold contamination of coffee and the formation of ochratoxin A. The difficulties lie in the identification of adequate measures, given the peculiarities of each situation and production system, which depend on the technical, social, and economic conditions of each region. By better understanding these conditions, adequate procedures can be developed that align with HACCP principles.

11.6 Post-Harvest Factors that Can Lead to the Occurrence of Toxigenic Fungi and Ochratoxin A in Coffee

While coffee is amply consumed in countries with temperate climates, the geographies where it is produced are subject to tropical conditions during harvest, processing, and storage for variable periods. It is also subject to sudden changes in humidity and temperature in transport to its final destination. These factors can trigger certain physical, chemical, and biological processes in the product that impact quality and compromise its consumability[54].

Various studies have looked at the incidence of ochratoxin A in coffee fruit and beans, with particular focus given to determining the stages of product processing where contamination and fungi colonization occur, as well as the actual syntheses of ochratoxin A.

Considering that little is known about the mechanisms that form ochratoxin A in coffee, defining and applying control measures has become an enormous challenge for producers, coffee brokers, and institutions responsible for guaranteeing product quality to the end consumer.

Differences have been observed in the OTA levels of samples coming from the two coffee species of major commercial importance, Arabica and Canephora. Canephora coffee has demonstrated higher susceptibility to contamination compared to Arabica. However, the major causes of variability in the levels of mycotoxins detected are principally due to differences in coffee farm management. In general, less care is exercised in the production of Canephora coffee during pre- and post-harvest, giving the coffee a higher risk of OTA contamination.

This re-emphasizes the importance of appropriate control measures such as Good Agricultural Practices, including proper post-harvest care in order to avoid OTA contamination. In implementing these control measures, the points considered to be most crucial include the period directly after harvest, when the coffee tree is subject to hydric stress and other injuries of biotical and abiotical origin that leave it more susceptible to colonization by fungi, and during the harvest, when direct contact with the soil can be a major source of contamination.

The post-harvest phase is of major importance for preserving product safety. Critical control points within the various post-harvest stages are listed below.

11.6.1 Coffee Processing

The various ways coffee can be processed are discussed in Chapter 5, Coffee Processing. The following are the aspects of processing that present contamination risks for fungi that are potential producers of OTA and other mycotoxins.

11.6.1.1 Hydraulic Separation

Hydraulic separation is a post-harvest stage recommended for all types of coffee processing. In the hydraulic separator, part of the coffee lot will float. These "floaters" comprise coffee that has dried on the tree; fruit that are over-ripe and almost-dry, called "raisins"; and underdeveloped coffee fruit; among other types.

All types of coffee coming from the hydraulic separator (floaters, ripes, and unripes) present different levels of risk exposure to ochratoxigenic fungi and to ochratoxin A.

Floaters are composed of less dense fruit that float in the hydraulic separator. There are many factors that result in lower fruit density. Among them are poor seed development, insect damage (normally from the coffee bean borer), damage caused by pathogenic fungi, and over-ripe fruit. Because of this, floaters may arrive from the field already vulnerable and at higher risk of exposure to OTA contamination in the processing phase.

The best quality portion is composed of the denser ripe, semi-ripe, and unripe fruit that do not float in the water. This portion is usually subject to stricter standards in order to preserve its sensorial qualities.

A study examining OTA levels in these two lots concluded that even though floaters, in general, have a higher risk of exposure to ochratoxin A, the samples analyzed presented a percentage that was not contaminated. Thus, whether or not the floaters become contaminated depends on various factors relative to their composition and ambient conditions[55]. Within a program to reduce the occurrence of ochratoxin A, floaters should be monitored and processed separately.

11.6.1.2 Dry Processing

Dry processing is the oldest processing method and is a simple and natural way to process recently harvested coffee fruit.

Research has shown the importance of the participation of microorganisms, including fungi that can potentially produce ochratoxin A, in compromising the quality and security of the final product. Dry processing involves a higher level of exposure to OTA contamination in the final product where ambient conditions are favorable to the development of these fungi and the synthesis of mycotoxins, as well as when drying is inadequately performed[56].

A series of experiments was conducted to determine how OTA developed under different conditions and stages of post-harvest processing. These experiments consistently found that OTA formed in the coffee fruit and skin during sun drying[57]. Over the course of two major harvest months, the dried coffee beans showed an average OTA concentration over 100 times lower than that found in the coffee fruit skins. OTA contamination occurred after five days of drying and frequently continued to increase until between 15 and 20 days of drying. The experiments were conducted in the humid weather conditions of southern Thailand (26–27 °C, 77%–82% RH).

Considering that experiments have demonstrated that the fruit skin is a major source of OTA contamination of the bean, dry processing should be performed with extreme care to prevent transmission of any fungi in the fruit skin or other tissues to the bean.

11.6.1.3 Wet Method

The wet process significantly reduces the risk of OTA contamination during both pulping/hull removal and during drying and storage. Since the mucilage constitutes an excellent substrate for the development of fungi that can potentially produce OTA, its removal decreases contamination risk[58].

11.6.2 Coffee Drying

Drying coffee fruit on patios may, under certain environmental conditions, allow for the development of fungi that can potentially produce OTA and mycotoxins. Coffee fruit with initial moisture content between 59% and 63%, containing a readily available carbon source via their free sugars, were dried in the high humidity of southern Thailand (26–27 °C, 77%–82% RH), thus presenting an excellent substrate for fungi. OTA contamination occurred within 5 days of drying and continued to increase until 15 to 20 days of drying. Additionally, experiments with re-humidifying done between 5 and 10 days of drying led to a vigorous synthesis of OTA in the fruit, thus showing the potentially negative impact of rain on drying[59].

In general, the drying of coffee fruit is a high-risk stage for the occurrence and development of fungi that can potentially cause OTA and produce mycotoxins.

Under unfavorable environmental conditions for sun drying, it is recommended to follow sun drying (pre-drying) with mechanical drying or to only use mechanical dryers, minimizing the exposure to OTA contamination in the final product.

11.6.3 Dry Milling and Sorting

Since the skin of coffee fruit is a significant source of ochratoxin A, and the defects caused by inadequate management of pre- and post-harvest stages can expose the fruit to high-risk contamination, the sorting and milling of coffee beans are effective methods of reducing ochratoxin A[60].

Studies conducted with Canephora coffee demonstrated that unripe and ripe fruit contained just traces of ochratoxin A in the beans (an average of 0.3 µg kg^{-1}), while their skins had an average of 0.4 µg kg^{-1}. Raisin fruit (fruit that are over-ripe and almost-dry) were the most contaminated, with levels ranging from undetectable to 31.8 µg kg^{-1}, with an average of 3.3 µg kg^{-1}, for the beans, and 0.3 to 38.6 µg kg^{-1}, with an average of 7.3 µg kg^{-1} [61], for the skins. In dried fruit, contamination levels varied greatly, principally in the skin, with levels from 5.9 to 1,206 µg kg^{-1}, with an average of 389 µg kg^{-1}. Contamination in the beans of dried fruit varied from 1.4 to 51.6 µg kg^{-1}, with an average of 16.5 µg kg^{-1}. This was the first experiment that demonstrated that ochratoxin A was concentrated in the coffee fruit skin, the part removed during pulping and milling.

Roasted coffee adulterated with the fruit skin presented higher levels of OTA contamination, around six times more than samples of pure soluble coffee[62]. This occurred because the coffee fruit skin is a substrate that is more favorable to the development of fungi and the production of OTA than the coffee bean.

A study of the relationship between the occurrence of ochratoxin A and size grading concluded that size grading reduces the risk of OTA contamination, despite the low levels of contamination encountered[63]. Of the twelve samples analyzed, seven showed levels of contamination that varied between 0.120 and 0.591µg kg^{-1}. Of the contaminated samples, three were the residuals left after sorting and one was an unsorted green sample.

11.6.4 Storage

During storage, spores can penetrate the fruit and beans and produce toxins if storage conditions are inadequate, principally in terms of the coffee's moisture content.

Research done in Thailand has yielded no evidence regarding an increase in OTA in Canephora coffee stored in tropical conditions[64], indicating that the toxin was already present in the warehouse. These results reinforce the possibility that pre-harvest, harvest, and processing factors are critical points in reducing contamination risk.

The ideal moisture level at the end of drying is between 11% and 12%, corresponding to 0.52 and 0.59 water activity (a$_{w}$). As shown in table 11.1, these levels are unfavorable for the development of fungi that can produce and synthesize ochratoxin A.

In the event contamination is suspected, the storage facility should by analyzed (looking for gutters or other possible sources of humidity in the air and in the bean), and any suspect lots inventoried with appropriate information noted (age, original classification by type and cup quality, lot origin). Once this inventory is completed, representative samples of the lots should be collected and sent to an appropriate lab to determine the level of contamination.

Where visible signs of deterioration occur in a product that has been exposed to conditions that are extremely favorable to product quality deterioration as well as conditions that compromise food safety, the lot should be deemed inappropriate for consumption and appropriate documentation should be generated. In these cases it is not necessary to invest in high-cost analysis[65]. It should be noted that these cases are very rare in Brazil as the coffee is, in most cases, stored in warehouses that are properly constructed, maintained, and monitored.

11.7 Roasting and Decaffeination

Although roasting is usually under the purview of industrialization rather than post-harvest, it is worth mentioning its role in the reduction of ochratoxin A in coffee.

Until the last decade of the 20th century, there was a consensus that ochratoxin A was degraded during the roasting process. However, after reports of the occurrence of OTA in samples of coffees from various origins, sold in Japan in 1988, various studies were undertaken regarding the presence of mycotoxins in roasted, ground, and brewed coffee.

The majority of the studies showed around a 75% decrease in OTA levels during roasting, partially removed physically with the residues and partially through mechanisms that were not elucidated[66]. The remaining quantity of OTA was principally extracted with water.

Since 1995, in a total of 985 results in samples taken of roasted and ground coffee, and 359 of instant coffee (regular and decaf), 83% of the samples of roasted and ground coffee and 71% of the samples of instant coffee had levels of ochratoxin A under 1 µg kg^{-1}. Some studies showed variations in the reduction of mycotoxin, according to the roast level and method of extraction. Therefore, roasting is an efficient means of reducing levels of ochratoxin A and its ingestion through coffee consumption.

The inhibiting effect of caffeine on the development of fungi that produce mycotoxins, including ochratoxin A, has been well-documented[67] (Figure 11.4).

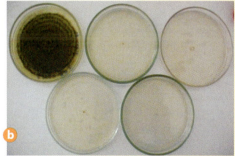

Based on these results, some authors have suggested that the decaffeination process poses some risk for fungi contamination and mycotoxin synthesis in coffee and cacao[68]. However, the decaffeination process occurs at the industrialization stage, after the critical phases of high contamination risk.

FIGURE 11.4 The effects of adding caffeine to a medium (0%, 0.5%, 0.8%, 1.5% and 2.0%) on the mycellium growth in the fungi ⓐ *A. ochraceus* and ⓑ *A. parasiticus*.

Recent genetic improvement work has targeted hybrids with reduced levels of caffeine. This poses a higher risk as it also decreases alkaloid levels, which, in turn, increases susceptibility of the fruit and beans to contamination by toxigenic fungi and mycotoxins. However, currently the preferred method to reduce caffeine levels is through an additional processing stage, appropriately called the decaffeination process.

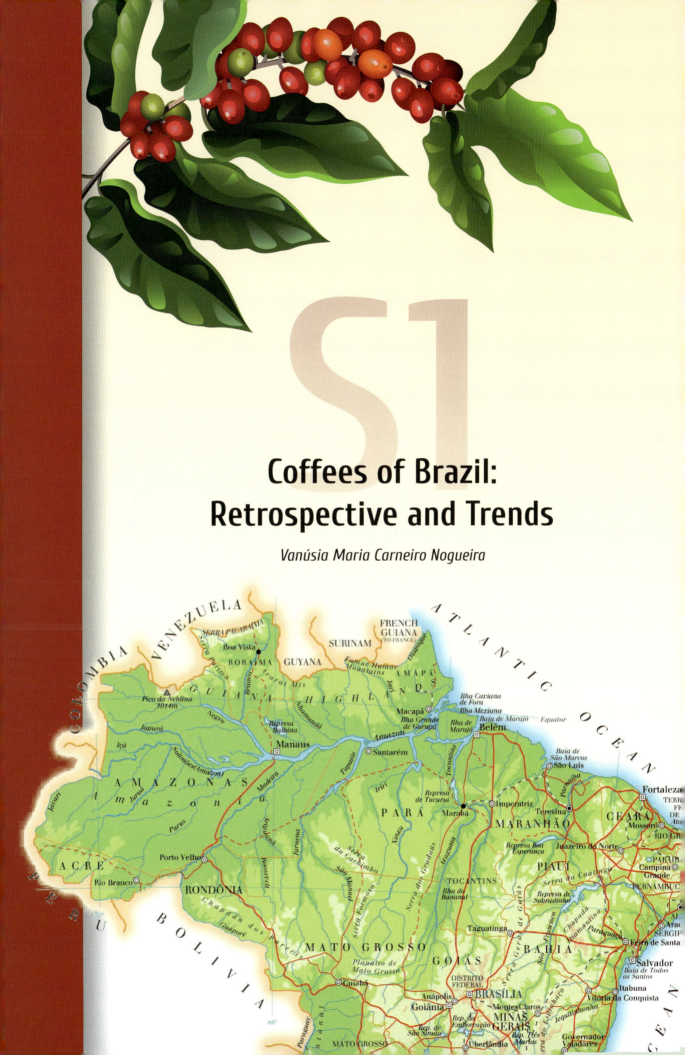

S1

Coffees of Brazil:
Retrospective and Trends

Vanúsia Maria Carneiro Nogueira

The Beginning of Everything

Coffee is a native product to regions of Africa such as Ethiopia, Sudan, and Kenya. Before entering the world stage as a commercial product, coffee was consumed as ripe fruit or juice, which was sometimes fermented to produce an alcoholic beverage.

The Arabs were first to commercially cultivate coffee and expand its consumption. There are no exact records of when coffee arrived in Arabia. Scholars estimate that it was in the millennium before Christ by way of merchants from the kingdom of Sheba, who brought seeds from their native land, Kaffa, in what is now Ethiopia.

Beginning in the 14th century, coffee was cultivated in Yemen, which tried to maintain a monopoly on its cultivation. But in the 15th and 16th centuries, the consumption of coffee spread throughout the Arabian Peninsula. Seeds of the fruit were transported to India and the drink also spread widely through the Ottoman Empire.

The first shipment of beans to Western Europe occurred in 1615, from the port of Mocha, in Yemen, to Venice. During this period Europe became familiar with cocoa, tea, and tobacco, and enthusiasm grew for exotic products from tropical countries. Coffee, with its stimulating properties, was appreciated by the European middle class from the beginning. In 1616, the Dutch began to distribute coffee seedlings from the Amsterdam Botanical Garden around the world. In 1706, the first seedlings arrived in Dutch Guyana (now Suriname), descendants of plants from the island of Java, Indonesia. In 1717, the first seedling arrived in French Guyana from Yemen. In 1727, a Brazilian official visiting French Guyana received a coffee seedling as a sign of friendship; it was from the wife of the governor of the colony, hidden in a bundle of flowers. After Martinique, Haiti (1725), and Brazil, coffee was brought to Jamaica (1730), Cuba, Puerto Rico (1755), Guatemala, and Costa Rica, finally arriving in Colombia in 1794 and in Mexico in 1796. In the 19th century, coffee reached Hawaii and other African countries such as Tanzania, and Uganda.

The Origin of the Word *Coffee*

There is disagreement about the origin of the word coffee, but three versions are most accepted. The first says that coffee comes from the Arabic *qalwah*, a term that was used by the Muslims to refer to wine, a drink prohibited by Islam. The second version states that the word comes from the Ethiopian region of Kaffa, whose name means "plant or land of God" in Egyptian. The third version suggests that the origin is the word *kahué*, a Turkish term that indicates strength.

The First Cafés

The first known coffee shop was opened in Istanbul in 1554. The first European cafés are dated 1650 and 1671, in Oxford, England and Marseille, France. By 1715, London had already registered two thousand establishments selling coffee. It is interesting to note that Café Florian, founded in Venice in 1720, remains in operation today.

Coffee houses were meeting places of merchants, intellectuals, and artists to discuss ideas that were not always in line with the rulers. Beginning in 1750, some English cafés began to allow women, who were supposed to drink their coffee separately from men. These were the first steps toward the emancipation of women.

Currently, it is estimated that more than one billion people are coffee consumers—close to 15% of the world's population.

Spotlight on Brazil

In 1720, world consumption of coffee totaled 90 tons, most of it coming from Java. Both the increase in consumption and in the price of the fruit encouraged Europeans to seek other tropical regions to increase coffee production. In 1770, consumption reached 320 tons, still coming mainly from Asia. By 1820, the coffee trade had reached 90 thousand tons, 50% of which came from Brazil.

The earliest available production statistics are from 1852, and show a global production of 4.6 million bags, to which Brazil contributed 2.4 million, Central America and the Caribbean 560 thousand, Asia 1.6 million, and Africa, the birthplace of coffee, 20 thousand bags. Between 1925 and 1929, Brazil came to be responsible for 70% of global production.

Coffee was very well adapted to the climate in Brazil and the country had much area for expansion. The plant was cultivated in Brazil beginning on large plantations in full sun, whereas in other regions it was often grown on small properties and in shady regions. In the middle of the 19th century, the main region of coffee production in Brazil was the Paraiba valley, near Brazil's capital at the time, Rio de Janeiro. At this time slave labor was used exclusively. The second largest region in the country was the state of São Paulo, with 12% of the total area and with labor coming predominantly from European immigrants and the settlement system (sistema colonato). After the abolition of slavery, this system came to be used in most of the country.

Since the arrival of coffee in Brazil, its cultivation has performed a fundamental role in the formation of the country. In the 19th and 20th centuries, the paths of both the economy and politics of Brazil were directly related to the production and commercialization of coffee. Brazilian "coffee culture" opened roads, created and populated cities, formed social classes, and made itself present in the socioeconomic life of the country. For around 100 years, Brazil captured close to 1% of international trade through coffee. The port of Rio de Janeiro was the main coffee port in the world until the end of the Brazilian Empire (1822–1889). After this period, Rio gave way to Santos, as coffee production in the country was already moving to the state of São Paulo.

Global quality references were created in Brazil due to the relevance of its production with respect to total global production. The coffee exported by Rio de Janeiro came to be known to the world as *Rio* coffee, an inferior quality coffee. By contrast, exportation by the port of Santos, and "Santos coffee," came to be a worldwide standard of quality for commercial coffees.

In the 20th century, São Paulo came to be the fastest growing state in the federation, and the capital accumulated by the coffee economy was responsible for the birth of industry in the country. This century was also marked by the beginning of over-production crises that led the government to intervene directly with acquisitions and/or warehousing of stocks and the prohibition of new planting. After World War II, coffee lost its status as the engine of the Brazilian economy. However, stimulated by successive price peaks and insufficient stocks of coffee, Brazilians began coffee plantations in the state of Paraná. For close to 30 years, Paraná production was responsible for a substantial part of Brazilian output, reaching periods of over-production that again led the government to intervene in the sector. For close to 30 years, Paraná, with its high quality, fertile soil, was responsible for a substantial part of Brazilian output. It is worth noting that the Tropic of Capricorn runs through Parana, and as coffee production expanded south through Paraná, coffee was grown outside the tropics for the first time. However, strong frosts occurred in the region in 1975 and 1977, indicating that the axis of domestic coffee production should be shifted to regions with milder climates. In 1980, the state of Minas Gerais assumed the lead in domestic production, where it currently remains, accounting for around 50% of Brazil's total production. In Minas Gerais, production occurs on properties of diverse sizes, spread over the entire state.

Currently, Brazil produces coffees with quality equal to or superior to those considered the best coffees in the world. It is the only country to offer the market diverse qualities and flavors of coffee. Although over time coffee's share of Brazilian exports has decreased due to the diversification of exported products, it is still a strong generator of foreign exchange. Brazil remains the leading exporter of green coffee and is the second largest consumer of roasted and ground coffee.

The Vanguard of Specialty Coffee in Brazil

Over the decades, the main attributes of Brazilian coffee were being lost mainly due to the popularization of blends to standardize the product. On one hand, standardization is seen in a very good light by the consumer for providing beverage consistency. On the other hand, unique and special characteristics of individual coffees can be lost through blending. In 1997, starting with a project entitled the "Gourmet Project," supported by ICO, the process of systematizing the various sensory attributes of coffee and of developing new methodologies of classifying and grading fine or gourmet coffees—or as they are known nowadays, *specialty coffees*—was begun. The sensory evaluation methodologies "Cup of Excellence," and later SCAA, originated from these works. Both are currently widely used in the market. Also as a result of the "Gourmet Project" project, the "Cup of Excellence" coffee quality contest was designed, with its first edition taking place in Brazil in 1999.

During this period of reevaluating both the strategies adopted up to that point and the model of governance of the sector, a group of 12 producers who were concerned about quality united and founded the BSCA, the Brazil Specialty Coffee Association.

The BSCA is a non-profit society that brings together people and corporations in the internal and external markets, seeking to disseminate information and stimulate technical improvement in the production, commercialization, and industrialization of specialty coffees. In addition, it seeks to promote environmental preservation and environmentally sustainable development. In order to do this it fosters programs, projects, and partnerships with public, private, domestic, and foreign entities.

The BSCA's objective is to elevate the standards of excellence of the Brazilian coffees offered on the internal and external markets through research, the publication of quality control techniques, and product promotion. It is the only Brazilian institution to certify lots and monitor quality control labels for specialty coffees, with total traceability via individual numbering, which can be consulted by the consumer through the site www.bsca.com.br.

For all the actions and initiatives adopted and for the success obtained, The BSCA is currently recognized internationally as the vanguard of fine coffee production in Brazil. An example of this is the BSCA's acting, in partnership with ACE, the Alliance for Coffee Excellence, on the organization and collaboration of coffee quality contests in 10 producer countries. The

Cup of Excellence is a great opportunity to show to the demanding consumer market the high quality of the best coffees in the world. It also enables winning producers to sell their coffees on an internet auction, with the participation of more than 50 consumer countries, at extremely remunerable prices with respect to the conventional market. Given the success of this initiative and the volume of excellent natural coffees produced in Brazil, in 2011, the BSCA promoted the first edition of the Cup of Excellence – Natural – Late Harvest, the only contest in the world specifically aimed at natural coffees produced by the dry method.

Specialty Coffees

Today the specialty coffee segment represents close to 12% of the international market for coffee. The quality attributes of coffee range from physical characteristics such as origin, varieties, fruit color, and bean size, to environmental and social concerns such as production systems and labor conditions of coffee workers.

Such differentiated coffees currently sell for a price premium of 30% to 40% over commercial grade coffee. In some cases this can surpass the 100% mark. The differentiation of specialty coffees should be based on physical and sensory attributes that contribute to a cup quality that is superior to the standard.

According to the BSCA, the main categories of specialty coffee are:

a. **Coffee from certified origins**
   Related to origin regions of the plantations, as some of the quality attributes of the product are inherent to the region where the plant is cultivated.
b. **Gourmet coffee**
   Arabica coffee beans with a screen size larger than 16 and of high quality. This is a differentiated product, almost devoid of defects.
c. **Organic coffee**
   Produced under the rules of organic agriculture. The coffee must be grown exclusively with organic fertilizers, and weed and pest control must be done biologically. Despite having higher commercial value, in order to be considered as belonging to the class of specialty coffees, organic coffee should have qualitative specifications that add value and strengthen it on the market.
d. **Fair trade coffee**
   Produced by small farmers and consumed by people concerned about the socio-environmental conditions under which the coffee is grown.

Coffee from Plant to Cup: How to Put Quality in the Cup with Pleasure and Knowledge

About two decades ago, international coffee specialists determined, by means of very well structured studies, that in order to increase global coffee consumption and also add value to the product, the concepts of quality needed to reach the consumer. Several questions were raised: How to put quality in the cup with pleasure and knowledge? How to get the best qualities of coffee to reach the end consumer, that is, to reach the cup? How to present this quality to the consumer? What would be the best way for the consumer to understand and recognize the quality of this product?

Many actions were taken from this determination. Among them we can cite the change in the approach to dealing with the end consumer. A cup of coffee would become a moment of pleasure, tasted calmly and with awareness. Coffeehouses began to be well decorated and began to offer more beverage options than the traditional European espresso or American drip coffee. Alternatives were included, such as offering entertainment, free Internet access, and a common space. For household consumption of the product, which accounts for the majority of the commercialized volume of coffee, several forms of extraction of single servings were developed. The pace of life nowadays demands a more practical style. Always seeking the need for convenience, single servings in any format are a solid consumption trend, provided that the product is of high quality. Countless initiatives can be evaluated at this moment in the market with this strong practicality trend demanded by the modern market.

Advancement in the World

In conclusion, we cannot avoid returning to our starting point. After being discovered in Africa, transformed in Middle East, and winning over the palates of the Europeans, Americans, and Japanese, coffee continues to conquer new markets. The creation of new consumption preferences in emerging markets such as South Korea, Oceania, and China is a clear demonstration of the globalization of the ancient custom of consuming coffee. People in other countries and continents are discovering the pleasure and stimulation of this widely consumed beverage. How about a tasty cup right now?

Brazilian Coffee
Classification

Joel Shuler

Brazilian Coffee Classification

There are two main ways to classify the quality of a coffee. One is through a physical analysis of the unroasted green coffee, the other through a sensorial tasting of the roasted beverage. The Official Brazilian Classification method, commonly known by its Portuguese acronym COB (Classificaçao Oficial Brasileira) has served as the basis for many other classification systems, and is still the principle classification system used for a large portion of the world's coffee. Currently this method is officially regulated by Federal Guideline No 8 of the Brazilian Ministry of Agriculture, Livestock, and Supply (MAPA). Understanding the COB is important for coffee growers as it enables them to understand how their product is evaluated and priced. Similarly, it is important for coffee buyers in that it contains the language of coffee commerce and the reasoning behind this language. Even micro-roasters, who tend to operate solely in the realm of Specialty Coffee and are more familiar with SCAA or COE sensorial analyses, can benefit by understanding the expanded classification of coffee through various grades. After all, the history of coffee classification is a moving intersection between suppliers' offerings and buyers' needs and tastes.

As technologies advance, consumer tastes change, and coffee offerings become more diverse, classification systems must adapt, and a better understanding of the complete gamut of classification will enable those in the industry to understand and even contribute to these changes. Given its importance both past and present, learning the COB is a good way to do this. Presented below is the Classificaçao Oficial Brasileiro, or COB, as codified by the Brazilian Ministry of Agriculture, Livestock, and Supply.

Technical Regulation of Identity and of Quality for the Classification of Green Coffee

Federal Guideline No 8, June 11, 2003, by MAPA

1   **Objective: the objective of this regulation is to define identifying characteristics for the quality classification of green coffee.**

2   **Definition of Product: green coffee is the endosperm of the fruit of various species of the genus *Coffea*, principally *Coffea arabica* L. and *Coffee canephora* Pierre (robusta or conilon).**

3   **Concepts: For purposes of this regulation, the following terms are defined:**

   3.1   Moisture Content: percent of water found in the product sample, which should be free of foreign materials and impurities.

   3.2   Foreign Material: vegetative remnants not coming from the product; grains or seeds from other species; and foreign bodies of any nature, such as stones or clods.
   3.2.1   Stone or clods: any stone or clod, of differing size, originating from harvesting coffee off of the ground or from drying patio fragments.

   3.3   Impurity: Hull, stick, or other remnants coming from the coffee itself.
   3.3.1   Hull: a fragment of the dried coffee fruit skin, of any size, originating from poor calibration of the milling machine.
   3.3.2   Stick: a fragment of a coffee shrub branch.

   3.4   Black Bean: a bean or piece of a bean with opaque black coloration.

3.5   Sour Bean: a bean or piece of a bean that is colored brown by the fermentation process.

3.6   Black-Green Bean: a black bean that shines due to the intact silver skin.

3.7   Immature bean: a bean that did not reach maturation, with its silverskin still intact, with a closed ventral furrow and varied green tones in its coloration.

3.8   Parchment: a bean in which the parchment was not removed during milling.

3.9   Broken: a piece of a bean, varying in size.

3.10   Shell: a bean in the form of a conch shell, resulting from the separation of intertwined beans originating from the fertilization of two ovules in one ovary locule.

3.11   Pod: a bean that did not have its hull removed in the milling process.

3.12   Shell core: a flat bean of little thickness, resulting from the separation of intertwined beans originating from the fertilization of two ovules in one ovary locule.

3.13   Malformed bean: a bean with an incomplete formation, presenting little mass and sometimes a wrinkled surface.

3.14   Flattened bean: Bean with an altered form due to flattening.

3.15   Coffee Borer Bean Damage: a bean damaged by the coffee bean borer; presenting one or more orifices that are either clean or dirty
    3.15.1   Coffee Bean Borer Damage – Dirty: a bean or piece of a bean that has been damaged by the coffee bean borer and that has developed black or blue colorations due to this damage.
    3.15.2   Coffee Bean Borer Damage – Lace-like: a bean or piece of a bean that has been damaged by the coffee bean borer and that has three or more perforations but no black colorations.
    3.15.3   Coffee Bean Borer Damage – Clean: a bean or piece of a bean that has been damaged by the coffee bean borer and that has up to two perforations and no black colorations.

3.16   Triangle Bean: a bean in triangular form caused by three or more seeds developing in a single fruit.

3.17   Grinder Bean: a broken bean that falls through screen 14 (14/64") but is at least 2/3 intact.

3.18   Head bean: A bean composed of two intertwined beans, originating from two beans developing in one ovary locule. It is not considered a defect until the two beans separate into shell and shell core, which are considered defects.

3.19   Fox Bean: a perfect bean having, however, the skin of the spermoderm intact due to climatic factors, and colored a reddish-brown hue.

3.20   Quaker: a bean that does not brown when roasting.

4 **Classification: green coffee will be classified into CATEGORY, SUBCATEGORY, GROUP, SUBGROUP, CLASS, AND TYPE, according to the species, bean form and size, aroma and flavor, cup quality, color, and quality, respectively.**

4.1 Category: according to the species to which it belongs, green coffee will be classified into 2 (two) categories:

4.1.1 Category 1: coffee belonging to the species *Coffea arabica*.

4.1.2 Category 2: coffee belonging to the species *Coffea canephora*.

4.2 Subcategory: green coffee, according to its form and size, will fall into one of 2 (two) categories:

4.2.1 Flat: a bean with a convex dorsal surface and a ventral plane that is flat or lightly curved, with a central furrow running in the longitudinal direction.

4.2.1.1 Green coffee of the Flat subcategory, according to bean size and the circular screen holes that retain the bean, will be classified into:

4.2.1.1.1 Flat large: screens 19/18 and 17.

4.2.1.1.2 Flat medium: screens 16 and 15.

4.2.1.1.3 Flat small: screens 14 and smaller

4.2.2 Peaberry: a bean in ovoid form, also with a furrow running down the center longitudinally.

4.2.2.1 Green coffees of the Peaberry subcategory, which are classified by bean size per the dimension of the largest oblong screen perforation that retains the bean, will be classified as:

4.2.2.1.1 Peaberry large: screens 13/12 and 11.

4.2.2.1.2 Peaberry medium: screen 10.

4.2.2.1.3 Peaberry small: screens 9 and below.

4.2.3 When green coffee is not subjected to size separation, or when subjected beans are divided into four or more screens, it will be considered "Unclassified Coffee."

4.2.4 In classifying by screen size, the maximum allowable variance for a coffee to be classified at a screen size is 10%. A coffee with a variance greater than 10% should be classified as the lower size.

4.3 Group: Green coffee, in accordance with its aroma and flavor, will be classified into one of 2 (two) groups. The flavor and aroma will be defined by means of a cupping.

4.3.1 Group I – Arabica

4.3.2 Group II – Robusta

4.4 Subgroup: green coffee, in accordance with its cup quality, will be classified into 7 (seven) Subgroups of Group I, and 4 (four) Subgroups of Group II, described as follows:

4.4.1 Fine Cup Quality of Group I – Arabica

4.4.1.1 Strictly soft: coffee that presents, on the whole, all of the requirements of the flavor "soft," but more accentuated.

4.4.1.2 Soft: coffee that presents pleasurable flavor and aroma and is sweet and gentle on the palate.

4.4.1.3   Slightly soft: coffee that presents a flavor with small amounts of sweetness and smoothness, but with no astringency or asperity (harshness) on the palate.

4.4.1.4   Hard: coffee that presents an acrid, astringent, or harsh flavor, but which does not present any off flavors.

4.4.2   Phenolic Beverages – Arabica

4.4.2.1   Rioy: coffee that presents a light iodine-like flavor.

4.4.2.2   Rio: coffee that presents a strong iodine-like flavor.

4.4.2.3   Rio Zona: coffee that presents a strong iodine or phenolic acid-like aroma and flavor, and is thus repulsive to the palate.

4.4.3   Beverages of Group II – Robusta

4.4.3.1   Excellent: coffee that presents a neutral flavor and medium acidity.

4.4.3.2   Good: coffee that presents a neutral flavor and light acidity.

4.4.3.3   Regular: coffee that presents a typical robusta flavor with no acidity.

4.4.3.4   Abnormal: coffee that presents a flavor that is not characteristic to robusta coffee.

4.5   Class: green coffee, in accordance with bean coloration will be classified into one of 8 (eight) classes:

4.5.1   Bluish green and cane green: colors characteristic of coffee that has been pulped and demucilaged.

4.5.2   Green: coffee in which the bean presents a green coloration.

4.5.3   Yellowed: coffee in which the bean presents a yellowed coloration, indicating product aging.

4.5.4   Yellow

4.5.5   Brown

4.5.6   Bluish-grey

4.5.7   Whitened

4.5.8   Discrepant: a mix of colors originating from the mixing of different lots.

4.6   Type: green coffee, in accordance with the percentage of defects, foreign material, and impurities, will be classified according to Tables 1,2, and 3 of this regulation.

5   Moisture Content

5.1   Independent of its classification, the moisture levels of green coffee cannot exceed the maximum tolerance of 12.5% (twelve and one-half percent).

6   Foreign Matter and Impurities

6.1   The maximum percent of foreign matter and impurities permitted in green coffee will be 1% (one percent). If it exceeds this value, the product will be temporarily disqualified and may not be commercialized until it is further sorted and re-classified.

7   Substandard: Green coffee will be classified as Substandard that presents:

7.1   Percentages of defects exceeding the maximum tolerance limits established in the tables of this regulation.

7.2   More than 50 black beans, more than 100 sour beans, or more than 100 black-green beans.

7.3   More than 300 defects, with the exception of broken beans, malformed beans, shells, shell cores, and coffee bean borer damage – clean.

7.4   Product classified as Substandard may not be sold and should be:

      7.4.1   Sorted, broken down, and reconstituted, so that it may fall into a specific Type.

      7.4.2   Re-bagged and re-marked so that it meets the demands of this Regulation

8   Disqualified Coffee

8.1   A green coffee that contains more than 1% (one percent) foreign matter and impurities will be disqualified and temporarily prohibited from sale until it is further sorted.

8.2   A green coffee that contains live insects will be disqualified until the insects have been exterminated.

8.3   A green coffee will be disqualified and prohibited from sale and from human and animal consumption if it contains one or more of the following characteristics:

      8.3.1   Mold

      8.3.2   Poor state of conservation

      8.3.3   A strange odor of any nature, inappropriate to the product

      8.3.4   Phytosanitary product residues, traces of mycotoxins or other contaminants, or noxious substances harmful to health above levels established by specific legislation.

      8.3.5   The presence of toxic seeds

      8.3.6   Product that has been disqualified may only be used for other means after a hearing with the Brazilian Ministry of Agriculture, Livestock, and Supply.

      8.3.7   The Classification Service should immediately communicate to the Brazilian Ministry of Agriculture, Livestock, and Supply the occurrence of the disqualification of any product, so that appropriate measures can be taken by the appropriate technical sector.

      8.3.8   It is up to the Brazilian Ministry of Agriculture, Livestock, and Supply to determine the destination of disqualified product, and it should, when possible, communicate with other official agencies.

      8.3.9   In a specific case where permission or authorization is granted to use disqualified product for other means, the Brazilian Ministry of Agriculture, Livestock, and Supply must establish all of the necessary procedures to accompany the product until its complete use, and it is up to the property owner, or his or her representative, to pay all costs pertinent to the operation and to be the trustee and responsible for the inviolability and indivisibility of the lot in all phases of manipulation, facing appropriate civil and penal punishment in cases of irregularities or unauthorized use of the product under these conditions.

9   Packaging

9.1   The packaging used in the storage of green coffee can be made of natural, synthetic, or other appropriate material.

9.2   Within the same lot it is obligatory that all packaging be of the same material and have the same storage capacity.

9.3   The specifications for the production and capacity of the packaging should be in accord with appropriate legislation.

10 Labeling

10.1   No specific marking or label relative to classification is required; however, the product must be accompanied by its respective Classification Report.

11 Samples

11.1   Before the sample is taken, the lot should be inspected to verify that there are no abnormalities, such as the presence of live insects or the existence of any conditions that would lead to disqualification of the lot (foul odors, poor state of conservation, mold, among others).

11.2   The sample should be taken using a grain probe in at least 10% of the lot, but always representing the average of a lot and a minimum sample size of 30g (thirty grams) of each bag.

11.3   The samples will be homogenized and divided into the minimum of 3 (three) separate samples, each with the minimum weight of 1 kg (one kilogram) and properly identified, sealed, and authenticated.

11.4   One sample will be delivered to the interested party and two will remain with the entity responsible for the classification. The rest of the sample will be put back into the lot or returned to the entity in possession of the product.

11.5   The sample for the classification will be 300 g (three hundred grams), obtained after homogenization and the quartering of one of the two samples destined for the entity responsible for the classification, with the other intact sample remaining for future testing.

12 Classification Procedures

12.1   Carefully verify if the sample contains evidence of live insects and if the green coffee presents one or more disqualifying characteristics. If such characteristics are found, the classifier should take the measures outlined in item 8.2 and 8.3 of this regulation.

12.2   If the product meets the appropriate conditions to be classified, the sample should be homogenized and reduced through the process of quartering until obtaining a working sample of 300 g (three hundred grams) as weighed on a digital scale previously calibrated.

12.3   Verify the physical aspect of the product, identifying the drying characteristic and the product class, noting the result on the report.
   12.3.1   The information regarding the processing method used to prepare the coffee should be given by the proper party and noted on the report.

12.4   Identify and separate any foreign material and impurities in the sample. Weigh them, noting on the report the weight and percentage found. If the percentage is greater than that established in item 6.1, the lot must be temporarily disqualified.

12.5 From the remainder of the product to be classified, one subsample should be obtained by quartering the product in order to determine the moisture content of the sample. Impurities and foreign matter should be removed when encountered. The weight of the subsample should be in accordance with the manufacturers recommendations of the equipment used to verify moisture levels. Once moisture level is obtained it should be noted on the report.

12.6 Once in possession of the sample, free of foreign matter and impurities, separate out the defects observing the following steps:

12.6.1 When verifying the incidence of two or more defects in the same bean, the more severe defect should prevail, in accordance with the following list in decreasing order: black, sour, black green, shell, underdeveloped, immature, broken.

12.7 Note on the report the defects found in the sample.

12.8 Group the defects into similar groups in accordance with the equivalence as established in the Tables of this regulation.

12.9 In the foreign matter and impurities, separate the hulls, sticks, stones, and clods. Count and note on the report and establish the equivalency of defects in accordance with the Tables of this regulations.

12.10 Sum the total number of defects and determine the product type in accordance with the parameters established in the Tables of this Regulation, noting the result on the report.

12.11 Determine the Category of the green coffee in accordance with the product species and note this in the respective field on the report.

12.12 Determine the Subcategory: to determine the subcategory of the product, use the following procedures:

12.12.1 Re-divide the working sample obtained after the separation of defects and of foreign matter and impurities, reducing the sample to 100g.

12.12.2 Pass the product through a series of screens, arranged in the following order:

Screen 19 = flat
Screen 13 = peaberry
Screen 18 = flat
Screen 12 = peaberrry
Screen 17 = flat
Screen 11 = peaberry
Screen 16 = flat
Screen 10 = peaberry
Screen 15 = flat
Screen 09 = peaberry
Screen 14 = flat
Screen 13 = flat
Screen 08 = peaberry
Screen 10 = flat
Bottom Pan

12.12.3 Weigh the quantities retained in each screen and note on report.

12.12.4 In determining the Subcategory, the weight is equal to the percentage.

12.13 Determination of Group and Subgroup: The determination of the Group and Subgroup of green coffee will be completed using the following cupping procedures:

12.13.1 Sanitize the grinder and other utensils (cups, kettles, spoons, etc.).

12.13.2 Set the grinder to a coarse grind and purge with a small amount of roasted coffee of the sample that is to be tested, discarding the ground product, thus ensuring the removal of possible cross contamination from previously tested product.

12.13.3 Divide the sample of roasted coffee into portions of approximately 8 to 10 g and put the proportioned coffee in the grinder, positioning each cup under the orifice of the grinder. Repeat this 7 times as 7 cups must be tested.

12.13.4 Put the cups on a round and rotating cupping table, grouped together by lot.

12.13.5 Using spring water or filtered water that has not been treated chemically, carefully pour the water over the product just after it first boils.

12.13.6 Mix the infusion with a spoon, allowing the cupper to smell it and obtain a preliminary judgment from the vapors released, and then remove the foam

12.13.7 From the released aromas, the classifier should establish an initial judgment of each coffee and separate out the beverages with favorable characteristics, which should be tasted first, from those less favorable.

12.13.8 In passing from one sample to another, with all samples the cupping spoon should be washed in cups placed on the fixed arm of the revolving table

12.13.9 Wait for the grinds to decant and then remove the foam and residue that remain on the coffee surface.

12.13.10 Wait for the beverage to cool. The decision of when to begin cupping is up to the cupper.

12.13.11 Initiate the cupping by dipping the spoon calmly in the cup, filling it with liquid. Lift the spoon to the mouth and strongly slurp so that a portion of the sprayed liquid goes to where the tongue meets the palate, with the beverage remaining in the mouth only long enough to evaluate flavor and aromas, then expel it in the spittoon.

12.13.12 Note on the report the cup quality presented by the sample and put the sample into the appropriate Group and Subgroup.

12.13.13 In cases where the caffeine content has been solicited by an interested party, the classifier should put the result of a credentialed laboratory, mentioning the name and the registration number and archive a copy of the document together with the Classification Report.

12.13.14 State on the report the reason(s) the product is considered outside of type or disqualified.

12.13.15 Revise, date, stamp, and sign the Classification Report. The stamp must have the classifier's name and registration number from the Brazilian Ministry of Agriculture, Livestock, and Supply.

13 Classification Report

13.1 The Classification Report will be emitted by the Brazilian Ministry of Agriculture, Livestock, and Supply or through the entities properly licensed by them through proper legislation.

13.2 The Classification Report is the valid document to prove the realization of a classification, corresponding to a determined lot of classified product.

13.3 The report will only be considered valid when it has the classifier ID (stamp and signature), and when the classifier is properly registered with the Brazilian Ministry of Agriculture, Livestock, and Supply.

13.4 The deadline to contest a classification result by means of arbitration is 45 (forty-five) days from the moment the Classification Report was emitted.

13.5 The Classification Report must state, in addition to the information already established in this regulation, the following:

    13.5.1 The discrimination of the results of each analysis made and of the percentages found for each quality determination as well as conclusive information (detailing Category, Subcategory, Group, Subgroup, Class, and Type) that will be transcribed on the respective Classification Report.

    13.5.2 The reasons a sample was classified as Substandard

    13.5.3 The reasons a sample was disqualified.

14 Fraud

14.1 The intentional alteration of any nature, conducted in the classification, packaging, or storage of the product, as well as the alteration of product documents

14.2 Also considered fraud will be the sale of green coffee that is not in accordance with this Regulation.

15 General Disposition

15.1 It is the exclusive purview of the Technical Organ of Brazilian Ministry of Agriculture, Livestock, and Supply to resolve cases arising from the use of this regulation.

16 Tables

16.1   Table 1: Classification of Green Coffee: Defect Equivalencies (Intrinsic)

Defects	Quantity	Equivalency
Black Beans	1	1
Sour Beans	2	1
Shell	3	1
Immature Beans	5	1
Broken Beans	5	1
Coffee Bean Borer Damage	2 to 5	1
Malformed Beans	5	1

Observations
1. Black Beans are considered the principal or capital defect
2. Sour Beans and Coffee Bean Borer Damage are considered secondary defects
3. The Black-Green defect "stinker" is considered the same as a sour bean defect

16.2   Table 2: Classification of Green Coffee: Impurity Equivalencies (extrinsic)

Defects	Quantity	Equivalency
Pod	1	1
Parchment	2	1
Large Stick, Stone, or Dirt Clod	1	5
Medium Stick, Stone, or Dirt Clod	1	2
Small Stick, Stone, or Dirt Clod	1	1
Large Hull	1	1
Small Hull	2 to 3	1

Observations
1. Large stones, clods, and sticks roughly correspond to the dimensions of flat bean screen sizes 18/19/20
2. Medium stones, clods, and sticks roughly correspond to the dimensions of flat bean screen sizes 15/16/17
3. Small stones, clods, and sticks roughly correspond to the dimensions of flat bean screen sizes 14 and lower
4. Hulls are related to roughly the size of a dried coffee pod

16.3 Classification of green coffee by defect/type

Defects	Type	Points	Defects	Type	Points
4	2	+100	46	5	50
4	2 – 05	+ 95	49	5 – 05	55
5	2 –10	+ 90	53	5 –10	60
6	2 –15	+ 85	57	5 –15	65
7	2 – 20	+ 80	61	5 – 20	70
8	2 – 25	+ 75	64	5 – 25	75
9	2 – 30	+ 70	68	5 – 30	80
10	2 – 35	+ 65	71	5 – 35	85
11	2 – 40	+ 60	75	5 – 40	90
11	2 – 45	+ 55	79	5 – 45	95
12	3	+ 50	86	6	100
13	3 – 05	+ 45	91	6 – 05	105
15	3 –10	+ 40	100	6 –10	110
17	3 –15	+ 35	108	6 –15	115
18	3 – 20	+ 30	115	6 – 20	120
19	3 – 25	+ 25	123	6 – 25	125
20	3 – 30	+ 20	130	6 – 30	130
22	3 – 35	+ 15	138	6 – 35	135
23	3 – 40	+ 10	145	6 – 40	140
25	3 – 45	+ 05	153	6 – 45	145
26	4	Base	160	7	150
28	4 – 05	05	180	7 – 05	155
30	4 –10	10	200	7 –10	160
32	4 –15	15	220	7 –15	165
34	4 – 20	20	240	7 – 20	170
36	4 – 25	25	260	7 – 25	175
38	4 – 30	30	280	7 – 30	180
40	4 – 35	35	300	7 – 35	185
42	4 – 40	40	320	7 – 40	190
44	4 – 45	45	340	7 – 45	195
			360	8	200
			> 360	Disqualified	

17 Example of a Classification Report

(IDENTIFICATION OF THE ORGAN RESPONSIBLE FOR THE CLASSIFICATION)

CLASSIFICATION REPORT FOR GREEN COFFEE # _____ / _____

Arabica Coffee ( )   Robusta Coffee ( )

INTERESTED PARTY:		
HARVEST:	SAMPLE #:	
WAREHOUSE:	CITY:	
PAVILION:	BLOCK:	CELL:
# OF BAGS:	LOT:	

PHYSICAL CLASSIFICATION

Defective Equivalencies

Description	# of Imperfect Beans	# of Defects	Description	# of Imperfect Beans	# of Defects
Black Bean			Pod		
Sour Bean			Parchment		
Shell			Flattened		
Malformed			Large Hull		
Immature			Medium/Small Hull		
Broken			Large sticks, stones or clods		
Coffee Bean Borer Damage			Medium sticke, stones or clods		
Coffee Bean Borer Damage - Dirty			Small sticks, stones or clods		
Coffee Bean Borer Damage - Lace- like			SUBTOTAL (2)		
Coffee Bean Borer Damage - Clean			TYPE DETERMINATION		
SUBTOTAL (1)					

TYPE DETERMINATION		
Subtotal (1)		
Subtotal (2)		
General Total (1) + (2)		
	TYPE	

CATEGORY

SUBCATEGORY – Screen Size ( ) 15 ( ) 18 ( ) 16 ( ) 19 ( ) 17 ( ) 20 ( ) Unclassified Green Coffee	( ) FLAT ( ) Large ( ) Medium ( ) Small	( ) PEABERRY ( ) Large ( ) Medium ( ) Small
GROUP 1 Arabica ( )		GROUP 2 Robusta ( )

SUBGROUP - ( ) Strictly Soft ( ) Soft ( ) Slightly Soft ( ) Hard ( ) Rio ( ) Rioy ( ) Rio Zona	( ) Excellent ( ) Good ( ) Regular ( ) Abnormal

CLASS
( ) Bluish Green   ( ) Yellow   ( ) Cane Green   ( ) Bluish Gray   ( ) Whitened
( ) Green   ( ) Greenish   ( ) Brown   ( ) Discrepant

CONCLUSION

MOISTURE CONTENT: _____   EQUIPMENT USED: _____

CATEGORY: _____   SUBCATEGORY: _____

GROUP: _____   SUBGROUP: _____

CLASS: _____   TYPE: _____

OBSERVATIONS: _____
_____
_____

Classified by _____ on (date) _____

CLASSIFIER'S LICENSE # _____

GREEN COFFEE REPORT # _____

Arabica ( )   Robusta ( )

ADDITIONAL QUALITY CHARACTERISTICS		
PROCESS ( ) Dry ( ) Wet	DRYING ( ) Dried Well ( ) Average Drying ( ) Dried Poorly	PHYSICAL APPEARANCE ( ) Good ( ) Average ( ) Poor
ROAST		
Coffea arábica ( ) Fine Roast ( ) Good Roast ( ) Average Roast ( ) Poor Roast	*Coffea canephora* ( ) Excellent Roast ( ) Almost Excellent Roast ( ) Very Good Roast ( ) Good Roast ( ) Average Roast ( ) Poor Roast	
CAFFEINE CONTENT		
( ) Normal Coffee	( ) Decaffeinated Coffee	

Classified by: _____ On (date) _____

Classification and Sensorial Analysis of Specialty Coffees

As the market for higher quality "specialty" coffees has grown, the need has arisen to better define coffees on this end of the quality spectrum. There are currently two main classification systems for specialty coffees: The Specialty Coffee Association of America (SCAA) analysis and the Cup of Excellence (COE) analysis. Both of these analysis systems focus on the sensorial "cupping" of the coffee in order to determine not only a coffee's numerical score, but also its specific flavor profile.

The next chapter will explore these forms and detailed classification criteria for specialty coffees.

Classification of
Specialty Coffees

Silvio Luis Leite

Introduction

With most traditional classification systems, the focus is on defect identification, and cuppers are trained to identify the presence of specific defects. When defects are not found, the coffee is considered "clean and uniform" and its score is "hard cup or higher." In Brazil, terms such as "Santos Fine Cup" and "Santos Good Cup" have developed to describe defect-free coffee.

However, coffee has a lot more to offer, and the practice of coffee tasting can lead to identifying not just the absence of defects, but flavor complexity and a wide array of flavors that vary with region, country, and other factors.

The available variety of flavors is extensive and changes according to the sensory preferences of each market. Germans buyers, for example, always rigorous in their analyses, evaluate lots considering physical aspects, roast function, and cup quality, and their standards often serve to establish references for their companies during the harvest year. Northern Europeans often opt for coffees with predominant acidity. Italians prefer balance, with a clean finish in a dryer cup, though they sometimes favor more acidity in the cup. Yet such generalizations are not always accurate, and there can be diversity and variation in quality criteria among companies from a single country.

Thus, although for a long time the training of new professionals, agronomists, and even producers, was done based on defects, the rise of new techniques for sensory analysis and for the preparation of the drink based on its positive qualities and attributes has become increasingly more important and necessary.

Cup of Excellence® Cupping Form

In 1999, through the Brazil Specialty Coffee Association (BSCA), new tools were created with the objective of recognizing and valuing the quality of coffee. The actions began with the program called "Gourmet Project," culminating with the creation of the "Cup of Excellence" (COE) program and its revolutionary testing methodology. The methodology currently used by the COE was created by George Howell and co-authored by Silvio Luis Leite. The Cup of Excellence, from its beginning, has been one of the most important programs in bringing recognition and appreciation to high quality coffees worldwide.

In evaluating specialty coffees, it is important to understand the diverse variables that compose flavor, that every defect has an origin, and that the beans transmit this to the professional. It is important to understand the "language of the beans" through classification and tasting. The coffees, through their aspect and their flavor, tell their processing history and level of care. Through their physical aspect, they allow many interpretations about drying, humidity, and uniformity. Via tasting, they allow the interpretation of the positive attributes without defects, or conversely, defective ones that had some problem at harvest and/or processing. These defects can have different intensities and affect the concept of the overall beverage in a distinct manner. Fermented beans, for example, are easily detected in the aspect and also in tasting. Although it is impossible to correct the error of a specific lot with this specific problem, the producer, upon receiving such information, can correct the preparation of new lots.

Several methodologies are available for evaluating specialty coffees, including the form widely used in contests carried out by COE, as well as those currently used by the SCAA. On the following pages the COE form is presented in its entirety, including an overview and detailed instruction on using the form, followed by the SCAA cupping form.

Cup of Excellence® Cupping Form

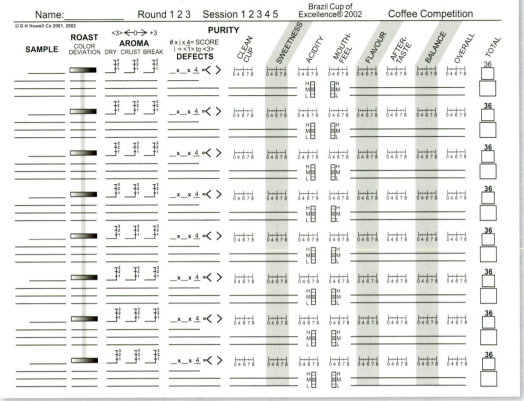

Cup of Excellence® Cupping Form

Overview of Form:

The Cup of Excellence® cupping form rates the cup quality of submitted coffees using multiple categories: absence of defects, cleanness of cup, sweetness, quality of acidity and of mouthfeel, flavor, aftertaste and balance.

The form can also provide private notes for each cupper's future reference, since all judges will receive their completed forms back at the end of the competition, and since a list of the actual farm names for each coded coffee will also be given out at that time.

The Cup of Excellence® cupping form is loosely based on the scoring system described in Winetaster's Secrets by the late Andrew Sharp and on the original SCAA Cupping form, no longer in use.

1. Filling out the form:

There are **six steps** in filling out the cupping form.

1.1 - General information: the cupper fills the top of the page with his name, the round number and the session number; in the far left column, titled *Sample*, the cupper writes the sample numbers, beginning with 1 and continuing up to 10 (eight samples can be entered on one sheet).

1.2 - Roast Color: the cupper may view a tray, provided at each session, showing all the ground coffee samples to be cupped in the session, to observe the uniformity of roasts. A sample that is more darkly or more lightly roasted than the norm can be singled out by drawing a line to the left or right of the center line:

Lighter or Darker

1.3 - Aroma: the third column notes the coffee's *aroma*, first the dry grounds, then the crust that forms after water is poured on the coffee, and lastly when the cupper breaks the crust and smells the vapors that are released. Values are as follows:

-3	-2	-1	0	+1	+2	+3
	unpleasant				pleasant	
very		slightly		slightly		very

The cupper writes the number 2 for moderately pleasant this way:

$$2 \begin{array}{l} ^{3} \\ ^{2} \\ _{1} \end{array}$$

The cupper can also note the intensity of the aroma by drawing a line through the ladder structure. Thus a moderately pleasurable aroma of high intensity would be notated this way:

$$2 \begin{array}{l} ^{3} \\ ^{2} \\ _{1} \end{array}$$

1.3.1 - Notations on the aroma <u>are not part of the score.</u> They are strictly for the cupper's future reference. The competition setting does not provide the adequate number of cups nor the proper amount of time to accurately measure the aroma of a rapidly cooling beverage.

1.4 - The Defects column records both the intensity of a defect and the number of cups (out of four glasses for each sample) on the judge's table which have that defect. The cupper must not include samples from other tables in an assessment. The intensity scale is:

1	2	3
Slight	Moderate	Intense

If more than one cup is defective, the cupper records the most defective cup's intensity. Then the cupper records the number of defective cups present on his table for the particular sample (a maximum of four). These two numbers are multiplied by a constant of 4 to derive the negative score.

Thus if sample #6 has two defective cups, one moderate, one intense, the cupper records:

$$\underline{\quad 3 \quad} \times \underline{\quad 2 \quad} \times \underline{\quad 4 \quad} = \langle 24 \rangle$$

intensity of defect **x** number of cups **x** constant = **negative score**

COE form courtesy of George Howell. Used with permission.

Cup of Excellence® Cupping Form

1.5 - Eight columns of cup quality criteria follow: Clean Cup, Sweetness, Acidity, Mouthfeel, Flavor, Aftertaste, Balance, Overall. The judge indicates a score by drawing a line across the appropriate value. The scale of values is as follows:

Unacceptable	Poor	Ordinary	Fine	Great
0	2	4	6	8

1.5.1 - A cupper may score half points. The following would be a score of 7½ points:

1.5.2 - The Cup of Excellence® cupping competition is primarily concerned with finding and rating coffee qualities from "fine" to "great". Therefore the quality "fine" is in the middle of each scale chart, allowing for greater differentiation beyond "fine".

1.5.3 - If a judge rates any category as 0 unacceptable, then the total score of that coffee will be 0. No more time need be wasted on it.

1.5.4 - Placing lines across a graded scale allows the cupper to record changing perception as the coffee cools. For instance, when the coffee is hot one might give a 7 score in sweetness but later reduce it to 6 when tasted at a cooler temperature. Do not erase. Simply place an arrow pointing from the 7 line to the 6 line on the grading scale:

1.5.5 - If the cupper has recorded a changing perception, it is recommended that the last perception value should be used in totaling. It is the decisive score. Nevertheless, if preferred, the cupper can choose the median between the two extremes. The cupper must then take care to consistently do this for all coffees.

1.5.6 - Before totaling, the cupper should write the score for each category and circle it just above its corresponding graded scale. This makes totaling far easier. It makes the clerical task of recording all scores on a computer much quicker and easier as well:

1.6 - Totaling: once all eight of the quality criteria have been scored to the judge's satisfaction, the scores are tallied. The scores are added, using the following formula:

Clean Cup
Sweetness
Acidity
Mouthfeel
Flavor
Aftertaste
Balance
+ Overall

Sub-total + Defect Score = Raw Score (small boxes on far right)
Positive!   Negative!

The 100 Basis Score is achieved by adding 36 to the Raw Score:

36 + Raw Score = 100 Basis Score (big boxes on right)

Cup of Excellence® Cupping Form

2. Cup quality criteria:

2.1 - Clean cup: this is the basic starting point for coffee quality. Clean cup is complete freedom from taints or faults. It is the transparency necessary for a coffee's terroir to shine through.

2.2 - Sweetness: the sensation of sweetness correlates directly with how uniformly ripe a coffee was when harvested. Sweetness is not entirely dependent on how much sugar is in the roasted coffee, but also on other components which combine to create the impression of sweetness.

2.3 - Acidity: this is what brightens a coffee. It gives life. In wine it is often referred to as nerve (nervosité in French), backbone or spine.

2.3.1 - Quantity of acidity is not directly related to quality. The judge must score the quality of the acidity, not how much acidity is in a particular coffee. As with wine, not all coffees should be notably acidic. It is rather the expression of that acidity, whether powerful or very mild, that is important: is the acidity harsh or overly tart? Is the acidity refined or tangy or does it have a pleasant snap? These are the kinds of questions the judge should ask when scoring a particular coffee's acidity.

2.3.2 - While not determining a coffee's quality it is nonetheless important to rank the amount of acidity in each coffee. The vertical ladder-like structures below the line graphs provide a place to record the acidity's intensity, irrespective of its quality, for future reference. **H represents high acidity, M is moderate, and L is light:**

H ☐
M ▓
L ☐

2.4 - Mouthfeel: the tactile sensation imparted by a coffee. Mouthfeel can include the perception of viscosity, density, weight, texture and astringency.

2.4.1 - As with acidity, the degree of mouthfeel presence is not the same thing as quality. The judge must score the quality of the mouthfeel.

2.4.2 - The cupper also wants to record how much body a coffee displays, and so a ladder-like graph has been provided for this purpose. **In the case of body: H represents heavy body, M is medium, and L is light.**

2.5 - Flavor: this is a combination of taste (sweet, sour, bitter, salty and pungent) and aroma - or nose. This is where a fine coffee can truly stand out as an elegant, and even forceful, expression of place terroir. The judge must determine whether a coffee's flavor profile is merely generic or a genuine expression of terroir brought out by the care of the harvester and the skill of the processor. For example, a slightly off - fruity flavor determined by the fermenting process may be pleasant for many but it is a reproducible flavor the world over. What is worse, such a flavor masks what terroir the coffee might otherwise express.

2.6 - Aftertaste: the lingering flavor after the coffee has been swallowed can either reinforce the pleasure derived from a coffee's other attributes or it can weaken and even sabotage it. Does the coffee sweetly disappear or is there a harshness that emerges?

2.7 - Balance: is the coffee harmonious? Is something excessive? Is the coffee missing something?

2.8 - Overall: does the coffee have an exciting complexity or is it a simple but very pleasing coffee? Does the cupper simply not like it? This category is the cupper's personal call.

Final points: One way the coffee experience is truly different from wine is the perception of how it changes from hot to cold over a considerable amount of time. Most judge a coffee by their first impression. The first step to leaving the commodity world, however, is for consumers to discover delight in the elegant and slowly evolving transformations of the rare best coffees. Judges should take care to explore each coffee presented at the Cup of Excellence® competitions in this way.

The complexity of this form pales in comparison with many forms used in wine competitions. The present Cup of Excellence® cupping form is an attempt to develop coffee cupping along the lines wine has so successfully pioneered. It is easy to cite comparisons as well as differences with wine. What is clear is that quality coffee is not merely a product of nature, but a complex achievement resulting from craftsmanship, care, discipline and love. Such coffees deserve our going out of our way to properly express those

Cup of Excellence® Scoring Categories

DEFECTS
Phenolic, rio, riado ➡ automatic disqualification
Ferment
Oniony, sweaty

CLEAN CUP
+ purity | free from measurable faults | clarity
- dirty | earthy | moldy | off-fruity

SWEETNESS (prevalence of…)
+ ripeness | sweet
- green | undeveloped | closed | tart

ACIDITY
+ lively | refined | firm | soft | having spine | crisp | structure | racy
- sharp | hard | thin | dull | acetic | sour | flabby | biting

MOUTHFEEL (texture, viscosity, sediment, weight, astringency)
+ buttery | creamy | round | smooth | cradling | rich | velvety | tightly knit
- astringent | rough | watery | thin | light | gritty

FLAVOR (nose + taste)
+ character | intensity | distinctiveness | pleasure | simple-complex | depth
(possible notations: nutty, chocolate, berry, fruit, caramel, floral, beefy, spicy, honey, smokey…)
- insipid | potato | peas | grassy | woody | bitter-salty-sour | gamey | baggy

AFTERTASTE
+ sweet | cleanly disappearing | pleasantly lingering
- bitter | harsh | astringent | cloying | dirty | unpleasant | metallic

BALANCE
+ harmony | equilibrium | stable-consistent (from hot to cold) | structure | tuning | acidity-body
- hollow | excessive | aggressive | inconsistent change in character

OVERALL (not a correction!)
+ complexity | dimension | uniformity | richness | (transformation from hot to cold…)
- simplistic | boring | do not like!

The SCAA cupping method was developed by the Specialty Coffee Association of America (SCAA) for the evaluation of specialty coffees. The methodology and protocols are defined by the Statistics and Standards Committee of the SCAA.

A key difference between the COE and SCAA forms is that the SCAA methodology awards points for fragrance and aroma while the COE does not. Also, the SCAA form views sweetness as a pass/fail attribute, while the COE methodology ranks sweetness on a scale as it does aftertaste, acidity, and other cup attributes.

Specialty Coffee Association of America Coffee Cupping Form

Name: _____

Date: _____

SPECIALTY
COFFEE ASSOCIATION
OF AMERICA

Quality scale:

6.00 - Good	7.00 - Very Good	8.00 - Excellent	9.00 - Outstanding
6.25	7.25	8.25	9.25
6.50	7.50	8.50	9.50
6.75	7.75	8.75	9.75

Sample

Roast Level of Sample

Fragrance/Aroma — Score: — Qualities: Dry / Break 6 7 8 9 10

Flavor — Score: 6 7 8 9 10

Aftertaste — Score: 6 7 8 9 10

Acidity — Score: 6 7 8 9 10 — Intensity High / Low

Body — Score: 6 7 8 9 10 — Level Heavy / Thin

Uniformity — Score:

Clean Cup — Score:

Balance — Score: 6 7 8 9 10

Sweetness — Score:

Overall — Score: 6 7 8 9 10

Defects (subtract) — Taint=2 / Fault=4 — # cups X Intensity =

Total Score

Final Score

Sample

Roast Level of Sample

Fragrance/Aroma — Score: — Qualities: Dry / Break 6 7 8 9 10

Flavor — Score: 6 7 8 9 10

Aftertaste — Score: 6 7 8 9 10

Acidity — Score: 6 7 8 9 10 — Intensity High / Low

Body — Score: 6 7 8 9 10 — Level Heavy / Thin

Uniformity — Score:

Clean Cup — Score:

Balance — Score: 6 7 8 9 10

Sweetness — Score:

Overall — Score: 6 7 8 9 10

Defects (subtract) — Taint=2 / Fault=4 — # cups X Intensity =

Total Score

Final Score

Sample

Roast Level of Sample

Fragrance/Aroma — Score: — Qualities: Dry / Break 6 7 8 9 10

Flavor — Score: 6 7 8 9 10

Aftertaste — Score: 6 7 8 9 10

Acidity — Score: 6 7 8 9 10 — Intensity High / Low

Body — Score: 6 7 8 9 10 — Level Heavy / Thin

Uniformity — Score:

Clean Cup — Score:

Balance — Score: 6 7 8 9 10

Sweetness — Score:

Overall — Score: 6 7 8 9 10

Defects (subtract) — Taint=2 / Fault=4 — # cups X Intensity =

Total Score

Final Score

Notes:

June 2003

Bibliography

Chapter 1

1. Ghosh, B. N., and Gacanja, W. (1970). A study of the shape and size of wet parchment coffee beans. *Journal of Agricultural Engineering Research*, 15 (2), 91-99.

   Houk, W. G. (1938). Endosperm and perisperm of coffee with notes on the morphology of the ovule and seed development. *American Journal of Botany*, 25 (1), 56-61.

   Mendes, A. J. T. (1941). Cytological observations in *Coffea*: embryo and endosperm development in *Coffea arabica* L. *American Journal of Botany*, 28 (9), 784-789.

2. Rodrigues, V. E. G. (2001). *Morfologia externa: organografia, organogenia vegetal*. Lavras: UFLA.

   Silva, E. A. A. (2002). *Coffee* (Coffea arabica *cv. Rubi*) *seed germination: mechanism and regulation*. Ph.D. Thesis, Wageningen University, Wageningen, Netherlands.

   Wilbaux. R. (1963). Agricultural engineering. Rome: FAO.

3. Arcila-Pulgarín, M. I., and Orozco-Castaño, F. J. Estudio morfologico del desarrollo del embrión en cafe. (1987). *Cenicafé*, Colombia, 38, 62-63.

   Ghosh, B. N., and Gacanja, W. (1970). A study of the shape and size of wet parchment coffee beans. *Journal of Agricultural Engineering Research*, 15 (2), 91-99.

   Salazar, G. M. R., Riaño, H. N. M., Arcila, P. J., and Ponced, C. A. (1994). Estudio morfológico, anatómico y ultraestrutural del fruto de café *Coffea arabica* L. Cenicafe, 45 (3), 93-105.

4. Salazar, G. M. R., Riaño, H. N. M., Arcila, P. J., and Ponced, C. A. (1994). Estudio morfológico, anatómico y ultraestrutural del fruto de café *Coffea arabica* L. Cenicafe, 45 (3), 93-105.

   Wilbaux. R. (1963). Agricultural engineering. Rome: FAO.

5. Marín-López, S. M., Arcila-Pulgarín, J., Montoya-Restrepo, E. C., and Oliveros-Tascón, C. E. (2003). Cambios físicos y químicos durante la maduración del fruto de café. *Cenicafé*, 54 (3), 208-225.

6. Lopes, C. R., Musche, R., and Hec, M. (1984). Identificação de pigmentos flavonóides e ácidos fenólicos nos cultivares Bourbon e Caturra de *C. arabica* L. *Revista Brasileira de Genética*, 7 (2), 657-669.

7. Elías, 1979, Cited by: Peñaloza, W., Molina, M. R., Brenes, R. G., and Bressani, R. (1985). Solid state fermentation: an alternative to improve the nutritive value of coffee pulp. *Applied and Environmental Microbiology*, 49 (2), 388-393..

8. Zuluaga, V. J. (1999). *Chemical properties of coffee*.

9. Rolz, C., Menchú, J. F., Espinosa, R., and García-Prendes, A. (1971). Coffee fermentation studies. *In 5 International Conference on Coffee Science*. 1971 (259-269).

10. Marín-López, S. M., Arcila-Pulgarín, J., Montoya-Restrepo, E. C., and Oliveros-Tascón, C. E. (2003). Cambios físicos y químicos durante la maduración del fruto de café. *Cenicafé*, 54 (3), 208-225.

11. Zuluaga, V. J. (1999). *Chemical properties of coffee*.

12. Wilbaux, R. (1961). *Le traitement du café*. (Bulletin de renseignements).

13. Carvalho, V. D. (1997). *Cafeicultura empresarial: produtividade e qualidade*. Monografia de Especialização, Universidade Federal de Lavras, Lavras, Brasil.

    Wilbaux. R. (1963). Agricultural engineering. Rome: FAO.

14. Rolz, C., Menchú, J. F., Espinosa, R., and García-Prendes, A. (1971). Coffee fermentation studies. *In 5 International Conference on Coffee Science*. 1971(259-269).

    Zuluaga, V. J. (1999). Chemical properties of coffee.

15. Avalone, S., Guiraud, J. P., Guyot, B., Olguin, E., and Brillouet, J. M. (2001). Fate of mucilage cell wall polysaccharides during coffee fermentation. *Journal of Agricultural and Food Chemistry*, 49, 5556-5559.

    Rolz, C., Menchú, J. F., Espinosa, R., and García-Prendes, A. (1971). Coffee fermentation studies. *In 5 International Conference on Coffee Science*. 1971(259-269).

16. Coleman et al., 1955; Rolz et al., 1971, Cited by: Menchú, J. F., and Rolz, C. (1973). Coffee fermentation technology. *Café Cacao Thé*, 17 (1), 53-60.

17. Salazar, G. M. R., Riaño, H. N. M., Arcila, P. J., and Ponced, C. A. (1994). Estudio morfológico, anatómico y ultraestrutural del fruto de café *Coffea arabica* L. Cenicafe, 45 (3), 93-105.

18. Salazar, G. M. R., Riaño, H. N. M., Arcila, P. J., and Ponced, C. A. (1994). Estudio morfológico, anatómico y ultraestrutural del fruto de café *Coffea arabica* L. Cenicafe, 45 (3), 93-105.

19. Wilbaux, R. (1961). *Le traitement du café*. (Bulletin de renseignements).

20. Ghosh, B. N.; Gacanja, W. (1970). A study of the shape and size of wet parchment coffee beans. *Journal of Agricultural Engineering Research*, 15 (2), 91-99.

21. Houk, W. G. (1938). Endosperm and perisperm of coffee with notes on the morphology of the ovule and seed development. *American Journal of Botany*, 25 (1), 56-61.

    Salazar, G. M. R., Riaño, H. N. M., Arcila, P. J., and Ponced, C. A. (1994). Estudio morfológico, anatómico y ultraestrutural del fruto de café *Coffea arabica* L. Cenicafe, 45 (3), 93-105.

    Wilbaux. R. (1963). Agricultural engineering. Rome: FAO.

22. Chin & Roberts (1980), Mendes (1940), cited by Silva, E. A. A. (2002). *Coffee* (Coffea arabica *cv. Rubi*) seed germination: mechanism and regulation. Ph.D. Thesis, Wageningen University, Wageningen, Netherlands.

23. Deddecca, D. M. (1957). Anatomia e desenvolvimento ontogenético de *Coffea arábica* L. var Typica Cramer. *Bragantia*, 16, 315-368.

24. Rena, A. B. and Maestri, M. (1986). Cultura do cafeeiro: fatores que afetam a produtividade. *In 1 Simpósio sobre fatores que afetam a produtividade do cafeeiro*. 1986 (447).

25. Feldman, J. R., Ryder, W. S., and Kune, J. T. (1969). Importance of nonvolatile compounds to the flavor of coffee. *Journal of Agricultural and Food Chemistry*, 17 (4), 733-739. Sivetz, M. (1963). Coffee processing technology. USA: AVI.

26. Bradbury, A. G. W. (2001). Carbohydrates. In Clarke, R. J., and Vitzthum, O. G. (Ed.), *Coffee: recent developments* (1-17). Oxford: Blackwell Science.

27. Navarini, L., Gilli, R., Gombac, V., Abatangelo, A., and Bosco, M. (1999). Polysaccharides from hot water extracts of *Coffea arabica* beans: isolation and characterization. *Carbohydrate Polymers*, 40, 71-81.

    Redgwell, R. J., Curti, D., Fisher, M., Nicolas, P., and Fay, L. B. (2002). Coffee bean arabinogalactans: acidic polymers covalently linked to protein. *Carbohydrate Research*, 337, 239-253.

28. Clifford, M. N. (1985). Chemical and physical aspects of green coffee and coffee products. In Clifford, M. N., and Willson, K. C. (Ed.), *Coffee: botany, biochemistry and production of beans and beverage* (305-374). Westport: Avi.

29. Fisher, M., Reimann, S., Trivato, V., and Redgwell, R. J. (2001). Polysaccharides of green Arabica and Robusta coffee beans. *Carbohydrate Research*, 330, 93-101.

30. Bradbury, A. G. W. and Halliday, D. J. (1987) Polysaccharides in green coffee beans. *In 12 International Conference on Coffee Science*. 2001 (265-269).

    Fisher, M., Reimann, S., Trivato, V., and Redgwell, R. J. (2001). Polysaccharides of green Arabica and Robusta coffee beans. *Carbohydrate Research*, 330, 93-101.

31. Bradbury, A. G. W. and Halliday, D. J. (1987) Polysaccharides in green coffee beans. *In 12 International Conference on Coffee Science*. 2001 (265-269).

32. Rogers, W. J., Michaux, S., Bastin, M., and Bucheli, P. (1999). Changes to the content of sugars, sugar alcohols, myo-inositol, carboxilic acids and inorganic anions in developing grains from different varieties of Robusta (*Coffea canephora*) and Arabica (*C. arabica*) coffees. *Plant Science*, 149, 115-123.

Silwar, R. and Lüllmann, C. (1988). The determination of mono-and disaccharides in green arabica and robusta coffees using high performance liquid chromatography. *Café Cacao Thé*, 32 (4), 319-322.

33. Bradbury, A. G. W. (2001). Carbohydrates in coffee. *In 19 International Conference on Coffee Science*.

34. Aguiar, A. T. da E., Fazuoli, L. C., Salva, T. J., and Favarin, J. L. (2005). Diversidade química de cafeeiros da espécie *Coffea canephora*. *Bragantia*, 64 (4), 577-582.

Clifford, M. N. (1975). The composition of green and roasted coffee beans. Process Biochemistry. Oxford, 13-19.

Mazzafera, P., Guerreiro Filho, O., and Carvalho, A. (1988). A cor verde do endosperma do café. *Bragantia*, 47 (2), 159-170.

35. Dentan, E. (1985). The microscopy structure of the coffee bean. In Clifford, M. N., and Willson, K. C. (Ed.), *Coffee: botany, biochemistry and production of beans and beverage* (284-304). New York: AVI.

36. Folstar, P. (1985). Lipids. In Clarke, R. J., and Macrae, R. (Ed.), Coffee (203-222). New York: Elsevier.

37. Maier, 1981 Cited by Speer, K., and Kölling-Speer, I. (2006). The lipid fraction of the coffee bean. *Brazilian Journal of Plant Physiology*, 18 (1), 201-216.

38. Speer, K., Kölling-Speer, I. (2006). The lipid fraction of the coffee bean. *Brazilian Journal of Plant Physiology*, 18 (1), 201-216.

39. Thaler & Gaigl, 1963, Cited by Macrae, R. (1985). Nitrogenous components. In Clarke, R. J., and Macrae, R. (Ed.), Coffee (115-152). New York: Elsevier Applied Science.

40. Redgwell, R. J., Curti, D., Fisher, M., Nicolas, P., and Fay, L. B. (2002). Coffee bean arabinogalactans: acidic polymers covalently linked to protein. *Carbohydrate Research*, 337, 239-253.

41. Ludwig, E., Lipke, U., Raczek, U., and Jäger, A. (2000). Investigations of peptides and proteases in green coffee beans. *European Food Research Technology*, 211, 111-116.

42. Thaler, 1975, Cited by Clifford, M. N. (1985). Chemical and physical aspects of green coffee and coffee products. In Clifford, M. N., and Willson, K. C. (Ed.), *Coffee: botany, biochemistry and production of beans and beverage* (305-374). Westport: Avi.

43. Mazzafera, P. and Robinson, S. P. (2000). Characterization of polyphenol oxidase in coffee. *Phytochemistry*, 55, 285-296.

44. Eskin, 1979; Thaler, 1975, Cited by Clifford, M. N. (1985). Chemical and physical aspects of green coffee and coffee products. In Clifford, M. N., and Willson, K. C. (Ed.), *Coffee: botany, biochemistry and production of beans and beverage* (305-374). Westport: Avi.

45. Clifford, M. N., Johnston, L. K., Knight, S., and Kuhnert, N. (2003). Hierarchical scheme for LC-MS identification of chlorogenic acids. *Journal of Agricultural and Food Chemistry*, 51, 2900-2911.

46. Clifford, M. N. and Kazi, T. (1987). The influence of coffee bean maturity on the content of chlorogenic acids, caffeine and trigonelline. *Food Chemistry*, 26, 59-69.

47. Ky, C. L., Louarn, J., Dussert, S., Guyot, B., Hamon, S, and Noirot, M. (2001). Caffeine, trigonelline, chlorogenic acids and sucrose diversity in wild *Coffea arabica* L. and *C. canephora* P. accessions. *Food Chemistry*, 75, 223-230.

48. Clifford, M. N., Kellard, B., and Birch, G. G. (1989). Characterization of caffeoylferuloylquinic acids by simultaneous isomerization and transesterification with tetramethylammonium hydroxide. *Food Chemistry*, 34, 81-89.

49. Clifford, M. N., and Wight, J. (1976). The measurement of feruloylquinic acids and caffeoylquinic acids in coffee beans: development of the technique and its preliminary application to green coffee beans. *Journal of Science of Food and Agriculture*, 27, 73-84.

Ky, C. L., Louarn, J., Dussert, S., Guyot, B., Hamon, S, and Noirot, M. (2001). Caffeine, trigonelline, chlorogenic acids and sucrose diversity in wild *Coffea arabica* L. and *C. canephora* P. accessions. *Food Chemistry*, 75, 223-230.

50. Clifford, M. N., Williams, T., and Bridson, D. (1989). Chlorogenic acids and caffeine as possible taxonomic criteria in *Coffea* and *Psilanthus*. *Phytochemistry*, 28 (3), 829-838.

Farah, A. and Donangelo, C. M. (2006). Phenolic compounds. Brazilian Journal of Plant Physiology, 18 (1), 23-36.

51. Dentan, E. (1985). The microscopy structure of the coffee bean. In Clifford, M. N., and Willson, K. C. (Ed.), *Coffee: botany, biochemistry and production of beans and beverage* (284-304). New York: AVI.

52. Clifford, M. N. and Kazi, T. (1987). The influence of coffee bean maturity on the content of chlorogenic acids, caffeine and trigonelline. *Food Chemistry*, 26, 59-69.

Clifford, M. N., and Ramirez-Martinez, J. R. (1991). Phenols and caffeine in wet-processed coffee beans and coffee pulp. *Food Chemistry*, 40, 35-42.

53. Aguiar, A. T. Da E., Fazuoli, L. C., Salva, T. J., and Favarin, J. L. (2005). Diversidade química de cafeeiros da espécie *Coffea canephora*. *Bragantia*, 64 (4), 577-582.

54. Clifford, M. N. (1985). Chemical and physical aspects of green coffee and coffee products. In Clifford, M. N., and Willson, K. C. (Ed.), *Coffee: botany, biochemistry and production of beans and beverage* (305-374). Westport: Avi.

55. Hughes & Smith, 1946, Cited by Macrae, R. (1985). Nitrogenous components. In Clarke, R. J., and Macrae, R. (Ed.), Coffee (115-152). New York: Elsevier Applied Science.

56. Casal, S., Oliveira, M. B., and Ferreira, M. A. (2000). HPLC/diode-array applied to the thermal degradation of trigonelline, nicotinic acid and caffeine in coffee. *Food Chemistry*, 68, 481-485.

57. Alcázar, A., Fernández-Cáceres, P. L., Martín, M. J., Pablos, F., and Gonzáles, A. G. (2003). Ion chromatography determination of some organic acids, chloride and phosphate in coffee and tea. *Talanta*, 61, 95-101.

Jham, G. N., Fernandes, S. A., Garcia, C. F., and Silva, A. A. (2002). Comparison of GC and HPLC for the quantification of organic acids in coffee. *Phytochemical Analysis*, 13, 99-104.

Rogers, W. J., Michaux, S., Bastin, M., and Bucheli, P. (1999). Changes to the content of sugars, sugar alcohols, myo-inositol, carboxilic acids and inorganic anions in developing grains from different varieties of Robusta (*Coffea canephora*) and Arabica (*C. arabica*) coffees. *Plant Science*, 149, 115-123.

Stegen, G. H. D. v.d. and Duijn, J. van. (1987). Analysis of normal organic acids in coffee. *In 12 International Conference on Coffee Science*. 1987 (238-246).

Weers, M., Balzer, H., Bradbury, A., and Vitzthum, O. G. (1995). Analysis of acids in coffee by capillary electrophoresis. *In 16 International Conference on Coffee Science*, 1995 (218-223).

58. Balzer, H. H. (2001). Acids in coffee. In Clarke, R. J., and Vitzthum, O. G. (Ed.), *Coffee: recent developments* (18-32). Oxford: Blackwell Science.

Weers, M., Balzer, H., Bradbury, A., and Vitzthum, O. G. (1995). Analysis of acids in coffee by capillary electrophoresis. *In 16 International Conference on Coffee Science*, 1995 (218-223).

59. Maier, H. G. (1987). The acids of coffee. In 12 International ASIC Conference. 1987 (229-237).

60. Clarke, R. J. (1985). Water and mineral contents. In Clarke, R. J., and Macrae, R. (Ed.), *Coffee: Water and Mineral Contents* (42-82). New York: Elsevier Applied Science.

61. Martín, M. J., Pablos, F., and Gozáles, A. G. (1998). Characterization of green coffee varieties according to their metal content. *Analytica Chimica Acta*, 358, 177-183.

62. Morgano, M. A., Pauluci, L. F., Mantovani, D. M., and Mori, E. E. M. (2002). Determinação de minerais em café cru. *Ciência e Tecnologia de Alimentos*, 22 (1), 19-23.

Chapter 2

1. Wilbaux. R. (1963). Agricultural engineering. Rome: FAO
2. Fennema, O. R. (1993). *Química de los alimentos*. Acribia: Zaragoza, Espanã.
   Multon, J. L. (2000). *Basics of moisture measurement in grain*. In Institut National de la Recherche Agronomique. Uniformity.
   Taiz, L., and Zeiger, E. (1991). *Plant physiology*. Redwood City: The Benjamin.
3. Fennema, O. R. (1993). *Química de los alimentos*. Acribia: Zaragoza, Espanã.
   Taiz, L., and Zeiger, E. (1991). *Plant physiology*. Redwood City: The Benjamin.
4. Campbell, M. K. (2000). *Bioquímica*. Porto Alegre: Artmed.
   Fennema, O. R. (1993). *Química de los alimentos*. Acribia: Zaragoza, Espanã.
   Taiz, L., and Zeiger, E. (1991). *Plant physiology*. Redwood City: The Benjamin.
5. Campbell, M. K. (2000). *Bioquímica*. Porto Alegre: Artmed.
6. Némethy, G. and Scheraga, H. A. (1962). The structure of water and hydrophobic bonding in proteins: III. the thermodynamic properties of hydrophobic bonds in proteins. *Journal Physical Chemistry*, 66, 1773-1789.
   Taiz, L., and Zeiger, E. (1991). *Plant physiology*. Redwood City: The Benjamin.
7. Campbell, M. K. (2000). *Bioquímica*. Porto Alegre: Artmed.
8. Fennema, O. R. (1993). *Química de los alimentos*. Acribia: Zaragoza, Espanã.
   Campbell, M. K. (2000). *Bioquímica*. Porto Alegre: Artmed.
9. Némethy, G., and Scheraga, H. A. (1962). The structure of water and hydrophobic bonding in proteins: III. the thermodynamic properties of hydrophobic bonds in proteins. *Journal Physical Chemistry*, 66, 1773-1789.
   Taiz, L., and Zeiger, E. (1991). *Plant physiology*. Redwood City: The Benjamin.
10. Incropera, F. P. and Dewitt, D. P. (1992) *Fundamentos de transferência de calor e de massa*. Rio de Janeiro: Guanabara Koogan.
11. Multon, J. L. (2000). *Basics of moisture measurement in grain*. In Institut National de la Recherche Agronomique. Uniformity.
12. Multon, J. L. (2000). *Basics of moisture measurement in grain*. In Institut National de la Recherche Agronomique. Uniformity.
13. Némethy, G., and Scheraga, H. A. (1962). The structure of water and hydrophobic bonding in proteins: III. the thermodynamic properties of hydrophobic bonds in proteins. *Journal Physical Chemistry*, 66, 1773-1789.
14. Reichardt, K. (1985). A água: absorção e translocação. In Ferri, M. G. (Coord.). *Fisiologia vegetal* (3-74). São Paulo: EPU.
15. Fennema, O. R. (1993). Química de los alimentos. Acribia: Zaragoza, Espanã.
16. Multon, J. L. (2000). *Basics of moisture measurement in grain*. In Institut National de la Recherche Agronomique. Uniformity.
17. Fennema, O. R. (1993). *Química de los alimentos*. Acribia: Zaragoza, Espanã.
    Multon, J. L. (2000). *Basics of moisture measurement in grain*. In Institut National de la Recherche Agronomique. Uniformity.
18. Fennema, O. R. (1993). *Química de los alimentos*. Acribia: Zaragoza, Espanã.
    Multon, J. L. (2000). *Basics of moisture measurement in grain*. In Institut National de la Recherche Agronomique. Uniformity.
19. Fennema, O. R. (1993). *Química de los alimentos*. Acribia: Zaragoza, Espanã.
    Multon, J. L. (2000). *Basics of moisture measurement in grain*. In Institut National de la Recherche Agronomique. Uniformity.
20. Multon, J. L. (2000). *Basics of moisture measurement in grain*. In Institut National de la Recherche Agronomique. Uniformity.
21. Multon, J. L. (2000). *Basics of moisture measurement in grain*. In Institut National de la Recherche Agronomique. Uniformity.
22. Labuza, T. P. (1970). Properties of water as related to the keeping quality of food. *In 3 International Congress of Food Science and Technology*. 1970 (618-635).
23. Reh, C. Y., Gerber, A., Prodolliet, J., and Vuataz, G. (2006). Water content determination in green coffee: method comparison to study specificity and accuracy. *Food Chemistry*, 96 (3), 423-430.
24. Brasil (1992). Ministério da Agricultura e Reforma Agrária. Secretária Nacional de Defesa Agropecuária. *Regras para análise de sementes*. Brasília, DF, Brasil.
25. Funk, D. B. (2000). *Uniformity in dielectric grain moisture measurement*. In Institut National de la Recherche Agronomique. Uniformity.
26. Funk, D. B. (2000). *Uniformity in dielectric grain moisture measurement*. In Institut National de la Recherche Agronomique. Uniformity.
27. Cabrera, H. A. P., and Taniwaki, M. H. (2003). Determinação do teor de umidade do café cru beneficiado: comparação entre diferentes metodologias. *Revista Brasileira de Armazenamento*, (6), 25-29.
I. ISO International Standard (1978). Green coffee: *determination of moisture (routine method)*. ISO 1447:1978.
II. ISO International Standard (1983). Green coffee: *determination of loss in mass at 105 ºC*. ISO 6673:1983.
III. ISO International Standard (2001). Green coffee: *determination of moisture (basic reference method)*. ISO 1446:2001.

Chapter 3

1. Abalone, R., Cassinera, A., Gastón, A., and Lara, M. A. (2004). Some physical properties of amaranth seeds. Biosystems Engineering, 89 (1), 109-117.
   Afonso Júnior, P. C. (2001) Aspectos físicos, fisiológicos e de qualidade do café em função da secagem e do armazenamento. Tese de Doutorado, Universidade Federal de Viçosa, Viçosa, Brasil.
   Pérez-Alegría, L. R., Ciro, H. J., and Abud, L. C. (2001). Physical and thermal properties of parchment coffee bean. Transactions of the ASAE, 44 (6), 1721-1726.
   Borém, F.M. (2004) *Pós-colheita do café*. Lavras: UFLA/FAEPE.
   Dursun, E., and Dursun, I. (2005). Some physical properties of caper seed. Biosystems Engineering, 92 (2), 237-245.

   Ghosh, B. N. (1968). Effect of moisture content on the static coefficient of friction of parchment coffee beans. Journal of Agricultural Engineering Research, 13 (3), 249-253.
   Ghosh, B. N., and Gacanja, W. (1970). A study of the shape and size of wet parchment coffee beans. Journal of Agricultural Engineering Research, 15 (2), 91-99.
   Kaleemullah, S., and Gunasekar, J. J. (2002). Moisture-dependent physical properties of arecanut kernels. Biosystems Engineering, 82 (3), 331-338.
   Mohsenin, N. N. (1986). Physical properties of plant and animal materials. New York: Gordon and Breach.
   Ozarslan, C. (2002). Physical properties of cotton seed. Biosystems Engineering, 83 (2), 169-174.

Kumar, V. A., and Mathew, S. (2003). A method for estimating the surface area of ellipsoidal food materials. *Biosystems Engineering*, 85 (1), 1-5.

2. Afonso Júnior, P. C. (2001) *Aspectos físicos, fisiológicos e de qualidade do café em função da secagem e do armazenamento*. Tese de Doutorado, Universidade Federal de Viçosa, Viçosa, Brasil.

Brooker, D. B., Bakker-Arkema, F. W., and Hall, C. W. (1992). *Drying and storage of grains and oilseeds*. Westport: AVI.

Mohsenin, N. N. (1986). *Physical properties of plant and animal materials*. New York: Gordon and Breach.

Pabis, S., Jayas, D. S., and Cenkowski, S. (1998). *Grain drying: theory and practice*. New York: J. Wiley & Sons.

3. Pabis, S., Jayas, D. S., and Cenkowski, S. (1998). *Grain drying: theory and practice*. New York: J. Wiley & Sons.

4. Afonso Júnior, P. C. (2001) *Aspectos físicos, fisiológicos e de qualidade do café em função da secagem e do armazenamento*. Tese de Doutorado, Universidade Federal de Viçosa, Viçosa, Brasil.

Couto, S. M., Magalhães, A. C., Queiroz, D. M., and Bastos, I. T. (1999). Massa específica aparente e real e porosidade de grãos de café em função do teor de umidade. *Revista Brasileira de Engenharia Agrícola e Ambiental*, 3 (1), 61-68.

Ribeiro, R. C. M. S. (2001). *Determinação das propriedades termofísicas de café cereja descascado*. Dissertação de Mestrado, Universidade Federal de Lavras, Lavras, Brasil.

5. Afonso Júnior, P. C. (2001) *Aspectos físicos, fisiológicos e de qualidade do café em função da secagem e do armazenamento*. Tese de Doutorado, Universidade Federal de Viçosa, Viçosa, Brasil.

6. Brooker, D. B., Bakker-Arkema, F. W., and Hall, C. W. (1992). *Drying and storage of grains and oilseeds*. Westport: AVI.

7. Mohsenin, N. N. (1986). *Physical properties of plant and animal materials*. New York: Gordon and Breach.

8. Couto, S. M., Magalhães, A. C., Queiroz, D. M., and Bastos, I. T. (1999). Massa específica aparente e real e porosidade de grãos de café em função do teor de umidade. *Revista Brasileira de Engenharia Agrícola e Ambiental*, 3 (1), 61-68.

Ruffato, S., Corrêa, P. C., Martins, J. H., Mantovani, B. H. M., and Silva, J. N. (1999). Influência do processo de secagem sobre a massa específica aparente, massa específica unitária e porosidade de milho-pipoca. *Revista Brasileira de Engenharia Agrícola e Ambiental*, 3 (1), 45-48.

9. Afonso Júnior, P. C. (2001) *Aspectos físicos, fisiológicos e de qualidade do café em função da secagem e do armazenamento*. Tese de Doutorado, Universidade Federal de Viçosa, Viçosa, Brasil.

10. Afonso Júnior, P. C. (2001) *Aspectos físicos, fisiológicos e de qualidade do café em função da secagem e do armazenamento*. Tese de Doutorado, Universidade Federal de Viçosa, Viçosa, Brasil.

11. Oliveros-Tascón, C. E., and Mejía, G. R. (1985). Coeficiente de friccion, angulo de reposo y densidades aparentes de granos de cafe *Coffea arabica* variedad caturra. *Cenicafé*, 36 (1), 22-38.

12. Agrawal, K. K., Clary, B. L., and Schroeder, E. W. (1972). *Matematical models of peanut pod geometry*. ASAE, Saint Joseph.

Mohsenin, N. N. (1986). *Physical properties of plant and animal materials*. New York: Gordon and Breach.

13. Corrêa, P. C., Afonso Júnior, P. C., Queiroz, D. M., Sampaio, C. P., and Cardoso, J. B. (2002). Variação das dimensões características e da forma dos frutos de café durante o processo de secagem. *Revista Brasileira de Engenharia Agrícola e Ambiental*, 6 (3), 466-470.

14. Fortes, M., and Okos, M. R. (1980). Changes in physical properties of corn during drying. *Transaction of the ASAE*, 23 (4), 1004-1008.

Mata, M. E. R. M. C., Aragão, R. F., Santana, E. F., and Silva, F. A. S. (1986). Estudo da morfologia geométrica em grãos. *Revista Nordestina de Armazenagem*, 3 (1), 3-30.

15. Mohsenin, N. N. (1986). *Physical properties of plant and animal materials*. New York: Gordon and Breach.

16. Oliveros-Tascón, C. E., and Mejía, G. R. (1985). Coeficiente de friccion, angulo de reposo y densidades aparentes de granos de cafe *Coffea arabica* variedad caturra. *Cenicafé*, 36 (1), 22-38.

Magalhães, A.C., Couto, S.M., Queiroz, D.M., Andrade, E.T. (2000) Dimensões Principais, Massa e Volume Unitários, Esfericidade e Ângulo de Repouso de Frutos de Café. *Revista Brasileira de Produtos Agroindustriais*, 2 (2), 39-56.

17. Afonso Júnior, P. C., and Corrêa, P. C. (2000). Propriedades térmicas dos grãos de café. *In 1Simpósio de Pesquisa dos Cafés do Brasil*, 2000 (1142-1146).

18. Sokhansanj, S., and Yang, W. (1996). Revision of the ASAE standard D245.4: moisture relationships of grains. *Transactions of the ASAE*, 39 (2), 639-642.

19. Hall, C. W. (1980). *Drying and storage of agricultural crops*. Westport: AVI.

20. Brunauer, S., Emmet, P. H., and Teller, E. (1938). Adsorption in multi-molecular levels. *Journal of American Chemical Society*, 60 (2), 309-319.

21. Anderson, R. B. (1946). Modifications of the Brunauer, Emmet and Teller equation. *Journal of American Chemical Society*, 68 (4), 686-691.

22. Harkins, W. D., and Jura, G. (1944). Vapor adsorption method for determination of the aerea of a solid without assumption of a molecular area. *Journal of American Chemical Society*, 66 (8), 1366-1371.

23. Smith, S. E. (1947). The sorption of water vapor by high polymers. *Journal of American Chemical Society*, 69 (4), 646-651.

24. Halsey, G. (1948). Physical adsorption on non-uniform surfaces. *The Journal of Chemical Physics*, 16 (8), 931-937.

25. Aguerre, R. J., Suárez, C., and Viollaz, P. E. (1989). Modeling temperature dependence of food sorption isotherms. *Lebensmittel-Wissenschaft und-Technologie*, 22 (1), 1-5.

26. Roa, Citado Por Rossi, S. J., and Roa, G. (1976). Aplicação de métodos de análise numérica e regressão não linear para estimação da condutividade térmica e difusividade térmica para cereais. *In 6 Congresso Brasileiro de Engenharia Agrícola*, 1976.

27. Corrêa, P. C., Martins, J. H., and Melo, E. C. (1995). *Umigrãos: programa para o cálculo do teor de umidade de equilíbrio para os principais produtos agrícolas*. Viçosa: Centreinar-UFV, Brasil.

28. Aguerre, R. J., Suárez, C., and Viollaz, P. E. (1989). Modeling temperature dependence of food sorption isotherms. *Lebensmittel-Wissenschaft und-Technologie*, 22 (1), 1-5.

29. Henderson, S. M. (1952). A basic concept of equilibrium moisture. *Agriculture Engineering*, 33 (1), 29-32**.**

30. Bach, D. B. (1979). *Curvas de equilíbrio higroscópico de feijão preto*. Dissertação de Mestrado, Universidade Federal de Viçosa, Viçosa, Brasil.

Hall, C. W., and Rodriguez-Arias, J. H. (1958). Equilibrium moisture content of shelled corn. *Agricultural Engineering*, 39 (8), 466-470.

31. Thompson, H. J., and Shedd, C. K. (1954). Equilibrium moisture content and heat of vaporization of shelled corn and wheat. *Agricultural Engineering*, 35 (11), 786-788.

32. Christ, D. (1996). *Curvas de equilíbrio higroscópico e de secagem da canola (Brassica napus L. var. oleifera), e efeito da temperatura e da umidade relativa do ar de secagem sobre a qualidade das sementes*. Dissertação de Mestrado, Universidade Federal de Viçosa, Viçosa, Brasil.

Corrêa, P. C., and Moure, J. (2000). Higroscopicidad y propiedades térmicas de semillas de sorgo. *Alimentacion Equipos y Tecnologia*, Madri, 29 (1), 149-153.

Pena, R. S., Ribeiro, C. C., and Grandi, J. G. (1997). Influência da temperatura nos parâmetros de modelos bi-paramétricos que predizem isotermas de adsorção de umidade do guaraná (*Paullinia cupana*) em pó. *Ciência e Tecnologia de Alimentos*, 17 (3), 229-232.

33. Chung, D. S., and Pfost, H. B. (1967). Adsorption and desorption of water vapor by cereal grains and their products. *Transactions of the ASAE*, 10 (4), 149-157.
34. Pfost, H. B., Maurer, S. G., Chung, D. S., and Milliken, G. A. (1976). *Summarizing and reporting equilibrium moisture data for grains.* Saint Joseph: ASAE.
35. Brooker, D. B., Bakker-Arkema, F. W., and Hall, C. W. (1992). *Drying and storage of grains and oilseeds.* Westport: AVI.
36. Oswin, C. R. (1946). The kinetics of package life: III. isotherms. *Journal of the Society Chemical Industry*, 65 (4), 419-421.
37. Chen, C., and Morey, R. V. (1989). Comparison of four EMC/ERH equations. *Transactions of the ASAE*, 32 (3), 983-990.
38. Chen, C. (2000). Factors which effect equilibrium relative humidity of agricultural products. *Transactions of the ASAE*, 43 (3), 673-683.
   Pena, R. S., Ribeiro, C. C., and Grandi, J. G. (1997). Influência da temperatura nos parâmetros de modelos bi-paramétricos que predizem isotermas de adsorção de umidade do guaraná (*Paullinia cupana*) em pó. *Ciência e Tecnologia de Alimentos*, 17 (3), 229-232.
   Sun, D. W., and Woods, J. L. (1994). The selection of sorption isotherm equations for wheat based on the fitting of available data. *Journal of Stored Products Research*, 30 (1), 27-43.
39. Harkins, W. D., and Jura, G. (1944). Vapor adsorption method for determination of the aerea of a solid without assumption of a molecular area. *Journal of American Chemical Society*, 66 (8), 1366-1371.
40. Iglesias, H. A., and Chirife, J. (1976). Prediction of the effect of temperature on water sorption isotherms of food material. *Journal of Food Technology*, 11 (2), 109-116.
41. Afonso Júnior, P. C. (2001) *Aspectos físicos, fisiológicos e de qualidade do café em função da secagem e do armazenamento.* Tese de Doutorado, Universidade Federal de Viçosa, Viçosa, Brasil.
42. Mohsenin, N. N. (1986). *Physical properties of plant and animal materials.* New York: Gordon and Breach.
43. Pereira, J. A. M., and Queiroz, D. M. (1987). *Higroscopia.* Viçosa: Centreinar-UFV, Brasil.
44. Carvalho, V. D., Chagas, S. J. R., Chalfoun, S. M., Botrel, N., and Juste Júnior, E. S. G. (1994). Relação entre a composição físico-química e química do grão beneficiado e a qualidade de bebida do café: I. atividades de polifenoloxidase e peroxidase, índice de coloração de acidez. *Pesquisa Agropecuária Brasileira*, 29 (3), 449-454.
45. Hall, C. W. (1980). *Drying and storage of agricultural crops.* Westport: AVI.
   Wylen, G. J. van, and Sonntag, R. E. (1976). *Fundamentos da termodinâmica clássica.* São Paulo: E. Blücher.
46. Tagawa, A., Murata, S., and Hayashi, H. (1993). Latent heat of vaporization in buckwheat using the data of equilibrium moisture content. *Transactions of the ASAE*, 36 (1), 113-118.
47. Othmer, D. F. (1940). Correlating vapour pressure and latent heat data. *Journal of Industrial Engineering Chemistry*, 32 (6), 841-856.
48. Brooker, D. B., Bakker-Arkema, F. W., and Hall, C. W. (1992). *Drying and storage of grains and oilseeds.* Westport: AVI.

49. Afonso Júnior, P. C. (2001) *Aspectos físicos, fisiológicos e de qualidade do café em função da secagem e do armazenamento.* Tese de Doutorado, Universidade Federal de Viçosa, Viçosa, Brasil.
50. Incropera, F. P., and Dewitt, D. P. (1992). *Fundamentos de transferência de calor e de massa.* Rio de Janeiro: Guanabara Koogan.
51. Moura, S. C. S. R., Germer, S. P. M., Jardim, D. C. P., and Sadahira, M. S. (1998). Thermophysical properties of tropical fruit juices. *Brazilian Journal of Food Technology*, 1 (1/2), 70-76.
   Sharma, D. K., and Thompson, T. L. (1973). Specific heat and thermal conductivity of sorghum. *Transactions of the ASAE*, 16 (1), 114-117.
52. Borém, F. M., Ribeiro, R. C. M. S., Correa, P. C., and Pereira, R. G. F. A. (2002). Propriedades térmicas de cinco variedades de café cereja descascado. *Revista Brasileira de Engenharia Agrícola e Ambiental*, 6 (3), 475-480.
   Corrêa, P. C., Sampaio, C. P., Regazzia, J., and Afonso Júnior, P. C. (2000). Calor específico dos frutos do café de diferentes cultivares em função do teor de umidade. *Revista Brasileira de Armazenamento*, (1), 18-22.
53. Chakrabarti, S. M., and Johnson, W. H. (1972). Specific heat of flue cured tobacco by differential scanning colorimetry. *Transactions of the ASAE*, 15 (5), 928-931.
   Wratten, F. T., Poole, W. D., Chesness, J. L., Ball, S., and Ramarao, V. (1969). Physical and thermal properties of corn cobs. *Transactions of the ASAE*, 12 (5), 801-803.
54. Drouzas, A. E., and Saravacos, G. D. (1988). Effective thermal conduntivity of granular starch materials. *Journal of Food Science*, 53 (6), 1795-1799.
55. Mohsenin, N. N. (1986). *Physical properties of plant and animal materials.* New York: Gordon and Breach.
56. Afonso Júnior, P. C., and Corrêa, P. C. (2000). Propriedades térmicas dos grãos de café. *In 1Simpósio de Pesquisa dos Cafés do Brasil*, 2000 (1142-1146).
57. Incropera, F. P., and Dewitt, D. P. (1992). *Fundamentos de transferência de calor e de massa.* Rio de Janeiro: Guanabara Koogan.
58. Incropera, F. P., and Dewitt, D. P. (1992). *Fundamentos de transferência de calor e de massa.* Rio de Janeiro: Guanabara Koogan.
59. Pabis, S., Jayas, D. S., and Cenkowski, S. (1998). *Grain drying: theory and practice.* New York: J. Wiley & Sons.
60. Kazarian, E. A., and Hall, C. W. (1965). Thermal properties of grains. *Transactions of the ASAE*, 8 (1), 33-37.
61. Park, K. J., Murr, F. E. X., and Salvadego, M. (1997). Medição da condutividade térmica de milho triturado pelo método da sonda. *Ciência e Tecnologia de Alimentos*, 17 (3), 242-247.
   Stolf, S. R. (1972). Medição da condutividade térmica dos alimentos. *Boletim do Instituto de Tecnologia de alimentos*, 29 (1), 67-79.
62. Park, K. J., Alonso, L. F. T., and Nunes, A. S. (1999). Determinação experimental da condutividade e difusividade térmica de grãos em regime permanente. *Ciência e Tecnologia de Alimentos*, 19 (2), 264-269.
63. Afonso Júnior, P. C., and Corrêa, P. C. (2000). Propriedades térmicas dos grãos de café. *In 1Simpósio de Pesquisa dos Cafés do Brasil*, 2000 (1142-1146).

Chapter 4

1. Brando, C. H. J. (2004). Harvesting and green coffee processing. In Coffee: *growing, processing, sustainable production* (605-714). Wiley.
2. Brando, C. H. J. (2004). Harvesting and green coffee processing. In Coffee: *growing, processing, sustainable production* (605-714). Wiley.
3. Brando, C. H. J. (2004). Harvesting and green coffee processing. In Coffee: *growing, processing, sustainable production* (605-714). Wiley.

   Puerta-Quintero, G. I. P. (1996). Evaluación de la calidad del café colombiano procesado por vía seca. *Cenicafé*, 47 (2), 85-90.
   Vincent, J. C. (1987). Green coffee processing. In Clarke, R. J., and Macrae, R. (Ed.). *Technology*. London: Elsevier.
   Wilbaux, R. (1963). *Agricultural engineering*. Rome: FAO.
4. Wilbaux, R. (1963). *Agricultural engineering*. Rome: FAO.
5. Vincent, J. C. (1987). Green coffee processing. In Clarke, R. J., and Macrae, R. (Ed.). *Technology*. London: Elsevier.

6. Puerta-Quintero, G. I. P. (1996). Evaluación de la calidade del café colombiano procesado por via seca. *Cenicafé*, 47 (2), 85-90.
   Wilbaux, R. (1963). *Agricultural engineering*. Rome: FAO.
7. Illy, A., Viani, R. (1995). *Espresso coffee: the chemistry of quality*. London: Academic.
   Puerta-Quintero, G. I. P. (1996). Evaluación de la calidade del café colombiano procesado por via seca. *Cenicafé*, 47 (2), 85-90.
   Toselo, A. (1957). *1º curso de cafeicultura do Instituto Agronômico do Estado de São Paulo*. Campinas: Instituto Agronômico, Brasil.
   Villela, T. C. (2002). *Qualidade de café despolpado, desmucilado, descascado e natural, durante o processo de secagem*. Dissertação de Mestrado, Universidade Federal de Lavras, Lavras, Brasil.
   Vincent, J. C. (1987). Green coffee processing. In Clarke, R. J., and Macrae, R. (Ed.). *Technology*. London: Elsevier.
   Wilbaux, R. (1963). *Agricultural engineering*. Rome: FAO.
8. Tosello, A. (1957). *1º curso de cafeicultura do Instituto Agronômico do Estado de São Paulo*. Campinas: Instituto Agronômico, Brasil.
9. Puerta-Quintero, G. I. P. (1996). Evaluación de la calidade del café colombiano procesado por via seca. *Cenicafé*, 47 (2), 85-90.
10. Silva, R. F. da. (2003). *Qualidade do café cereja descascado produzido na região Sul de Minas Gerais*. Dissertação de Mestrado, Universidade Federal de Lavras, Lavras, Brasil.
11. Wilbaux, R. (1963). *Agricultural engineering*. Rome: FAO.
12. Vincent, J. C. (1987). Green coffee processing. In Clarke, R. J., and Macrae, R. (Ed.). *Technology*. London: Elsevier.
13. Vincent, J. C. (1987). Green coffee processing. In Clarke, R. J., and Macrae, R. (Ed.). *Technology*. London: Elsevier.
14. Vincent, J. C. (1987). Green coffee processing. In Clarke, R. J., and Macrae, R. (Ed.). *Technology*. London: Elsevier.
15. Puerta-Quintero, G. I. P. (2000). Influencia de los granos de café cosechados verdes, en la calidad física y organoléptica de la bebida. *Cenicafé*, 51 (2), 136-159.
16. Borem, F. M. , Reinato, C. H. R. , Silva, P., and Faria, L. F. (2006). Processamento e secagem dos frutos verdes do cafeeiro. *Revista Brasileira de Armazenamento*, (1), 19-24.
17. Borem, F. M., Reinato, C. H. R., Candiano, C. A., Faria, L. F., and Silva, P. J. (2004). Processamento do café verde descascado I: aspectos técnicos e econômicos. *In 30 Congresso Brasileiro de Pesquisas Cafeeiras*, 2004 (254-255).
18. Borem, F. M., Reinato, C. H. R., Candiano, C. A., Faria, L. F., and Silva, P. J. (2004). Processamento do café verde descascado I: aspectos técnicos e econômicos. *In 30 Congresso Brasileiro de Pesquisas Cafeeiras*, 2004 (254-255).
19. Puerta-Quintero, G. I. P. (1996). Evaluación de la calidade del café colombiano procesado por via seca. *Cenicafé*, 47 (2), 85-90.

Vilela, T. C. (2002). *Qualidade de café despolpado, desmucilado, descascado e natural, durante o processo de secagem*. Dissertação de Mestrado, Universidade Federal de Lavras, Lavras, Brasil.
   Vincent, J. C. (1987). Green coffee processing. In Clarke, R. J., and Macrae, R. (Ed.). *Technology*. London: Elsevier.
   Wilbaux, R. (1963). *Agricultural engineering*. Rome: FAO.
20. Clifford, M. N., and Ramirez-Martinez, J. R. (1991). Phenols and caffeine in wet-processed coffee beans and coffee pulp. *Food Chemistry*, 40, 35-42.
   Clifford, M. N., and Ramirez-Martinez, J. R. (1991). Tannins in wet-processed coffee beans and coffee pulp. *Food Chemistry*, 40, 191-200.
   Guimarães, R. M., Vieira, M. G. G. C., Fraga, A. C., Pinho, E. V. R. V., and Ferraz, V. P. (2002). Tolerância à dessecação em sementes de cafeeiro (*Coffea arabica* L.). *Ciência e Agrotecnologia*, 26 (1), 128-139.
   Leloup, V., Gancel, C., Liardon, R., Rytz, A., and Pithon, A. (2004) Impact of wet and dry process on green coffee composition and sensory characteristics. *In 20 International Conference in Coffee Science*. 2004 (93-101).
   Mazzafera, P., and Purcino, R. P. (2004). Post harvest processing methods and physiological alterations in the coffee fruit. *In 20 International Conference in Coffee Science*, 2004.
21. Leloup, V., Gancel, C., Liardon, R., Rytz, A., and Pithon, A. (2004) Impact of wet and dry process on green coffee composition and sensory characteristics. *In 20 International Conference in Coffee Science*. 2004 (93-101).
22. Selmar, D., Bytof, G., Knopp, S. E., Bradbury, A., Wilkens, J., and Becker, R. (2004) Biochemical insights into coffee processing: quality and nature of green coffee are interconnected with an active seed metabolism. *In 20 International Conference in Coffee Science*, 2004.
23. Bytof, G., Knopp, S. E., Schieberle, P., Teutsch, I., and Selmar, D. (2005). Influence of processing on the generation of g-aminobutyric acid in green coffee beans. *European Food Research Technology*, 220, 245-250.
24. Selmar, D., Bytof, G., Knopp, S. E., Bradbury, A., Wilkens, J., and Becker, R. (2004) Biochemical insights into coffee processing: quality and nature of green coffee are interconnected with an active seed metabolism. *In 20 International Conference in Coffee Science*, 2004.
25. Bytof, G., Knopp, S. E., Schieberle, P., Teutsch, I., and Selmar, D. (2005). Influence of processing on the generation of g-aminobutyric acid in green coffee beans. *European Food Research Technology*, 220, 245-250.
26. Leloup, V., Gancel, C., Liardon, R., Rytz, A., and Pithon, A. (2004) Impact of wet and dry process on green coffee composition and sensory characteristics. *In 20 International Conference in Coffee Science*. 2004 (93-101).
27. Borem, F. M., Vos, C. H. R., and Bino, R. J. (2006). Non-targeted metabolomics as a novel approach in coffee research. *In 21International Conference in Coffee Science*. 2006.

Chapter 5

1. Matos, A. T., Febrer, M. C. A., Silva, F.V (2000). Características químicas de composto orgânico produzido com casca de frutos de cafeeiro e águas residuárias da suinocultura. *In 1 Simpósio de Pesquisa dos Cafés do Brasil*, 2000. (975-978).
2. Matos, A. T., and Lo Monaco, P. A. (2003). Tratamento e aproveitamento agrícola de resíduos sólidos e líquidos da lavagem e despolpa dos frutos do cafeeiro. *Engenharia na Agricultura. Boletim técnico*, 7, 68.
3. Santos, J. H., and Matos, A. T. (2000). Contaminação do solo em áreas de depósito de cascas de frutos de cafeeiro. *In 2 Simpósio de Pesquisa dos Cafés do Brasil*, 2000, (981-984).
4. Matos, A. T., Vidigal, S. M., Sediyama, M. A. N., Garcia, N. P., and Ribeiro, M. F. (1998). Compostagem de alguns resíduos orgânicos utilizando-se águas residuárias da suinocultura

como fonte de nitrogênio. *Engenharia Agrícola e Ambiental*, Campina Grande, 2(2), 119-246
5. Vasco, J. Z. (1999). Procesamiento de frutos de café por vía humeda y generación de subproductos. *In 3 International Seminar on Biotechnology in the Coffeee Agroindustry*, 1999. (345-355).
6. Brandão, V. S. (1999). *Tratamento de águas residuárias de suinocultura utilizando-se filtros orgânicos*. Dissertação Mestrado, Universidade Federal de Viçosa, Viçosa, Brasil.
7. Fazenaro, F. L., Lo Monaco, P. A., Matos, A. T., Pereira, R. R. A., and Silva, N. C. L. (2004). Caracterização física da escuma de água residuária da despolpa do fruto do cafeeiro. *In 1 Simpósio de Iniciação Científica*, 2004.

8. Lima, C. R., Lo Monaco, P. A., Matos, A. T., Moreira, R. M. G., and Fazenaro, F. L. (2004). Caracterização química da escuma de água residuária da despolpa do fruto do cafeeiro. *In 9 Simpósio de Iniciação Científica,* 2004

9. Delgado Delgado, E. A.; Barois, I. (2000) Lombricompostaje de la pulpa de café em México. In: *Seminário Internacional sobre Biotecnologia na Agroindústria Cafeeira,* (335-343).

10. Zambrano-Franco, D. A.; Isaza-Hinestroza, J. D. I. (1998) Demanda química de oxigênio y nitrógeno total de los subproductos del processo tradicional de beneficio húmedo del café. *Cenicafé,* Chinchina, v. 49, n. 4, p. 279-289.

11. Matos, A. T. (2003). Tratamento e destinação final dos resíduos gerados no beneficiamento do fruto do cafeeiro. In ZAMBOLIM, L. (Ed.). *Produção Integrada de café.* (647-705).

12. Rigueira, R. (2005). *Avaliação da qualidade do café processado por via úmida, durante as operações de secagem e armazenagem.* Tese Doutorado, Universidade Federal de Viçosa, Viçosa, Brasil.

13. Kiehl, J. E. (1985) Fertilizantes orgânicos. *Agronômica Ceres,* 492.

14. Matos, A. T., and Lo Monaco, P. A. (2003). Tratamento e aproveitamento agrícola de resíduos sólidos e líquidos da lavagem e despolpa dos frutos do cafeeiro. *Engenharia na Agricultura. Boletim técnico, AEAGRI,* 7, 68.

15. Matos, A. T., and Febrer, M. C. A. (2000). Características químicas de composto orgânico produzido com casca de frutos de cafeeiro e águas residuárias da suinocultura. *In 1 Simpósio de Pesquisa dos Cafés do Brasil,* 2000. (975-978).

16. Pereira, R. R. D. A., Matos, A. T., Magalhães, M. A., Lo Monaco, P. A., and Fazenaro, F. L. (2004). Compostagem de casca e pergaminho de grãos de café utilizados na filtragem da agua residuária da despolpa dos frutos do cafeeiro. *In 9 Simpósio de Iniciação Científica,* 2004.

17. Gonçalves, M. S. J. (1997). Composto de resíduos sólido urbano: qualidade e utilização. *In 1 Seminário de Produção de Corretivos Orgânicos a Partir de Resíduos Sólidos Urbanos: Sua Importância, para a Agricultura Nacional e Meio Ambiente,* 1997.

18. Brasil, Ministério da Agricultura e do Abastecimento. (1993) *Legislação fertilizantes, corretivos e inoculantes. Brasília,* DF, 143p

19. Pavanelli, G. (2001). *Eficiência de diferentes tipos de coagulantes na coagulação, floculação e sedimentação de água com cor ou turbidez elevada.* Dissertação Mestrado, Escola de Engenharia de São Carlos, Universidade de São Paulo, São Carlos, Brasil.

20. Matos, A. T.; Eustáquio Júnior, V.; Pereira, P. A.; Matos, M. P. (2007) Tratamento da água para reuso no descascamento/despolpa dos frutos do cafeeiro. *Engenharia na Agricultura,* 15(2), 173-178.

21. Chagas, R.C.; Saraiva, C.B.; Moreira, D.A.; Silva, D.J.P.; Matos, A.T.; Farage, J.A. (2009) Uso do extrato de moringa como agente coagulante no tratamento de águas residuárias de laticínios. In: 26 *Congresso Nacional de Laticínios,* EPAMIG, 2009.

22. Silva, F.J.A.; Silveira Neto, J.W.; Mota, F.S.B.; Santos, G.P. (2001) Descolorização de efluente da indústria têxtil utilizando coagulante natural (Moringa oleifera e quitosana). In: Congresso Brasileiro de Engenharia Sanitária e Ambiental, 21, ABES, 2001.

23. Abdulsalam, S.; Gital, A.A.; Misau, I.M.; Suleiman, M.S. (2007) Water clarification using Moring oleifera seed coagulant: Maiduguri raw water as a case study. Journal of Food, Agriculture and Environment, Helsinki, 5 (1), 302-306.

24. Matos, A. T.; Cabanellas, C. F. G.; Ceccon, P.; Brasil, M. S.; Silva, C. M. (2007) Efeito da concentração de coagulantes e do pH da solução na turbidez da água, em recirculação, utilizada no processamento dos frutos do cafeeiro. *Engenharia Agrícola,* 27, 544-551.

25. Matos CABANELLAS, C. F. G. *Tratamento da água sob recirculação, em escala laboratorial, na despolpa dos frutos do cafeeiro.* (2004). Tese Mestrado, Universidade Federal de Viçosa, Brasil.

26. Brandão, V.S.; Matos, A.T.; Martinez, M.A.; Fontes, M.P.F. (2000) Tratamento de águas residuárias de suinocultura utilizando-se filtros orgânicos. *Revista Brasileira de Engenharia Agrícola e Ambiental,* 4(3), 327-33.

27. Magalhães, M.A.; Matos, A.T.; Azevedo, R.F.; Deniculi, W. (2005) Influência da compressão no desempenho de filtros orgânicos para tratamento de águas residuárias da suinocultura. *Engenharia na Agricultura,* 13(1), 26-32.

28. Matos, A. T. D., Magalhães, M. A., and Fukunaga, D. C. (2006). Remoção de sólidos em suspensão na água residuária da despolpa de frutos do cafeeiro em filtros constituídos por pergaminho de grãos de café submetido a compressões. *Engenharia Agrícola,* 26(2), 610-616.

29. Matos, A. T., and Lo Monaco, P. A. (2003). Tratamento e aproveitamento agrícola de resíduos sólidos e líquidos da lavagem e despolpa dos frutos do cafeeiro. *Engenharia na Agricultura. Boletim técnico, AEAGRI,* 7, 68.

30. Lo Monaco, P. A., Matos, A. T., Martinez, M. A., and Jordão, C. P. (2002) Eficiência de materiais orgânicos filtrantes no tratamento de águas residuárias da lavagem e despolpa dos frutos do cafeeiro. *Engenharia na Agricultura,* 10 (4), 40-47

31. Magalhães, M.A.; Matos, A.T.; Denículi, W.; Tinoco, I.F.F. Compostagem de bagaço de cana-de-açúcar triturado utilizado como material filtrante de águas residuárias da suinocultura. *Revista Brasileira de Engenharia Agrícola e Ambiental,* 10(2) 466–471.

32. Lo Monaco, P.A.; Matos, A.T.; Jordão, C.P.; Cecon, P.C.; Martinez, M.A. (2004) Influência da granulometria da serragem de madeira como material filtrante no tratamento de águas residuárias. *Revista Brasileira de Engenharia Agrícola e Ambiental,* 8 (1), 116-119.

33. Batista, R. O, Matos, A. T., Cunha, F. F., and Monaco, P. A. (2005). Obstrução de gotejadores utilizados para a aplicação de água residuária da despolpa dos frutos do cafeeiro. *Irriga,* 10 (3), 299-305.

34. Von Sperling (1996). *Introdução a qualidade das aguas e ao tratamento de esgoto.* Belo Horizonte, Editora UFMG. 452p.

35. Ortega, F. S.; Rocha, K. M.; Zaiat, M.; Pandolfelli, V. C. (2001) Aplicação de espumas cerâmicas produzidas via "gelcasting" em biorreator para tratamento anaeróbio de águas residuárias. *Cerâmica,* 47 (304), 199-203.

36. Young, C.J. (1991). Factors affecting the design and performance of upflow anaerobic filter. *Water Science & Technology.* 24 (8), 133–155.

37. Ortega, F. S.; Rocha, K. M.; Zaiat, M.; Pandolfelli, V. C. (2001) Aplicação de espumas cerâmicas produzidas via "gelcasting" em biorreator para tratamento anaeróbio de águas residuárias. *Cerâmica,* 47 (304), 199-203.

38. Couto, L. C. C.; Figueiredo, R. F. (1993) Filtro anaeróbio com bambu para tratamento de esgotos domésticos. In *Congresso Interamericano De Ingenieria Sanitaria Y Ambiental,* 23, La Habana, Cuba. Anales... La Habana: AIDIS. 2, 329-340.
Nour, E. A. A.; Coraucci Filho, B; Figueiredo, R. F.; Stefanutti, R.; Camargo, S. A. R. (2000) Tratamento de esgoto sanitário por filtro anaeróbio utilizando o bambu como meio suporte. In: Campos, J. R. *Tratamento de esgotos sanitários por processo anaeróbio e disposição controlada no solo - Coletânea de artigos técnicos.* São Carlos: ABES, 1, 210-231.

39. Torres, P.; Rodríguez, J. A.; Uribe, I. E. (2003) Tratamiento de aguas residuales del proceso de extracción de almidón de yuca en filtro anaerobio: influencia del medio soporte. *Scientia et Technica,* 23, 75-80.

40. Pinto, J. D. S.; Chernicharo, C. A. L. (1996) Escória de altoforno: uma nova alternativa de meio suporte para filtros anaeróbios. In: Simpósio Ítalo Brasileiro De Engenharia Sanitária E Ambiental, 3., Gramado. *Anais...* Rio de Janeiro: ABES/ANDIS, 1996. I-006. 10p.

41. Andrade Neto, C. O.; Pereira, M. G.; Santos, H. R.; Melo, H. N. S. (1999) Filtros anaeróbios de fluxo descendente afoga-

dos, com diferentes enchimentos. In: Congresso Brasileiro De Engenharia Sanitária E Ambiental, 20, Rio de Janeiro, 1999. *Anais...* Rio de Janeiro: ABES, 27-36.

42. Zellner, G.; Vogel, P.; Kneifel, H.; Winter, J. (1987) Anaerobic digestion of whey and permeate with suspended and immobilized complex and defined consortia. *Applied Microbiology and Biotechnology*, 27, 306-314.

 Kawasc, M.; Nomura, T.; Najima, T. (1989) Anaerobic fixed bed reactor with a porous ceramic carrier. *Water Science and Technology*, (21) 4-5, 77-86.

 Gourari, S.; Achkari-Begdouri, A. (1997) Use of baked clay media as biomass supports for anaerobic filters. *Applied Clay Science*, 12, 365-375.

 Ortega, F. S.; Rocha, K. M.; Zaiat, M.; Pandolfelli, V. C. (2001) Aplicação de espumas cerâmicas produzidas via "gelcasting" em biorreator para tratamento anaeróbio de águas residuárias. *Cerâmica*, (47) 304, 199-203.

43. Huysman, P.; Van Meenen, P.; Van Assche, P.; Verstraete, W. (1983) Factors affecting the colonization of non porous and porous packing materials in model upflow methane reactors. *Biotechnology Letters*, 5, 643-648.

 Fynn, G. H.; Whitmore, T. N. (1984) Retention of methanogens in colonized reticulades polyurethane foam biomass support particle. *Biotechnology Letters*, 6, 81-86.

 Gijzen, H. J.; Schoenmakers, T. J. M.; Caerteling, C. G. M.; Vogels, G. D. (1988) Anaerobic degradation of papermill sludge in a two-phase digester containing rumen microorganisms and colonized polyurethane foam. *Biotechnology Letters*, 10, 61-66.

 Zaiat, M.; Cabral, A. K. A.; Foresti, E. (1996) Cell wash-out and external mass transfer resistance in horizontal-flow anaerobic immobilized sludge reactor. *Water Research*, 30, 2435-2439.

 Ribeiro, R.; Varesche, M. B. A.; Foresti, E.; Zaiat, M. (2005) Influence of the carbon source on the anaerobic biomass adhesion on polyurethane foam matrices. *Journal of Environmental Management*, 74, 187-194.

44. Chaiprasert, P.; Suvajittanont, W.; Suraraksac, B.; Tanticharoend, M.; Bhumiratana, S. (2003) Nylon fibers as supporting media in anaerobic hybrid reactors: it's effects on system's performance and microbial distribution. *Water Research*, 37, 4605-4612.

45. Ruiz, I.; Veiga, M. C.; Santiago, P.; Blázquez, R. (1997) Treatment of slaughterhouse wastewater in a UASB reactor and an anaerobic filter. *Bioresource Technology*, 60, 251-258.

 Passig, F. H. (1997) *Estudo do desenvolvimento do biofilme e dos grânulos formados no filtro biológico anaeróbio*. 128p. Dissertação Mestrado, Escola de Engenharia de São Carlos, Universidade de São Paulo, Brasil.

 Show, K. Y.; Tay, J. H. (1999) Influence of support media on biomass growth and retention in anaerobic filters. *Water Research*, (33) 6, 1471-1481.

46. Show, K. Y.; Tay, J. H. (1999) Influence of support media on biomass growth and retention in anaerobic filters. *Water Research*, (33) 6, 1471-1481.

47. Luiz, F. A. R. (2007) *Desempenho de reatores anaeróbios de leito fixo no tratamento de águas residuárias da lavagem e descascamento/despolpa dos frutos do cafeeiro*. Universidade Federal de Viçosa, Brasil, 132p.

48. Luiz F. A. R. (2007) *Desempenho de reatores anaeróbios de leito fixo no tratamento de águas residuárias da lavagem e descascamento/despolpa dos frutos do cafeeiro*. Universidade Federal de Viçosa, Brasil, 132p.

49. Bae, B.; Autenrieth, R. L.; Bonner, J. S. (1995) Kinetics of multiple phenolic compounds degradation with a mixed culture in a continuous-flow reactor. *Water Environment Research*, 67, 215-223

50. Adak, A.; Pal, A. (2006) Removal of phenol from aquatic environment by SDS-modified alumina: Batch and fixed bed studies. *Separation and Purification Technology*, 50, 256-262.

51. Sperling, M. Von. (1996). Introdução à qualidade das águas e ao tratamento de esgotos. *Princípios do tratamento biológico de águas residuárias*. 1, 243 p.

 Fia, R.; Matos, A. T.; Luiz, F. A. R.; Pereira, P. A. (2007) Coeficientes de degradação da matéria orgânica de água residuária da lavagem e despolpa dos frutos do cafeeiro em condições anóxica e aeróbia. *Engenharia na Agricultura*, 15(1), 45-54.

 Matos, A. T., and Gomes Filho, R. R. (2001). Cinética da degradação do material orgânico de águas residuárias da lavagem e despolpa de frutos do cafeeiro. *In 2 Simpósio de Pesquisa dos Cafés do Brasil*, 2001.

52. Matos, A. T., and Gomes Filho, R. R. (2001). Cinética da degradação do material orgânico de águas residuárias da lavagem e despolpa de frutos do cafeeiro. *In 2 Simpósio de Pesquisa dos Cafés do Brasil*, 2001.

53. Matos, A. T., and Gomes Filho, R. R. (2001). Cinética da degradação do material orgânico de águas residuárias da lavagem e despolpa de frutos do cafeeiro. *In 2 Simpósio de Pesquisa dos Cafés do Brasil*, 2001.

54. Matos, A. T., Emmerich, I. N., and Russo, J. R. (2001). Tratamento de águas residuárias da lavagem e despolpa dos frutos do cafeeiro em rampas cultivadas com azevém. *In 2 Simpósio de Pesquisa Dos Cafés Do Brasil*, 2001.

55. Matos, A. T., Fukunaga, D. C., Pinto, A. B., and Russo, J. R. (2001). Remoção de DBO e DQO em sistemas de tratamento de águas residuárias da lavagem e despolpa dos frutos do cafeeiro com rampas cultivadas com aveia. *In 2 Simpósio de Pesquisa dos Cafés do Brasil*, 2001.

56. Matos, A. T.; Pinto, A. B.; Soares, A. A.; Pereira, O. G.; Lo Monaco, P. A. (2003) Produtividade de forrageiras utilizadas em rampas de tratamento de águas residuárias da lavagem e despolpa dos frutos do cafeeiro. *Revista Brasileira de Engenharia Agrícola e Ambiental*, Campina Grande, 7, 154-158.

57. Matos, A. T., Emmerich, I. N., and Russo, J. R. (2001). Tratamento de águas residuárias da lavagem e despolpa dos frutos do cafeeiro em rampas cultivadas com azevém. *In 2 Simpósio de Pesquisa Dos Cafés Do Brasil*, 2001.

 Matos, A. T., Fukunaga, D. C., Pinto, A. B., and Russo, J. R. (2001). Remoção de DBO e DQO em sistemas de tratamento de águas residuárias da lavagem e despolpa dos frutos do cafeeiro com rampas cultivadas com aveia. *In 2 Simpósio de Pesquisa dos Cafés do Brasil*, 2001.

58. Loehr E Oliveira, cited by Matos, A.T.; Sediyama, M.A.N. (1996) Riscos potenciais ao ambiente pela aplicação de dejeto líquido de suínos ou compostos orgânicos no solo. In: *Freitas, R.T.F. e Viana, C.F.A. I Seminário mineiro sobre manejo e utilização de dejetos de suínos, Anais...* 45-54.

59. Lo Monaco, P. A. (2005*). Fertirrigação do cafeeiro com águas residuárias da lavagem e descascamento de seus frutos*. Tese Doutorado, Universidade Federal de Viçosa, Viçosa, Brasil

60. Lo Monaco, P. A.; Matos, A. T.; Martinez, H. E. P.; Ferreira, P. A.; Ramos, M. M. (2009) Avaliação das características químicas do solo após a fertirrigação do cafeeiro com águas residuárias da lavagem e descascamento de seus frutos. *Revista Irriga*, Botucatu, (14) 3, 348-364.

61. Matos A. T.; Pinto, A. B.; Soares, A. A.; Pereira, O. G.; Lo Monaco, P. A. (2003) Produtividade de forrageiras utilizadas em rampas de tratamento de águas residuárias da lavagem e despolpa dos frutos do cafeeiro. *Revista Brasileira de Engenharia Agrícola e Ambiental*, Campina Grande - PA, 7, 154-158.

 Lo Monaco, P. A.; Matos, A. T.; Martinez, H. E. P.; Ferreira, P. A.; Ramos, M. M. (2009) Avaliação das características químicas do solo após a fertirrigação do cafeeiro com águas residuárias da lavagem e descascamento de seus frutos. *Revista Irriga*, Botucatu, (14) 3, 348-364.

62. Lo Monaco, P. A. (2005*). Fertirrigação do cafeeiro com águas residuárias da lavagem e descascamento de seus frutos*. Tese Doutorado, Universidade Federal de Viçosa, Viçosa, Brasil.

63. Moreira, R. M. G., Matos, A. T., Lo Monaco, P. A., Eustáquio Júnior, W., and Gutierrez, K. G. (2005). Produtividade do cafeeiro fertirrigado com águas residuárias geradas na despolpa de seus frutos. *In 15 Congresso Nacional de Irrigação e Drenagem*, 2005

64. Matos, A. T., Emmerich, I. N., and Russo, J. R. (2001). Tratamento de águas residuárias da lavagem e despolpa dos frutos do cafeeiro em rampas cultivadas com azevém. *In 2 Simpósio de Pesquisa Dos Cafés Do Brasil*, 2001

65. Moreira, R. M. G., Matos, A. T., Lo Monaco, P. A., Eustáquio Júnior, W., and Gutierrez, K. G. (2005). Produtividade do cafeeiro fertirrigado com águas residuárias geradas na despolpa de seus frutos. *In 15 Congresso Nacional de Irrigação e Drenagem*, 2005.

66. Moreira, R. M. G., Matos, A. T., Lo Monaco, P. A., Pereira, R. R. A., and Climaco, R. C. (2004). Fertirrigação do cafeeiro com aplicação, por aspersão, de águas residuárias da despolpa dos seus frutos. *In 9 Simpósio de Iniciação Científica*, 2004.

67. Leon, S. G., and Cavallini, J. M. (1999). Tratamento e uso de águas residuárias. *Campina Grande: UFPB*, 110.

68. Magalhães, M. A.; Matos, A. T.; Denículi, W.; Tinoco, I. F.F. (2006) de filtros orgânicos utilizados no tratamento de águas residuárias da suinocultura. *Revista Brasileira de Engenharia Agrícola e Ambiental*, (10) 2, 472–478.

Lo Monaco, P. A.; Matos, A. T.; Martinez, H. E. P.; Ferreira, P. A.; Ramos, M. M. (2009) Avaliação das características químicas do solo após a fertirrigação do cafeeiro com águas residuárias da lavagem e descascamento de seus frutos. *Revista Irriga*, Botucatu, (14) 3, 348-364.

69. Batista, R. O, Matos, A. T., Cunha, F. F., and Monaco, P. A. (2005). Obstrução de gotejadores utilizados para a aplicação de água residuária da despolpa dos frutos do cafeeiro. *Irriga*, 10 (3), 299-305

70. Fia, R., and Matos, A. T. (2001). Avaliação da eficiência e impactos ambientais causados pelo tratamento de águas residuárias da lavagem e despolpa dos frutos do cafeeiro em áreas alagadas. *In: 2 Simpósio de Pesquisa dos Cafés Do Brasil*, 2001.

71. Trotter, E. A.; Thomson, B.; Coleman, R. (1994) Evaluation of a subsurface flow wetland processing sewage from the Sevilleta LTER field station. Las Cruces: New Mexico Water Resources Research Institute. 52 p. (WRRI Report, 287).

72. Valentim, M. A. A. (1999) *Uso de Leitos Cultivados no Tratamento de Efluente de Tanque Séptico Modificado*. 113p. (Dissertação de Mestrado – Universidade Estadual de Campinas, 1999).

73. Kadlec & Knight, 1996; cited by Tobias, A. C. T. (2002) *Tratamento de resíduos da suinocultura: uso de reatores anaeróbios seqüenciais seguido de leitos cultivados*. Tese de Doutorado – Universidade Estadual de Campinas, Brasil, 125p.

74. Cooper, 1993; cited by Tobias, A. C. T. (2002) *Tratamento de resíduos da suinocultura: uso de reatores anaeróbios seqüenciais seguido de leitos cultivados*. Tese de Doutorado – Universidade Estadual de Campinas, Brasil, 125 p.

75. Fia, R. (2008) *Desempenho de sistemas alagados construí- dos no tratamento de águas residuárias da lavagem e descascamento/despolpa dos frutos do cafeeiro*. Tese Doutorado, Universidade Federal de Viçosa, Brasil, 181 p.

Chapter 6

1. Brooker, D. B., Bakker-Arema, F. W., and Hall, C. W. (1978). *Drying cereal grains*. Connecticut: AVI.

Hall, C. W. (1980). Drying and storage of agricultural crops. Connecticut: AVI.

2. Bala, B. K. (1997) Dying and storage of cereal grains. New Hampshire: *Science*, 302 p.

Brooker, D. B.; Bakker-Arkema, F. W.; Hall, C. W. (1992) *Drying and storage of grains and oilseeds*. Westport: AVI, 450 p.

Jayas, D. S.; Cenkowski, S.; Pabis, S.; Muir, W. (1991) Review of thin-layer drying and wetting equations. *Drying Technology*, 9 (3), p. 551-588.

Thompson, T. L.; Peart, R. M.; Foster, G. H. (1968) *Mathematical simulation of corn drying: a new model*. Transaction of the ASAE, Saint Joseph, 11 (4), 582-586.

3. McLoy, J. F. (1979) Mechanical drying of Arabica Coffee. *Kenya Coffee*, 44 (516), 13-26.

4. Reinato, C.H.R, Borem, F.M, Cirillo, M.A., Oliveira, E.C. (2012) Qualidade do café secado em terreiros com diferentes pavimentações e espessuras de camada. *Coffee Science*, 7 (3), 223-237.

5. Andrade, E. T., Hardoim, P. R., Borém, F. M., and Hardoim, P. C. (2001). Cinética de secagem de café cereja, bóia e cereja desmucilado em quatro diferentes tipos de terreiro. *In 27 Congresso Brasileiro de Pesquisas Cafeeiras*, 2001, (386-388).

Lacerda Filho, A. F. (1986). *Avaliação de diferentes sistemas de secagem e suas influências na qualidade de café (Coffea arabica L.)*. Dissertação Mestrado, Universidade Federal de Viçosa, Viçosa, Brasil.

Vilela, R. V. (1997). Qualidade do café: secagem e qualidade do café. *Informe Agropecuário*, 18 (187), 55-63.

6. Raposo, H. (1959). Café fino e seu preparo. *Rio de Janeiro: Serviço de Informação Agrícola*. 55.

7. Reinato, C. H. R. ; Borem, F. M. ; Cirillo, M. A. ; Oliveira, E. C. (2012) Qualidade do café secado em terreiros com diferentes pavimentações e espessura de camada. *Coffee Science*, 7, 223-237.

Santinato, R., and Teixeira, A. A. (1977). Estudos preliminares sobre tipos de terreiros para seca de café. *In Congresso Brasileiro de Pesquisas Cafeeiras*, 1977 (257-259).

8. Reinato, C. H. R. ; Borem, F. M. ; Cirillo, M. A. ; Oliveira, E. C. (2012) Qualidade do café secado em terreiros com diferentes pavimentações e espessura de camada. *Coffee Science*, 7, 223-237.

9. Borém, F. M. (2004). *Cafeicultura empresarial: produtividade e qualidade: pós-colheita do café*. Lavras,MG: UFLA.

10. Abrahão, E. J. (2001). Difusão de tecnologia: terreiro de baixo custo: lama asfáltica. *In Simpósio Brasileiro de Pesquisa dos Cafés do Brasil*, 2001, (2613-2616).

Abrahão, E. J, Ferreira, L. F., Felipe, M. P., Fabri Júnior, M. A., Regina, S. B., and Lasmar, W. M. (2002). Qualidade do café: terreiro pavimentação com lama asfáltica. *Cafeicultura*, 1 (4), 6-7.

11. Dafert (1896), cited by Silva, J. S., and Machado, M. C. (2002). Estado da arte da secagem de café no Brasil. In Zambolim, L. (Ed.). *O estado da arte de tecnologias na produção de café*. Viçosa, 521-558.

12. Micheli, G. (2000). La seca del café como factor de calidad café descascado y secado en parihuela. *In Simposio Latinoamericano de Caficultora*, 2000 (55-60).

13. Hardoim, P. C., Borém, F. M., Hardoim, P. R., and Abrahão, E. J. (2001). Secagem de café cereja, bóia e cereja desmucilado em terreiro de concreto, de lama asfáltica, de chão batido e de leito suspenso em Lavras. *In Congresso Brasileiro de Pesquisas Cafeeiras*, 2001, (126-128).

14. Borém, F. M.; Reinato, C. H. R.; Andrade, E. T. (2008) Secagem do Café. In: Flávio Meira Borém. (Org.). *Pós-Colheita do Café*. 1ed.Lavras: Editora UFLA, 1, p. 203-240.

15. Silva, J. S. (2000). Secagem e armazenagem do café. Viçosa: UFV.

16. Borém, F. M., Ribeiro, D. M., Pereira, R. G. F. A., Rosa, S. D. V. F. da, Morais, A. R. de. (2006). Qualidade do café submetido a diferentes temperaturas, fluxos de ar e períodos de pré-secagem. *Coffee Science*,1 (1), 55-63

17. Menchú, J. F., García, R., Rolz, C., and Calzada, J. F. (1981). El beneficiado de café y el aprovechamiento de sus sub-productos. *In Simposio Latinoamericano Sobre Caficultora*, 1981 (247-249).
18. Silva, J. S., Pinto, F. A. C., Machado, M. C., and Melo, E. C. (2001). Projeto, construção e avaliação de um secador de fluxos (concorrentes/contracorrentes) para secagem de café. *In Simpósio Brasileiro de Pesquisa dos Cafés do Brasil*, 2001 (964-980).
19. Reinato, C. H. R., and Borém, F. M. (2006). Variação da temperatura e do teor de água do café em secador rotativo usando lenha e GLP como combustíveis. Engenharia Agrícola, 26, 561-569.
20. Pinto Filho, A. C. (1993). *Projeto de um secador de fluxos contracorrentes/concorrentes e análise de seu desempenho na secagem de café (Coffea arabica L.).* Dissertação Mestrado, Universidade Federal de Viçosa, Viçosa, Brasil.
21. Silva, J. S. (2000). Secagem e armazenagem do café. Viçosa: UFV.
22. Berbert, P. A. (1991). *Secagem de café (Coffea arabica L.), em camada fixa, com inversão de sentido de fluxo de ar.* Dissertação Mestrado, Universidade Federal de Viçosa, Viçosa, Brasil.
23. Silva, J. S. (2001) Secagem e armazenagem do café. Viçosa: UFV, 162 p.
24. Borém, F. M. (2004*). Cafeicultura empresarial: produtividade e qualidade: pós-colheita do café.* Lavras, MG: UFLA.
25. Silva, J. S. (2000). Secagem e armazenagem do café. Viçosa: UFV.
26. Reinato, C. H. R.; Borém, F. M. (2006) Variação da temperatura e do teor de água do café em secador rotativo usando lenha e GLP como combustíveis. *Engenharia Agrícola*, 26, 561-569.
27. Borém, F.M. *Pos-colheita do Cafe.* Lavras, 2008. 631p.
28. Menchú, E. F. (1967) La determinación de la calidad del café: parte I: características, color y aspecto. *Agricultura de las Américas*, 16, (5),18-21.
29. Menchú, E. F. (1967). La determinación de la calidad del café: parte I: características, color y aspecto. *Agricultura de las Américas,* 16 (5), 18-21.
30. Borém, F. M.; Marque, E. R.; Alves, E. (2008) Ultrastructural analysis of drying damagein in parchment Arabica coffee endosperm cells. *Biosystems Engineering*, 99, 62-66.
31. Saath, R.; Borém, F. M.; Alves, E.; Taveira, J. H. S.; Medice, R.; Coradi, P. C. (2010) Microscopia eletrônica de varredura do endosperma de café (*Coffea arabica* L.) Durante o processo de secagem. *Ciência e Agrotecnologia*, 34, 196-203.
32. Taveira, J. H. S.; Rosa, S. D. V. F. da; Borém, F. M.; Giomo, G. S.; Saath, R.(2012) Perfis proteicos e desempenho fisiológico de sementes de café submetidas a diferentes métodos de processamento e secagem. *Pesquisa Agropecuária Brasileira*, 47, 1511-1517.
33. Saath, R.; Borém, F. M.; Alves, E.; Taveira, J. H. S.; Medice, R.; Coradi, P. C. (2010) Microscopia eletrônica de varredura do endosperma de café (*Coffea arabica* L.) Durante o processo de secagem. *Ciência e Agrotecnologia*, 34, 196-203.
34. Taveira, J.H.S. *Aspectos fisiologicos e bioquimicos associados à qualidade da bebida de café submetido a diferentes*
métodos de processamento e secagem. (2009). Dissertacao Mestrado, Universidade Federal de Lavras, Brasil.
35. Oliveira, P.D. (2010) *Metaboloma e fisiologia da pós-colheita: nova abordagem para o processamento e qualidade do café.* Dissertacao Mestrado, Universidade Federal de Lavras, Brasil.
36. Taveira, J.H.S. *Aspectos fisiologicos e bioquimicos associados à qualidade da bebida de café submetido a diferentes métodos de processamento e secagem.* (2009). Dissertacao Mestrado, Universidade Federal de Lavras, Brasil.
37. Prete, C. E. C. (1992) *Condutividade elétrica do exsudato de grãos de café (*Coffea arabica *L.) e sua relação com a qualidade da bebida.* Tese Doutorado, Escola Superior de Agricultura Luiz de Queiroz, Brasil.
38. Borém, F. M.; Reinato, C. H. R.; Andrade, E. T. (2008) Secagem do Café. In: Flávio Meira Borém. (Org.). *Pós-Colheita do Café.* 1ed. Lavras: Editora UFLA, 1, p. 203-240.
39. Taveira, J. H. S.; Rosa, S. D. V. F. da; Borém, F. M.; Giomo, G. S.; Saath, R.(2012) Perfis proteicos e desempenho fisiológico de sementes de café submetidas a diferentes métodos de processamento e secagem. *Pesquisa Agropecuária Brasileira*, 47, 1511-1517.
40. Saath, R.; Borém, F. M.; Alves, E.; Taveira, J. H. S.; Medice, R.; Coradi, P. C. (2010) Microscopia eletrônica de varredura do endosperma de café (*Coffea arabica* L.) durante o processo de secagem. Ciência e Agrotecnologia, 34, 196-203.
41. Oliveira, P. D.; Borém, F. M.; Isquierdo, E. P.; Giomo, G. S.; Lima, R. R.; Cardoso, R. A. (2013). Aspectos fisiológicos de grãos de café, processados e secados de diferentes métodos, associados à qualidade sensorial. *Coffee Science*, 8, 211-220.
42. Isquierdo, E. P.; Borém, F. M.; Andrade, E. T.; Correa, J. L. G.; Oliveira, P. D.; Alves, G. E. (2013). Drying Kinetics and quality of natural coffee. American Society of Agricultural and Biological Engineers. Transactions, 56, 1003-1010

Alves, G. E.; Isquierdo, E. P.; Borém, F.M.; Siqueira, V. C.; Oliveira, P. D.; Andrade, E.T. (2013). Cinética de secagem de café natural para diferentes temperaturas e baixa umidade relativa.. Coffee Science, 8, 238-247

Isquierdo, E. P. (2011) Cinética de secagem de café natural e suas relações com a qualidade para diferentes temperaturas e umidades relativas do ar. Tese de Doutorado, Universidade Federal de Lavras, Brasil.
43. Marques, E. R.; Borém , F. M.; Pereira, R. G. F. A.; Biaggioni, M. A. M. (2008) Eficácia do Teste de Acidez Graxa na Avaliação da Qualidade do Café Arábica (*Coffea arabica* L.) Submetido a Diferentes Período e Temperatura de Secagem. *Ciência e Agrotecnologia* (UFLA), 32, 1557-1562.
44. Isquierdo, E. P.; Borém, F. M.; Cirillo, M. A.; Oliveira, P. D.; Cardoso, R. A.; Fortunato, V. A. (2011) Qualidade do café cereja desmucilado submetido ao parcelamento de secagem. *Coffee Science*, 6, 83-90.
45. Martin, S., Donzeles, S.M.L., Silva, J.N, Zanatta, F.L., Cecon, P.R.(2009) Qualidade do café cereja descascado submetido a secagem contínuae intermitente, em secador de camada fixa. *Revista Brasileira de Armazenamento*, 11, 30-36.
46. Isquierdo, E. P.; Borém, F. M.; Oliveira, P. D.; Siqueira, V. C.; Alves, G. E.(2012) Quality of natural coffee subjected to different rest periods during the drying process. *Ciência e Agrotecnologia*, 36,439-445.

Chapter 7

1. Afonso Júnior, P. C. (2001). *Aspectos físicos, fisiológicos e de qualidade do café em função da secagem e do armazenamento.* Tese Doutorado, Universidade Federal de Viçosa, Viçosa, Brasil.
2. Afonso Júnior, P. C. (2001). *Aspectos físicos, fisiológicos e de qualidade do café em função da secagem e do armazenamento.* Tese Doutorado, Universidade Federal de Viçosa, Viçosa, Brasil.
3. Castro, L. H. (1991). *Efeito do despolpamento, em secador de leito fixo sob altas temperaturas, no consumo de energia e na qualidade do café.* Dissertação Mestrado, Universidade Federal de Viçosa, Viçosa, Brasil.
4. Lopes, R. P., Afonso, A. D. L., and Silva, J. de S. (2000). Energia no pré-processamento de produtos agrícolas. In SILVA, J. de S. (Ed). *Secagem e armazenagem de produtos agrícolas.* Viçosa: Aprenda Fácil. 191-219.

5. Reinato, C. H. R., Borém, F. M., Pereira, R. G. F. A., and Carvalho, F. M. (2003). Avaliação técnica, econômica e qualitativa do uso da lenha e do GLP na secagem do café. *Revista Brasileira de Armazenamento*, 27 (7), 3-13.
6. Cardoso Sobrinho, J. (2001). *Simulação e avaliação de sistemas de secagem de café*. Tese Doutorado, Universidade Federal de Viçosa, Viçosa, Brasil.
7. Reinato, C. H. R., Borém, F. M., Pereira, R. G. F. A., and Carvalho, F. M. (2003). Avaliação técnica, econômica e qualitativa do uso da lenha e do GLP na secagem do café. *Revista Brasileira de Armazenamento*, 27 (7), 3-13.
8. Cardoso Sobrinho, J. (2001). *Simulação e avaliação de sistemas de secagem de café*. Tese Doutorado, Universidade Federal de Viçosa, Viçosa, Brasil.
9. Lacerda Filho, A. F. (1986). *Avaliação de diferentes sistemas de secagem e suas influências na qualidade do café*. Dissertação Mestrado, Universidade Federal de Viçosa, Viçosa, Brasil.
10. Campos, A. T. (1988). *Desenvolvimento e análise de um protótipo de secador de camada fixa para café (Coffea arabica L.) com sistema de revolvimento mecânico*. Dissertação Mestrado, Universidade Federal de Viçosa, Viçosa, Brasil.
11. Pinto Filho, G. L. (1994). *Desenvolvimento de um secador de fluxos cruzados com reversão do fluxo de ar de resfriamento, para a secagem de café (Coffea arabica L.)*. Dissertação Mestrado, Universidade Federal de Viçosa, Viçosa, Brasil.
12. Cardoso Sobrinho, J. (2001). *Simulação e avaliação de sistemas de secagem de café*. Tese Doutorado, Universidade Federal de Viçosa, Viçosa, Brasil.
13. Osório, A. G. S.(1982). *Projeto e construção de um secador intermitente de fluxos concorrentes e sua avaliação na secagem do café*. Dissertação Mestrado, Universidade Federal de Viçosa, Viçosa, Brasil.
14. Lacerda Filho, A. F. (1986). *Avaliação de diferentes sistemas de secagem e suas influências na qualidade do café*. Dissertação Mestrado, Universidade Federal de Viçosa, Viçosa, Brasil.
15. Silva, L. C. (1991). *Desenvolvimento e avaliação de um secador de café intermitente de fluxos contracorrentes*. Dissertação Mestrado, Universidade Federal de Viçosa, Viçosa, Brasil.
16. Pinto, F. A. C. (1993). *Projeto de um secador de fluxos contracorrentes/concorrentes e análise de seu desempenho na secagem do café*. Dissertação Mestrado, Universidade Federal de Viçosa, Viçosa, Brasil.
17. Silva, J. N., Cardoso Sobrinho, J., Lacerda Filho, A. F. De, and Silva, J. de S. (2000). Vapor d'água para aquecimento do ar de secagem de café: custo e consumo de energia. *In 1 Simpósio de Pesquisa dos Cafés do Brasil*, 1151-1154.
18. Lora, E. S., Happ, J. F., and Cortez, L. A. B. (1997). Caracterização e disponibilidade da biomassa. In CORTEZ, L. A. B., LORA, E. S. *Tecnologias de conversão energética da biomassa*. Manaus: EDUA/EFEI, 5-37.
19. Bazzo, E. (1995). Geração de vapor. Florianópolis: UFSC.
20. Camargo, C. A. (1990). Conservação de energia na indústria do açúcar e do álcool. São Paulo: Instituto de Pesquisas Tecnológicas.
21. Camargo, C. A. (1990). Conservação de energia na indústria do açúcar e do álcool. São Paulo: Instituto de Pesquisas Tecnológicas.
    Ocón, J. (1967).Producción industrial del calor: combustibles, combustión y gasificación. In VIAN, A., AGUILAR, J. O. *Elementos de ingeniería química*. Madri, 141-181.
22. Andrade, E. B. (1982). Combustíveis e fornalhas. Viçosa: Centreinar.
    Souza, Z. De. (1980). Elementos de máquinas térmicas. Rio de Janeiro: Campus/EFEI.
23. Barros, D. M., Vasconcelos, E. C. (2001). Termelétricas a lenha. In MELLO, M. G. *Biomassa: energia dos trópicos em Minas Gerais*. Belo Horizonte: LabMídia/FAFICH/UFMG, 221-241.
24. Andrade, E. B. (1982). Combustíveis e fornalhas. Viçosa: Centreinar.
25. Vann & Claar, cited by Saglietti, J. R. C. (1991). *Rendimento térmico de fornalha a len.ha de fluxos cruzados*. Tese Doutorado, Universidade Estadual Paulista, Botucatu, Brasil.
26. Rocha, J. D., Silva, O. C., Paletta, C. E. M., and Coelho, J. T. (2001). Banco de dados de biomassa online *In 1 Congresso Internacional de Uso da Biomassa Plantada para Produção de Metais e Geração de Eletricidade*. 2001.
27. Rocha, J. D., Silva, O. C., Paletta, C. E. M., and Coelho, J. T. (2001). Banco de dados de biomassa online *In 1 Congresso Internacional de Uso da Biomassa Plantada para Produção de Metais e Geração de Eletricidade*. 2001.
28. Andrade, E. B. (1982). Combustíveis e fornalhas. Viçosa: Centreinar.
29. Vlassov, D. (2001) Combustíveis, combustão e câmaras de combustão. Curitiba: UFPR, 2001. 185 p.
30. Diniz, V. Y. (1981). Caldeiras a lenha. In PENEDO, W. R. (Comp.). *Gaseificação de madeira e carvão vegetal.* Belo Horizonte: CETEC, 113-131.
31. Andrade, E. B., Sasseron, J. L., and Oliveira Filho, D. (1984). Princípios sobre combustíveis, combustão e fornalhas. Viçosa: Centreinar.
32. Diniz, V. Y. (1981). Caldeiras a lenha. In PENEDO, W. R. (Comp.). *Gaseificação de madeira e carvão vegetal.* Belo Horizonte: CETEC, 113-131.
33. Mitre, M. N. (1982). Caldeiras para a queima de madeira e bagaço. In PENEDO, W. R. (Comp.). *Produção e utilização de carvão vegetal.* Belo Horizonte: CETEC, 319-348.
34. Pinheiro, P. C. Da C., Sampaio, R. S., and Bastos Filho, J. G. (2001). Fornos de carbonização utilizados no Brasil. *In 1 Congresso Internacional de Uso da Biomassa Plantada para Produção de Metais e Geração de Eletricidade*.
35. Almeida, M. R. (1982). Recuperação de alcatrão em fornos de alvenaria. In PENEDO, W. R. (Comp.). *Produção e utilização de carvão vegetal*. Belo Horizonte: Fundação Centro Tecnológico de Minas Gerais/CETEC. 175-180.
36. Mendes, M. G., Gomes, P. A., and Oliveira, J. B. (1982). Propriedades e controle de qualidade do carvão vegetal. In PENEDO, W. R. (Comp.). *Produção e utilização de carvão vegetal.* Belo Horizonte: CETEC, 75-89.
37. Oliveira, J. B., Gomes, P. A., and Almeida, M. R. de. (1982). Estudos preliminares de normalização de testes de controle de qualidade do carvão. In: _____. *Carvão vegetal: destilação, carvoejamento, propriedades, controle de qualidade.* Belo Horizonte: CETEC, 7-38.
38. Pimenta, A. S., and Barcellos, D. C. (2000). Como produzir carvão para churrasco. Viçosa: CPT, 76.
39. Silva, J. S., Precci, R. L., Machado, M. C. (2000). Fornalha a carvão vegetal para secagem de produtos agrícolas. *Associação dos Engenheiros Agrícolas de Minas Gerais/UFV*, 25.
40. Carioca, J. O. B., and Arora, H. L. (1984). Biomassa: fundamentos e aplicações tecnológicas. Fortaleza: UFCE.
41. Rodrigues, L. D., Silva, I. T. Da, and Rocha, B. R. P. da. (2002). Uso de briquetes compostos para produção de energia no Estado do Pará. *In 4 Encontro de Energia no Meio Rural*.2002.
42. Gomes Filho, A. (2000). Resíduos florestais. Disponível em: <http://www.ambiental.com.br/Cenbio/biomassa/floresta.htm>. Acessed 30 March 2000.
43. Lopes, R. P., Cardoso Sobrinho, J., Silva, J. de S., and Silva, J. N. (2001). Fontes de energia para secagem de café. *Associação dos Engenheiros Agrícolas de Minas Gerais*, 26.
44. Batista, F. L. (1981). Manual técnico construção e operação de biodigestores modelo indiano. Brasília, DF: Embrater.
45. Reis, B. De O, Silva, I. T. De, Silva, I. M. O. De, and Rocha, B. R. P. da. (2002). Produção de briquetes energéticos a partir de coroços de açaí. *In 4 Encontro de Energia no Meio Rural*. 2002.

46. Silva, J. S., Precci, R. L., Machado, M. C. (2000a). Fornalha a carvão vegetal para secagem de produtos agrícolas. *Associação dos Engenheiros Agrícolas de Minas Gerais/UFV*, 25.

47. Lopes, R. P., Silva, J. J., Ruffato, S., and Sena Júnior, D. G. (2002). Consumo de energia em dois sistemas de café. *Ciência e Agrotecnologia*, 26.

48. Donzeles, S. M. L. (2002). *Desenvolvimento e avaliação de um sistema híbrido, solar e biomassa, para secagem de café*. Tese Doutorado, Universidade Federal de Viçosa, Viçosa, Brasil.

49. Silva, J. S., and Lacerda Filho, A. F. Adaptação de fornalha de fogo direto na secagem de grãos. *Informe Agropecuário*, 9(99), 12-16.

50. Oliveira, G. A. de. (1996). *Desenvolvimento e teste de uma fornalha com aquecimento indireto e autocontrole da temperatura máxima do ar para secagem de produtos agrícolas*. Dissertação Mestrado, Universidade Federal de Viçosa, Viçosa, Brasil.

51. Valarelli, I. D. (1991) Desenvolvimento e teste de uma fornalha a resíduos agrícolas, de fogo indireto para secadores de produtos agrícolas. Tese Doutorado, Universidade Estadual Paulista, Botucatu, Brasil.

52. Torreira, R. P. (1995) Geradores de vapor. São Paulo: Exlibris. 710.

53. Torreira, R. P. (19--). Fluidos térmicos: água, vapor, óleos térmicos. São Paulo: Hemus.

54. Torreira, R. P. (19--). Fluidos térmicos: água, vapor, óleos térmicos. São Paulo: Hemus.

55. Araújo, C. (1982). Transmissão de calor. Rio de Janeiro: LTC.

56. Pera, H. Geradores de vapor: um compêndio sobre conversão de energia com vistas à preservação da ecologia. São Paulo: Fama, 1990. paginação irregular.

57. Nogueira, V. S., and Fioravante, N. (1987). Teste comparativo de secagem do café entre o sistema de vapor e o sistema de fornalha. *In 1 Congresso Brasileiro Sobre Pesquisas Cafeeiras*, 1987. 254-255.

58. Lopes, R. P. (2002). *Desenvolvimento de um sistema gerador de calor com opção para aquecimento direto e indireto de ar*. Tese Doutorado, Universidade Federal de Viçosa, Viçosa, Brasil.

59. Lopes, R. P. (2002). *Desenvolvimento de um sistema gerador de calor com opção para aquecimento direto e indireto de ar*. Tese Doutorado, Universidade Federal de Viçosa, Viçosa, Brasil.

60. Parodi, F. A., and Sánchez, L. G. (2002). Aspectos da co-gaseificação de resíduos agroindustriais e municipais. In *Encontro de Energia no Meio Rural*, 2002.

61. Silva, J. N. (1984*). Tar formation in corncob gasification*. Thesis (Ph.D.), Purdue University, Purdue.

62. Silva, J. N. (1984*). Tar formation in corncob gasification*. Thesis (Ph.D.), Purdue University, Purdue.

63. Saiki, E. T., Silva, J. N., Vilarinho, M. C., Cardoso Sobrinho, J. (2000). Viabilidade de secagem de café despolpado usando-se um gasificador/combustor a cavacos de lenha. *In 1 Simpósio de Pesquisa dos Cafés do Brasil*, 1151-1154.

64. Silva, I. S. (2004). *Reator de gaseificação de biomassa em fluxo contracorrente para aquecimento de ar de secagem*. Dissertação Mestrado, Universidade Federal de Viçosa, Viçosa, Brasil.

65. Martin, S. (2005). *Gasificador de biomassas de fluxo concorrente*. Dissertação Mestrado, Universidade Federal de Viçosa, Viçosa, Brasil.

66. Cardoso Sobrinho, J., Silva, J. N. da, Lacerda Filho, A. F. de, and Silva, J. de S. (2000). Custos comparativos de secagem de café usando-se lenha de eucalipto e gás liquefeito de petróleo. *In 1 Simpósio de Pesquisa dos Cafés do Brasil*, 2000 (1132-1137).

Chapter 8

1. Dias, C. A. (2003). *Logística e comercialização de cafés especiais no estado do espírito santo*. Dissertação Mestrado, Universidade Federal de Lavras, Lavras, Brasil.

2. Borém, F. M.; Ribeiro, F. C.; Figueiredo, L. P.; Giomo, G. S.; Fortunato, V. A.; Isquierdo, E. P. (2013). Evaluation of the sensory and color quality of coffee beans stored in hermetic packaging. Journal of Stored Products Research, 52, 1-6.

   Nobre, G. W.(2005). *Alterações qualitativas do café cereja descascado durante o armazenamento*. Dissertação Mestrado, Universidade Federal de Lavras, Lavras, Brasil.

   Ribeiro Ribeiro, F. C.; Borém, F. M.; Giomo, G. S.; Lima, R. R.; Malta, M. R.; Figueiredo, L. P.(2011) Storage of green coffee in hermetic packaging injected with CO2. Journal of Stored Products Research, 47, 341-348.

3. Trubey Trubey, R.; Raudales, R.; Morales, A. (2005) Café britt hermetic cocoon storage trial II report. Beneficio Pataliyo, Costa Rica: Mesoamerican Development Institute Corp, 14 p.

4. Harris, R. L.; Miller, A. (2009) Storing & preserving green coffee – part 2. Roast Magazine, New York, 31-38.

5. Ribeiro, F. C.; Borém, F. M.; Giomo, G. S.; Lima, R. R.; Malta, M. R.; Figueiredo, L. P.(2011) Storage of green coffee in hermetic packaging injected with CO2. Journal of Stored Products Research, 47, 341-348.

6. Brandão, F. (1989) Manual do armazenista.(2 ed). Viçosa: UFV.

7. Brandão, F. (1989) Manual do armazenista.(2 ed). Viçosa: UFV.

8. Afonso Júnior, P. C. (2001). *Aspectos físicos, fisiológicos e da qualidade do café em função da secagem e do armazenamento*. Tese Doutorado, Universidade Federal de Lavras, Lavras, Brasil.

   Arêdes, E. M. (2002). *Avaliação das perdas de matéria seca e de qualidade do café (*Coffea arabica *L.) beneficiado e armazenado em importantes municípios produtores da zona da mata mineira e em Alegre*. Dissertação Mestrado, Universidade Federal de Viçosa, Viçosa, Brasil.

   Bacchi, O. (1962). O branqueamento dos grãos de café. *Bragantia*, 21(28), 467-468.

   Carvalho, V. D., Chagas, S. J. R., Souza, S. M. C. (1997). Fatores que afetam a qualidade do café. *Informe Agropecuário*, 18 (187), 5-20.

   Coelho, K. F., Pereira, R. G. F. A., Vilela, E. R. (2001). Qualidade do café beneficiado em função do tempo de armazenamento e de diferentes tipos de embalagens. *Revista Brasileira de Armazenamento*, 25 (2), 22-27.

   Leite, R. A., Corrêa, P. C., Oliveira, M. G. A., Reis, F. P., Oliveira, T. T. (1999). Avaliação por métodos físicos da qualidade do café (*Coffea arabica* L.) pré-processado por "via seca" e "via úmida" durante dez meses de armazenamento. *Engenharia na Agricultura*, 7 (2), 106-115.

   Lopes, R. P., Hara, T., Silva, J. S., Riedel, B. (2000). Efeito da luz na qualidade (cor e bebida) de grãos de café beneficiados (*Coffea arabica* L.) durante a armazenagem. *Revista Brasileira de Armazenamento*, 25, 9-17.

   Melo, M., Fazuoli, L. C., Teixeira, A. A., Amorim, H. V. (1980). Alterações fisicas, químicas e organolépticas em grãos de café armazenados. *Ciência e Cultura*, 32 (4), 468-471.

   Nobre, G. W. (2005). *Alterações qualitativas do café cereja descascado durante o armazenamento*. Dissertação Mestrado, Universidade Federal de Lavras, Lavras, Brasil.

Oliveira, M. V. de. (1995). *Efeito do armazenamento no branqueamento de grãos de café beneficiado: modelagem matemática do processo*. Dissertação Mestrado, Universidade Federal de Lavras, Lavras, Brasil.

Vilela, E. R., Chandra, P. K., Oliveira, G. (2000). Efeito da temperatura e umidade relativa no branqueamento de grãos de café. *Revista Brasileira de Armazenamento*, 1, 31-37.

9. Afonso Júnior, P. C. (2001). *Aspectos físicos, fisiológicos e da qualidade do café em função da secagem e do armazenamento*. Tese Doutorado, Universidade Federal de Lavras, Lavras, Brasil.

Amorim, H. V. (1972). *Relação entre alguns compostos orgânicos do grão de café verde com a qualidade da bebida*. Tese Doutorado, Escola Superior de Agricultura de Luiz de Queiroz, Piracicaba, Brasil.

Amorim, H. V. (1978) Aspectos bioquímicos e histoquímicos do grão de café verde relacionados com a deterioração de qualidade. Tese de Livre Docência, Escola Superior de Agricultura Luiz de Queiroz, Brasil.

Amorim, H. V., Cruz, A. R., Dias, R. M., Gutierrez, L.E., Teixeira, A. A., Melo, M., and Oliveira, G. D. (1977). Transformações físicas, químicas e biológicas do grão do café e sua qualidade. In 5 Congresso Brasileiro de Pesquisas Cafeeiras,1977 (45-48).

Antunes, P., Sgarbigri, V. C. (1979). Influence of time and conditions of storage on technological and nutritional proprieties of a dry bean (*Phaseolus vulgaris* L.) variety rosinha G2. *Journal of Food Science*, 44 (1), 1704.

Carvalho et al., 1997.

Carvalho, V. D., Chagas, S. J. R., Souza, S. M. C. (1997). Fatores que afetam a qualidade do café. *Informe Agropecuário*, 18 (187), 5-20.

Corrêa, P. C., Afonso Júnior, P. C., Silva, F. S., Ribeiro, D. M. (2003). Qualidade dos grãos de café (*Coffea arabica* L.) durante o armazenamento em condições diversas. *Revista Brasileira de Armazenamento*, 7, 137-147.

Leite, R. A., Corrêa, P. C., Oliveira, M. G. A., Reis, F. P., Oliveira, T. T. (1999). Avaliação por métodos físicos da qualidade do café (*Coffea arabica* L.) pré-processado por "via seca" e "via úmida" durante dez meses de armazenamento. *Engenharia na Agricultura*, 7 (2), 106-115

Mazzafera, P., Guerreiro, F. O., Carvalho, A. (1984). Estudo de coloração verde do grão de café: determinação de flavonóides e clorofllas. In 11 *Congresso Brasileiro de Pesquisas Cafeeiras*,1984 (178-181).

Melo, M., Fazuoli, L. C., Teixeira, A. A., Amorim, H. V. (1980). Alterações físicas, químicas e organolépticas em grãos de café armazenados. Ciência e Cultura, 32 (4), 468-471.

Nobre, G. W. (2005). *Alterações qualitativas do café cereja descascado durante o armazenamento*. Dissertação Mestrado, Universidade Federal de Lavras, Lavras, Brasil

Northmore, J. M. (1968). *Raw bean colors and the quality of Kenya arabica coffee. Turrialba*, 31 (368), 339-341.

Oliveira, M. V. de. (1995). *Efeito do armazenamento no branqueamento de grãos de café beneficiado: modelagem matemática do processo*. Dissertação Mestrado, Universidade Federal de Lavras, Lavras, Brasil.

Vilela, E. R., Chandra, P. K., Oliveira, G. (2000). Efeito da temperatura e umidade relativa no branqueamento de grãos de café. *Revista Brasileira de Armazenamento*, 1, 31-37.

10. Rabechault, 1962, reported by Bacchi, O. (1962). O branqueamento dos grãos de café. *Bragantia*, 21(28), 467-468.

11. Chassevent, F. (1987). Coloque scientific international sur le café, rapport de aynthése agronomic. *Café Cacao Thé*, 31 (3), 219-221.

12. Northmore, J. M. (1968). *Raw bean colors and the quality of Kenya arabica coffee. Turrialba*, 31 (368), 339-341.

13. Mazzafera, P., Guerreiro, F. O., Carvalho, A. (1984). Estudo de coloração verde do grão de café: determinação de flavonóides e clorofllas. *In 11 Congresso Brasileiro de Pesquisas Cafeeiras*,1984 (178-181).

14. Graner, E. A., Godoy, J. C. (1967). Manual do cafeicultor. São Paulo: *Melhoramentos*.

15. AFONSO JR 2001: NOT LISTED IN NEW
Afonso Júnior, P. C. (2001). *Aspectos físicos, fisiológicos e da qualidade do café em função da secagem e do armazenamento*. Tese Doutorado, Universidade Federal de Lavras, Lavras, Brasil

Bacchi, O. (1962). O branqueamento dos grãos de café. *Bragantia*, 21(28), 467-468.

Godinho, R. P., Vilela, E. R., Oliveira, G. A., Chagas, S. J. R. (2000). Variações na cor e na composição química do café (*Coffea arabica* L.) armazenado em côco e beneficiado. *Revista Brasileira de Armazenamento*, 1, 38-43.

Hara, T. (1972). *Storage factors affecting coffee quality*. Dissertation Master, Purdue University, Purdue.

Nobre, G. W. (2005). *Alterações qualitativas do café cereja descascado durante o armazenamento*. Dissertação Mestrado, Universidade Federal de Lavras, Lavras, Brasil.

Vilela, E. R., Chandra, P. K., Oliveira, G. (2000). Efeito da temperatura e umidade relativa no branqueamento de grãos de café. *Revista Brasileira de Armazenamento*, 1, 31-37.

16. Bacchi, O. (1962). O branqueamento dos grãos de café. *Bragantia*, 21(28), 467-468.

17. Lopes, R. P., Hara, T., Silva, J. S., Riedel, B. (2000). Efeito da luz na qualidade (cor e bebida) de grãos de café beneficiados *(Coffea arabica* L.) durante a armazenagem. *Revista Brasileira de Armazenamento*, 25, 9-17.

18. Bacchi, O. (1962). O branqueamento dos grãos de café. *Bragantia*, 21(28), 467-468.

Hara, T. (1972). *Storage factors affecting coffee quality*. Dissertation Master, Purdue University, Purdue.

Oliveira, M. V. de. (1995). *Efeito do armazenamento no branqueamento de grãos de café beneficiado: modelagem matemática do processo*. Dissertação Mestrado, Universidade Federal de Lavras, Lavras, Brasil.

Texeira, A. A., Fazuoli, L. C., Carvalho, A. (1977). Qualidade da bebida do café: efeito do acondicionamento e do tempo de conservação. *Bragantia*, 36 (7), 103-108.

19. Bacchi, O. (1962). O branqueamento dos grãos de café. *Bragantia*, 21(28), 467-468.

20. Oliveira, M. V. de. (1995). *Efeito do armazenamento no branqueamento de grãos de café beneficiado: modelagem matemática do processo*. Dissertação Mestrado, Universidade Federal de Lavras, Lavras, Brasil.

21. Subrahmanyan, V., Bathia, D. S., Natarajan, C. P., Majunder, S. K. (1961). Storage of coffee beans. *Indian Coffee*, 25 (1), 26-36.

22. Bacchi, O. (1962). O branqueamento dos grãos de café. *Bragantia*, 21(28), 467-468.

Hara, T. (1972). *Storage factors affecting coffee quality*. Dissertation Master, Purdue University, Purdue.

Oliveira, M. V. de. (1995). *Efeito do armazenamento no branqueamento de grãos de café beneficiado: modelagem matemática do processo*. Dissertação Mestrado, Universidade Federal de Lavras, Lavras, Brasil.

23. Vilela, E. R., Chandra, P. K., Oliveira, G. (2000). Efeito da temperatura e umidade relativa no branqueamento de grãos de café. *Revista Brasileira de Armazenamento*, 1, 31-37.

24. Oliveira, M. V. de. (1995). *Efeito do armazenamento no branqueamento de grãos de café beneficiado: modelagem matemática do processo*. Dissertação Mestrado, Universidade Federal de Lavras, Lavras, Brasil.

Stirling, H. G. (1975). Further experiments on factors affecting quality loss in stored arabica coffee. *Kenya Coffee*, 40 (466), 28-35.

Subrahmanyan, V., Bathia, D. S., Natarajan, C. P., Majunder, S. K. (1961). *Storage of coffee beans*. Indian Coffee, 25 (1), 26-36.

25. Buchelli, P., Meyer, I., Pittet, A., Vuataz, G., Viani, R.(1998). Industrial storage of green robusta coffee under tropical conditions and its impact on raw material quality and ochratoxin. *Journal of Agricultural and Food Chemistry*, 46 (11), 4507-4511.

26. Nobre, , G. W. (2005). *Alterações qualitativas do café cereja descascado durante o armazenamento.* Dissertação Mestrado, Universidade Federal de Lavras, Lavras, Brasil.

27. Ribeiro, F. C.; Borém, F. M.; Giomo, G. S.; Lima, R. R.; Malta, M. R.; Figueiredo, L. P.(2011) Storage of green coffee in hermetic packaging injected with CO2. Journal of Stored Products Research, 47, 341-348.

28. Afonso Júnior, P. C., Corrêa, P. C., Goneli, A. L. D., Silva, F. S. da. (2004). Contribuição das etapas do pré-processamento para a qualidade do café. *Revista Brasileira de Armazenamento*, 29 (8), 6-53.

Alves, W. M., Faroni, L. R. D., Corrêa, P. C., Parizzi, F., Pimentel, M. A. G (2003). Influência do pré-processamento e do período de armazenamento na perda de matéria seca em café (*Coffea arabica* L.) beneficiado. *Revista Brasileira de Armazenamento*, 7, 122-127.

Arêdes, E. M. (2002). *Avaliação das perdas de matéria seca e de qualidade do café (*Coffea arabica *L.) beneficiado e armazenado em importantes municípios produtores da zona da mata mineira e em Alegre.* Dissertação Mestrado, Universidade Federal de Viçosa, Viçosa, Brasil.

29. Alves, W. M., Faroni, L. R. D., Corrêa, P. C., Parizzi, F., Pimentel, M. A. G (2003). Influência do pré-processamento e do período de armazenamento na perda de matéria seca em café (*Coffea arabica* L.) beneficiado. *Revista Brasileira de Armazenamento*, 7, 122-127.

Arêdes, E. M. (2002). *Avaliação das perdas de matéria seca e de qualidade do café (*Coffea arabica *L.) beneficiado e armazenado em importantes municípios produtores da zona da mata mineira e em Alegre.* Dissertação Mestrado, Universidade Federal de Viçosa, Viçosa, Brasil.

30. Alves, W. M., Faroni, L. R. D., Corrêa, P. C., Parizzi, F., Pimentel, M. A. G (2003). Influência do pré-processamento e do período de armazenamento na perda de matéria seca em café (*Coffea arabica* L.) beneficiado. *Revista Brasileira de Armazenamento*, 7, 122-127.

31. Alves, W. M., Faroni, L. R. D., Corrêa, P. C., Parizzi, F., Pimentel, M. A. G (2003). Influência do pré-processamento e do período de armazenamento na perda de matéria seca em café (*Coffea arabica* L.) beneficiado. *Revista Brasileira de Armazenamento*, 7, 122-127.

32. Afonso Júnior, P. C. (2001). *Aspectos físicos, fisiológicos e da qualidade do café em função da secagem e do armazenamento.* Tese Doutorado, Universidade Federal de Lavras, Lavras, Brasil.

Amorim, H. V. (1978). *Aspectos bioquímicos e histoquímicos do grão de café verde relacionados com a deterioração de qualidade.* Tese, Escola Superior de Agricultura Luiz de Queiroz, Piracicaba, Brasil.

Bacchi, O. (1962). O branqueamento dos grãos de café. *Bragantia*, 21(28), 467-468.

Bártholo, G. F., Guimarães, P. T. G. (1997). Cuidados na colheita e preparo do café. Informe Agropecuário, 18 (187), 33-42.

Borém, F. M., Reinato, C. H. R., and Pereira, R. G. F. A. (2003) Alterações na bebida do café despolpado secado em terreiro de concreto, lama asfáltica, terra, leito suspenso e em secadores rotativos. *In 3 Simpósio de Pesquisa dos Cafés do Brasil e Workshop Internacional de Café & Saúde*, 155, 447.

Coelho, K. F., Pereira, R. G. F. A., Vilela, E. R. (2001). Qualidade do café beneficiado em função do tempo de armazenamento e de diferentes tipos de embalagens. *Revista Brasileira de Armazenamento*, 25 (2), 22-27.

Coradi, P. C. (2006). *Alterações na qualidade do café cereja natural despolpado submetidos a diferentes condições de secagem e armazenamento.* Dissertação Mestrado, Universidade Federal de Lavras, Lavras, Brasil.

Costé, R. (1969). El cafe. Barcelona: Blume.

Esteves, A. B. (1960). Acidificação ao longo do tempo da gordura do grão de café cru. *Estudos Agronômicos*, 1 (4), 297-317.

Fourny, G., Cros, E., Vicent, J. C. (1982). Etude préliminaire de Loxydation de L huile de café. *Proc 10 ASIC*, 235-246.

Godinho Silva, R., Vilela, E. R., Pereira, R. G. G. A., Borém, F. M. (2001). Qualidade de grãos de café (*Coffea arabica* L.) armazenados em coco com diferentes níveis de umidade. *Revista Brasileira de Armazenamento*, 3, 3-10.

Jordão, B. A., Garruti, R. S., Angelucci, E., Tango, J. S., and Toselo, Y.(1974). Armazenamento de café a granel em silo com ventilação natural. *In 2 Congresso Brasileiro de Pesquisas Cafeeiras*, 1974 (385).

Kurzrock, T.; Kolling-Speer, I.; Speer, K.(2004) Effects of controlled sto- rage on the lipid fraction of green Arabica Coffee Beans. Food Chemistry, 66, 161-168.

Leite, I. P., Vilela, E. R., Carvalho, V. D.(1996). Efeito do armazenamento na composição física e química do grão de café em diferentes processamentos. *Pesquisa Agropecuária Brasileira*, 31 (3), 159-163.

Leite, R. A., Corrêa, P. C., Oliveira, M. G. A., Reis, F. P., Oliveira, T. T. (1999). Avaliação por métodos físicos da qualidade do café (*Coffea arabica* L.) pré-processado por "via seca" e "via úmida" durante dez meses de armazenamento. *Engenharia na Agricultura*, 7 (2), 106-115.

Nobre, G. W. (2005). *Alterações qualitativas do café cereja descascado durante o armazenamento.* Dissertação Mestrado, Universidade Federal de Lavras, Lavras, Brasil.

Pinto, N. A. V. D., Pereira, R. G. F. A., Fernandes, S. M., Thé, P. M., Carvalho, V. D. (2002). Caracterização dos teores de polifenóis e açúcares em padrões de bebida do café (*Coffea arabica* L.) cru e torrado do sul de Minas Gerais. *Revista Brasileira de Armazenamento*, 4, 52-58.

Prete, C. E. C. (1992). *Condutividade elétrica do exsudado de grãos de café (*Coffea arabica *L.) e sua relação com a qualidade da bebida.* Tese Doutorado, Escola Superior de Agricultura Luiz de Queiroz, Piracicaba, Brasil.

Reinato, C. H. R. (2006). *Secagem e armazenamento do café: aspectos qualitativos e sanitários.* Tese Doutorado, Universidade Federal de Lavras, Lavras, Brasil.

Ribeiro, D. M. (2003). *Qualidade do café cereja descascado submetido a diferentes temperaturas, fluxos de ar e períodos de pré-secagem.* Dissertação Mestrado, Universidade Federal de Lavras, Lavras, Brasil.

Vidal, H. M. (2001). *Composição lipídica do café (*Coffea arabica *L.) durante o armazenamento.* Dissertação Mestrado, Universidade Federal de Viçosa, Viçosa, Brasil.

Wajda, P.; Walczyk, D. (1978) Relationship between acid value of extracted fatty matter and age of green coffee bean. Journal of the Science of Food and Agriculture, 29 (4), 377-380.

33. Afonso Júnior, P. C. (2001). *Aspectos físicos, fisiológicos e da qualidade do café em função da secagem e do armazenamento.* Tese Doutorado, Universidade Federal de Lavras, Lavras, Brasil.

Bacchi, O. (1962). O branqueamento dos grãos de café. *Bragantia*, 21(28), 467-468.

Coradi, P. C. (2006). *Alterações na qualidade do café cereja natural despolpado submetidos a diferentes condições de secagem e armazenamento.* Dissertação Mestrado, Universidade Federal de Lavras, Lavras, Brasil.

Reinato, C. H. R. (2006). *Secagem e armazenamento do café: aspectos qualitativos e sanitários.* Tese Doutorado, Universidade Federal de Lavras, Lavras, Brasil.

34. Afonso Júnior, P. C. (2001). *Aspectos físicos, fisiológicos e da qualidade do café em função da secagem e do armazenamento*. Tese Doutorado, Universidade Federal de Lavras, Lavras, Brasil.

Reinato, C. H. R. (2006). *Secagem e armazenamento do café: aspectos qualitativos e sanitários*. Tese Doutorado, Universidade Federal de Lavras, Lavras, Brasil.

35. Fourny, G., Cros, E., Vicent, J. C. (1982). Etude préliminaire de Loxydation de L huile de café. *Proc 10 ASIC*, 235-246.

Wajada, P., Walczyz, D.(1980). Relationship between acid value of extracted fatty matter and age of green coffee bean. *Journal Science Food Agriculture*, 29 (1), 377-380.

36. Afonso Júnior, P. C. (2001). *Aspectos físicos, fisiológicos e da qualidade do café em função da secagem e do armazenamento*. Tese Doutorado, Universidade Federal de Lavras, Lavras, Brasil.

Coradi, P. C. (2006). *Alterações na qualidade do café cereja natural despolpado submetidos a diferentes condições de secagem e armazenamento*. Dissertação Mestrado, Universidade Federal de Lavras, Lavras, Brasil.

37. Nobre, G. W. (2005). *Alterações qualitativas do café cereja descascado durante o armazenamento*. Dissertação Mestrado, Universidade Federal de Lavras, Lavras, Brasil.

38. Ribeiro, F. C.; Borém, F. M.; Giomo, G. S.; Lima, R. R.; Malta, M. R.; Figueiredo, L. P.(2011) Storage of green coffee in hermetic packaging injected with CO2. Journal of Stored Products Research, 47, 341-348

Chapter 9

1. Gomes, F. C., and Teixeira, V. H. (2000). *Instalações para o processamento do café*. Lavras, MG: UFLA.

2. Bueno, C. F. H. (1998). Instalações para o beneficiamento do café. In *Encontro Nacional de Técnicos, Pesquisadores e Educadores de Construções Rurais. Congresso Brasileiro de Engenharia Agrícola*, 1998 (87-148).

3. Portland, A. B. C. (1984). *Prospecção de jazidas e coleta de amostras de solos para solo-cimento: misturas de dois solos*. São Paulo, SP: ABCP.

4. Bueno, C. F. H. (1998). Instalações para o beneficiamento do café. In *Encontro Nacional de Técnicos, Pesquisadores e Educadores de Construções Rurais. Congresso Brasileiro de Engenharia Agrícola*, 1998 (87-148).

5. Souza Júnior, T. F. (2004). *Tecnologia e qualidade do material concreto nas construções agroindustriais*. Dissertação de Mestrado, Universidade Federal de Lavras, Lavras, Brasil.

6. Borém, F. M. (2004). *Cafeicultura empresarial: produtividade e qualidade: pós-colheita do café*. Lavras, MG: UFLA.

7. Gomes, F. C., and Teixeira, V. H. (2000). *Instalações para o processamento do café*. Lavras, MG: UFLA.

Chapter 10

1. Reis, P. R., Souza, J. C., and Venzon, M. (2002). Manejo ecológico das principais pragas do cafeeiro. *Informe Agropecuário*, 23 (214/215), 83-99.

2. Souza, J. C., and Reis, P. R. (1997). Broca-do-café: histórico, reconhecimento, biologia, prejuízos, monitoramento e controle. *Boletim Técnico 50*. Belo Horizonte, MG: Epamig.

3. Nakano, O., Costa, J. D., Bertoloti, S. J., and Olivetti, C. M. (1976). Revisão sobre o conceito de controle químico da broca do café Hypothenemus hampei (Coleoptera: Scolytidae). In *Congresso Brasileiro de Pesquisas Cafeeiras*, 1976 (8-10).

Yokoyama, M., Nakano, O., Costa, J. D., Nakayama, K., and Perez, C. A. (1978). Avaliação de danos causados pela broca do café, *Hypothenemus hampei* (Coleoptera: Scolytidae). In *Congresso Brasileiro de Pesquisas Cafeeiras*, 1978 (26-27).

4. Paulini, A. E., Paulino, A. I. (1979). Evolução de *Hypothenemus hampei* em café conilon armazenado e influência da infestação na queda de frutos. In *Congresso Brasileiro de Pesquisas Cafeeiras*, 1979 (285-287).

5. Reis, P. R., Souza, J. C., and Melles, C. C. A. (1984). Pragas do cafeeiro. *Informe Agropecuário*, 10 (109), 3-57.

Reis, P. R., and Souza, J. C. (1986). Pragas do cafeeiro. In Rena, A. B., Malavolta, E., Rocha, M., Yamada, T. (Eds.). *Cultura do cafeeiro: fatores que afetam a produtividade* (323-378). Piracicaba, SP: POTAFOS.

Souza, J. C., and Reis, P. R. (1980). Efeito da broca-do-café, *Hypothenemus hampei* (Coleoptera: Scolytidae), na produção e qualidade do grão de café. In *Congresso Brasileiro de Pesquisas Cafeeiras*, 1980 (281-283).

6. Lucas, M. B., Salgado, L. O., Reis, P. R., and Souza, J. C. (1989). Perdas de peso no processo de beneficiamento do café em consequência do ataque da broca do café *Hypothenemus hampei* (Coleoptera: Scolytidae). *Ciência e Prática*, 13 (3), 314-324.

7. Chalfoun, S. M., Souza, J. C., and Carvalho, V. D. (1984). Relação entre a incidência de broca, *Hypothenemus hampei* (Coleoptera: Scolytidae) e microorganismos em grãos de café. In *Congresso Brasileiro de Pesquisas Cafeeiras*, 1984, 149-150.

8. Calafiori, M. H., Maluf, H., Silva, P. S., and Dias, J. A. C. S. (1978). Influência da broca-do-café, *Hypothenemus hampei* na bebida e sua associação com fungo. *Ecossistema*, 3, 80-81.

9. Reis, P. R., Souza, J. C., and Melles, C. C. A. (1984). Pragas do cafeeiro. *Informe Agropecuário*, 10 (109), 3-57.

10. Paulini, A. E., Paulino, A. I. (1979). Evolução de *Hypothenemus hampei* em café conilon armazenado e influência da infestação na queda de frutos. In *Congresso Brasileiro de Pesquisas Cafeeiras*, 1979 (285-287).

11. Ferreira, A. J., Paulini, A. E., and D'antonio, A. M. (1983). Tempo de exposição letal a três fases da broca do café *Hypothenemus hampei*: à temperatura de 45º centígrados. In *Congresso Brasileiro De Pesquisas Cafeeiras*, 1983, 230-232.

12. Benassi, V. L. R. M. (1996) Desenvolvimento da broca-do-café, *Hypothenemus hampei* (Coleoptera: Scolytidae), em frutos de diferentes teores de umidade de *Coffea canephora* e de *C. arabica*. In *Congresso Brasileiro de Pesquisas Cafeeiras*, 1996 (102).

13. Matiello, J. B., Barros, U. V., and Barbosa, C. M. (1999). Controle da broca do café em sementes. In *Congresso Brasileiro de Pesquisas Cafeeiras*, 1999 (10-11).

14. Souza, J. C., and Reis, P. R. (1997). Broca-do-café: histórico, reconhecimento, biologia, prejuízos, monitoramento e controle. *Boletim Técnico 50*. Belo Horizonte, MG: Epamig.

15. Lucas, M. B., Salgado, L. O., Reis, P. R., and Souza, J. C. (1989). Perdas de peso no processo de beneficiamento do café em consequência do ataque da broca do café *Hypothenemus hampei* (Coleoptera: Scolytidae). *Ciência e Prática*, 13 (3), 314-324.

16. Autuori (1931), cited by Lima, A. C. (1956). *Insetos do Brasil: coleópteros: 4º parte*. Rio de Janeiro, RJ: Escola Nacional de Agronomia.

17. Kingsolver (1991)2, cited by Pacheco, I. A., and Paula, D. C. (1995). *Insetos de grãos armazenados: identificação e biologia*. Campinas, SP: Fundação Cargill.

18. Halstead (1986), cited by Pacheco, I. A., and Paula, D. C. (1995). *Insetos de grãos armazenados: identificação e biologia*. Campinas, SP: Fundação Cargill.

19. Halstead (1963), cited by Pacheco, I. A., and Paula, D. C. (1995). *Insetos de grãos armazenados: identificação e biologia*. Campinas, SP: Fundação Cargill.
20. Gonçalves, L. I., Bitran, H. V., and Bitran, E. A. Estudos sobre a biologia do caruncho-do-café Araecerus fasciculatus (De Geer, 1775) (Coleoptera, Anthribidae). In *Congresso Brasileiro de Pesquisas Cafeeiras*, 1976, 23-24.
21. Cotton (1921) cited by Lima, A. C. (1956). *Insetos do Brasil: coleópteros: 4º parte*. Rio de Janeiro, RJ: Escola Nacional de Agronomia.
22. Anderson (1991), cited by Pacheco, I. A., and Paula, D. C. (1995). *Insetos de grãos armazenados: identificação e biologia*. Campinas, SP: Fundação Cargill.
23. Figueiredo Júnior, E. R. (1957). O controle do caruncho-das-tulhas. *O Biológico*, 23 (10), 197-200.
24. Stresser, R. (2005). *Tratamentos preventivos e curativos no controle de insetos e pragas em silos e armazéns graneleiros*. Disponível em: <http://www.tecnigran. com.br/html/tratamentos_preventivos_ e_curativos.htm>. Acesso em: 24 jun. 2005.
25. Bell, C. H. (2000). Fumigation in the 21st century. *Crop Protection*, 19, 563-569.
26. Busck e Oliveira Filho (1925), cited by Graner, E. A., and Godoy Júnior, C. (1997). *Manual do cafeicultor*. São Paulo, SP: São Paulo Melhoramentos.
27. Damon, A. (2000). A review of the biology and control of the coffee berry borer, *Hypothenemus hampei* (Coleoptera: Scolytidae). *Bulletin of Entomological Research*, 90 (6), 453-465.
28. Bitran, E. A. (1972). *Contribuição ao conhecimento de "traças" que ocorrem em café armazenado, danos ocasionados e controle*. Dissertação de Mestrado, Escola Superior de Agricultura Luiz de Queiroz, Universidade de São Paulo, São Paulo, Brasil.
29. Bitran, E. A. (1972). *Contribuição ao conhecimento de "traças" que ocorrem em café armazenado, danos ocasiona-*

dos e controle. Dissertação de Mestrado, Escola Superior de Agricultura Luiz de Queiroz, Universidade de São Paulo, São Paulo, Brasil.
30. Ferguson (1991), cited by Pacheco, I. A., and Paula, D. C. (1995). *Insetos de grãos armazenados: identificação e biologia*. Campinas, SP: Fundação Cargill.
31. Cruz (1980), cited by Pacheco, I. A., and Paula, D. C. (1995). *Insetos de grãos armazenados: identificação e biologia*. Campinas, SP: Fundação Cargill.
32. Pacheco, I. A., and Paula, D. C. (1995). *Insetos de grãos armazenados: identificação e biologia*. Campinas, SP: Fundação Cargill.
33. AGROFIT (2005). *Sistema de agrotóxicos fitossanitários*. Disponível em: <http://extranet.agricultura.gov.br/agrofit_cons/ principal_agrofit _cons>. Acesso em: 12 ago. 2005.
34. Puzzi, D. (1977). Manual de armazenamento de grãos: armazéns e silos. São Paulo, SP: Ceres.
35. Matias, R. S. (2005). Olho nos roedores. *Cultivar Grandes Culturas*, 7 (75), 40-42.
36. Bernard, E. (2005). Morcegos vampiros: sangue, raiva e preconceito. *Ciência Hoje*, 36 (214), 44-49.
37. Bernard, E. (2003). Ecos na escuridão: o fascinante sistema de orientação dos morcegos. *Ciência Hoje*, 32 (190), 14-20.
38. Bernard, E. (2005). Morcegos vampiros: sangue, raiva e preconceito. *Ciência Hoje*, 36 (214), 44-49.
39. Batista, L. R., and Chalfoun, S. M. (2003). Biodiversidade de fungos associados a frutos e grãos de café gêneros Aspergillus e Penicillium. In *Simpósio de Pesquisa dos Cafés do Brasil*, 2003 (161-162).
40. Vargas, E. A., Santos, E. A., França, R. C. A., Amorim, S. S., Faustino, L. C. S., Bezerra, E. E. D., Aquino, V. C., Lima, F. B., Whitaker, and T. B., Slate, A. (2003). Delineamento, elaboração e implementação de plano amostral oficial para determinação de ocratoxina A em café beneficiado. In *Simpósio de Pesquisa dos Cafés do Brasil*, 2003 (166).

Chapter 11

1. Buchanan, R. L., Fletcher, A. M. (1978). Methylxanthine inhibition of aflatoxin production. *Journal of Food Science*, 43 (654-655).
   Chalfoun, S. M., Pereira, M. C., and Angélico, C. L. (2000). Efeito da cafeína (1,3,7 - trimethylxchanthina) sobre o crescimento micelial de fungos associados ao café. *Revista Brasileira de Armazenamento*, 1, 50-53.
   Chalfoun, S. M., and Batista, L. R. (2006). Incidência de ocratoxina A em diferentes frações de grãos de café (*Coffea arabica* L.). *Coffee Science*, 1(1), 28-35.
   Nartowicz, V. B., Buchanan, R. L., and Segal, S. (1979). A flatoxin production in regular and decaffeinated coffee beans. *Journal of Food Science*, 44 (2), 446-448.
2. ICO (2006). *International coffee organization annual review 2003/2004*. Available on: <http:// www.ico.org/annual_review.asp>. Access in: 2 Jan. 2006.
3. World Health Organization. (2001). Safety evaluation of certin mycotoxin in food. Geneva, Rome: JECFA.
4. Abouzied, M. M., Horvath, A. D., Podlesny, P. M., Regina, N. P., Metodiev, V. D., Kamenova-Tozeva, R. M., Niagolova, N. D., Stein, A. D., Petropoulos, E. A., and Ganev, V. S. (2002). Ochratoxina A concentration in food and feed for a region with Nalkan Endemic Nephropathy. Food Additives and Contaminants, 19 (8), 755-764.
   Peraica, M., Radic, B., Lucic, A., and Pavlovic, M. (1999). Toxic effects of mycotoxins in humans. *Bulletin of WHO*, 77, p. 754-766.
5. Moss, M. O. (1996). Mode of formation of ochratoxin A. *Food Additive and Contaminants*, 13, 5-9.

6. Chalfoun, S. M., and Batista, L. R. (2003). *Fungos associados a frutos e grãos de café: Aspergillus e Penicillium*. Brasília, DF: Embrapa.
7. Christensen, M. (1982). The Aspergillus ochraceus Group: two new species from western soils and a synoptc key. *Mycologia*, 74 (2), 210-225.
   Raper, K. B., and Fennell, D. I. (1965). *The genus Aspergillus*. Baltimore, MD: Williams and Wilkins.
8. Merwe Merwe, K. J. Van Der, Steyn, P. S., Fourie, L., Scott, D. B., and Theron, J. J. (1965). Ochratoxin A, a toxic metabolite produced by Aspergillus ochraceus Wilh. *Nature*, 205 (4976), 1112-1113.
9. Abarca, M. L., Bragulat, M. R., Castellá, G., and Cabanes, F. J. (1997). New ochratoxigenic species in the genus Aspergillus. *Journal of Food Protection*, 60 (2), 1580-1582.
10. Bayman, P., Baker, J. L., Doster, M. A., Michailides, T. J., and Mahoney, N. E. (2002). Ochratoxin production by the Aspergillus ochraceus group and A. alliaceus. *Applied and Environmental Microbiology*, 68(5), 2326-2329.
11. Varga, J., Kevel, E., Rinyu, E., Téren, J., and Kozakiewicz, Z. (1996). Ochratoxin Production by Aspergillus Species. *Applied and Environmental Microbiology*, 62 (12), 4461-4464.
12. Batista, L. R., Chalfoun, S. M., Prado, G., Schwan, R. F., and Wheals, A. E. (2003). Toxigenic fungi associated with processed (green) coffee beans (*Coffea arabica* L.). *International Journal fo Food Microbiology*, 85(3), 293-300.
13. Accensi, F., Abarca, M. L., Cabañes, F. J. (2004) Occurrence of Aspergillus species in mixed feeds and component raw materials and their ability to produce ochratoxin A. *Food Microbiology*, 21 (5), 623-627.

14. Nasser, P. P. (2001). *Influência da separação de grãos de café (*Coffea arabica *L.) por tamanho na qualidade e ocorrência de ocratoxina A*. 2001. Dissertação de Mestrado, Universidade Federal de Lavras, Lavras, Brasil.

15. Prado, E., Marín, S., Ramos, A. J., and Sanchis, V. (2004). Occurrence of ochratoxigenic fungi and ochratoxina A in green coffee from different origins. *Food Science and Technology International*, 10, (1), 45-49.

16. Silva, C. F. (2004). *Sucessão microbiana e caracterização enzimática da microbiota associada aos frutos e grãos de café (*Coffea arabica *L.) do município de Lavras-MG*. Tese de doutorado, Universidade Federal de Lavras, Lavras, Brasil.

17. Urbano, G. R., Taniwaki, M. H., Leitão, M. F. E., and Vicentini, M. C. (2001). Occurrence of ochratoxin A- producing fungi in raw brazilian coffee. *Journal of Food Protection*, 64 (8), 1226-1230.

18. Pitt, J. I. (2000) Toxigenic fungi: which are important. *Medical Mycology*, 38, 17-22.

19. Schuster, E., Dunn-Coleman, N., Frisvad, J. C., and Dijck, P. W. M. Van. (2002). On the safety of Aspergillus niger: a review. *Applied Microbiology and Biotechnology*, 59 (4/5), 426-435.

20. Pitt, J. I., and Hockking, A. D. (1997) *Fungi and Food Spoilage*. London: Blackie Academic & Professional.

21. Cabañes, F. J., Accensi, F., Bragulat, M. R., Abarca, M. L., Castellá, G., Minguez, S., and Pons, A. (2002) What is source of ochratoxin A in wine? *International Journal of Food Microbiology*, 79 (3), 213-215.

22. Heenan, C. N., Shaw, K. J., and Pitt, J. I. (1998) Ochratoxin A production by Aspergillus carbonarius ans A. niger isolates and detection using coconut cream Ágar. *Journal of Food Mycology*, 1, 67-72.

23. Joosten, H. M. L. J., Goetz, J., Pittet, A., Schellenberg, M., and Bucheli, P. (2001). Production of ochratoxin A by Aspergillus carbonarius on coffee cherries. *International Journal of Food Microbiology*, 65, 39-44.

24. Téren, J., Varga, J., Hamari, Z., Rinyu, E., and Kevei, F. (1996). Immunochemical detection of ochratoxin A im black Aspergillus strains. *Mycopathologia*, 134 (3), 171-176.

25. Abarca, M. L., Bragulat, M. R., Castellá, G., and Cabañes, F. J. (1994). Ochratoxin production by strain of Aspergillus niger var niger. *Applied and Environmental Microbiology*, 60 (7), 2650-2652.

26. Urbano, G. R., Taniwaki, M. H., Leitão, M. F. E., and Vicentini, M. C. (2001). Occurrence of ochratoxin A- producing fungi in raw brazilian coffee. *Journal of Food Protection*, 64 (8), 1226-1230.

27. Nakajima, M., Tsubouchi, H., Miyabe, M., and Ueno, Y. (1997). Survey of Aflatoxin B 1 and Ochratoxin A in commercial green coffee beans by highperformance liquid chromatography linked with immunoaffmity chromatography. *Food and Agricultural Immunology*, 9 (2), 77-83.

28. Prado E., Marín, S., Ramos, A. J., and Sanchis, V. (2004). Occurrence of ochratoxigenic fungi and ochratoxina A in green coffee from different origins. *Food Science and Technology International*, 10, (1), 45-49.

29. Samson, R. A., Hoekstra, E. S., Frisvad, J. C., and Filtenborg, O. (2000). *Introdution to food-borne fungi (4th Ed.)*. Centraalbureau Voor: Schimmelcultures Baarn Delft, 2000.

30. Levi, C. P., Trenk, H. L., and Mohr, H. K. (1974). Study of the occurrence of ochratoxin A in green coffee beans. *Journal of the Association Official Analytical Chemists*, 57 (4), 866-870.

31. Abarca, M. L., Accensi, F., Bragulat, M. R., and Cabañes, F. J. (2001). Current importance of ochratoxin A: producing Aspergillus spp. *Journal of Food Protection*, 64 (6), 903-906.

Bucheli, P., Meyer, I., Pittet, A., Vuataz, G., and Viani, R. (1998). Industrial storage of green robusta under tropical conditions and its impact on raw material quality and ochratoxin A content. *Journal of Agricultural and Food Chemistry*, 46 (11), 4507-4511.

Bucheli, P., Meyer, I., Pittet, A., Vuataz, G., and Viani, R. (2000). Development of ochratoxin A during robusta (*Coffea canephora*) coffee cherry drying. *Journal of Agricultural and Food Chemistry*, 48 (4), 1358-1362.

Duris, D. Coffee and ochratoxin contamination. In *Food Safety Management in Developing Countries*, 2000, 1-5.

Heilmann, W., Rehfeldt, A. G., and Rotzoll, F. (1999). Behavior and reduction of ochratoxin A in green coffee beans in response to various processing methods. *European Food Research Technology*, 209, 297- 300.

Joosten, H. M. L. J., Goetz, J., Pittet, A., Schellenberg, M., and Bucheli, P. (2001). Production of ochratoxin A by Aspergillus carbonarius on coffee cherries. *International Journal of Food Microbiology*, 65, 39-44.

Leoni, L. A. B., Furlani, R. P. Z., Soares, L. M. V. S., and Oliveira, P. L. C. (1980). Ochratoxin in Brazilian green coffee. *Ciência e Tecnologia de Alimentos*, 21 (1), 105-107.

Levi, C. P. (1980). Mycotoxin in coffee. *Journal of the AOAC*, 63 (6), 1282-1285

Mantle, P. G., and Chow, A. M. (2000). Ochratoxin formation in Aspergillus ochraceus with particular reference to spoilage of coffee. *International Journal of Food Microbiology*, 56, 105-109.

Micco, M., Grossi, M., Miraglia, M., and Brera, C. (1989). A study of the contamination by ochratoxin A of green e roasted coffee beans. *Food Additives and Contaminants*, 6 (3), 333-339.

Moraes, M. H. P., and Luchese, R. H. (2003). Ochratoxin A on green coffee: influence of harvest and drying processing procedures. *Journal of Agriculture and Food Chemistry*, 51 (19), 5824-5828.

Romani, S., Sacchetti, G., López, C. C., Pinnavaia, G. C., and Rosa, M. D. (2000). Screening on the occurrence of ochratoxin A in green coffee beans of different origins and types. *Journal Agriculture Food Chemistry*, 48 (8), 3616-3619.

Studer-Rohor, I., Dietrich, D. R., Schlatter, J., and Schlatter, C. (1995). *The occurrence of ochratoxin A in coffee*. *Food Chemistry and Toxicology*, 33 (5), 341-355.

Taniwaki, M. H., Pitt, J. I., Teixeira, A. A., and Iamanaka, B. T. (2003). The source of ochratoxin A in Brazilian coffee and its formation in relation to processing methods. *International Journal of Food Microbiology*, 82 (2), 173-179.

Urbano, G. R., Taniwaki, M. H., Leitão, M. F. E., and Vicentini, M. C. (2001). Occurrence of ochratoxin A- producing fungi in raw brazilian coffee. *Journal of Food Protection*, 64 (8), 1226-1230.

32. Stegen, G. Van Der, Jörissen, U., Pittet, A., Saccon, M., Steiner, W., Vicenzi, M., Winkler, M., Zapp, J., and Schlatter, C. (1997). Screening of european coffee final products for ocurrence of ochratoxin A (OTA). Food Additives and Contaminants, 14 (3), 211-216.

33. World Health Organization (2001) *Safety Evaluation of Certain Mycotoxins in Food*. Geneva, 701 p.

34. Stegen, G. Van Der, Jörissen, U., Pittet, A., Saccon, M., Steiner, W., Vicenzi, M., Winkler, M., Zapp, J., and Schlatter, C. (1997). Screening of european coffee final products for ocurrence of ochratoxin A (OTA). Food Additives and Contaminants, 14 (3), 211-216.

35. FAO/Codex, Alimentarius Commission. (2004). *Report of the 36th session of the CODEX Committee on Food Additives and Contaminants*. Wasghington, DC

36. World Health Organization (2001) *Safety Evaluation of Certain Mycotoxins in Food*. Rome.

37. Levi, C. P., Trenk, H. L., and Mohr, H. K. (1974). Study of the occurrence of ochratoxin A in green coffee beans. *Journal of the Association Official Analytical Chemists*, 57 (4), 866-870.

38. Cantáfora, A., Grossi, M., Miraglia, M., and Benelli, L. (1983). Determination of ochratoxin A in coffee beans using reversed-phase high performance liquid chromatography. *La Rivista della Societá Italiana diSienza dell'Alimentazione*, 1 (12), 103-108.

39. Tsubouchi, H., Yamamoto, K., Hisada, K., and Sakabe, Y. (1984) A Survey occurence of mycotoxins and toxigenic fungi in imported green coffee beans. *Japanese Association of Mycotoxicology*, 19, 16-21.

40. Micco, M., Grossi, M., Miraglia, M., and Brera, C. (1989). A study of the contamination by ochratoxin A of green e roasted coffee beans. *Food Additives and Contaminants*, 6 (3), 333-339.

41. Studer-Rohor, I., Dietrich, D. R., Schlatter, J., and Schlatter, C. (1995). *The occurrence of ochratoxin A in coffee. Food Chemistry and Toxicology*, 33 (5), 341-355.

42. Nakajima, M., Tsubouchi, H., Miyabe, M., and Ueno, Y. (1997). Survey of Aflatoxin B 1 and Ochratoxin A in commercial green coffee beans by highperformance liquid chromatography linked with immunoaffmity chromatography. *Food and Agricultural Immunology*, 9 (2), 77-83.

43. Trucksess, M. W. (1999). Committee on natural toxins: mycotoxins. *Journal of the Association of Official Analytical Chemists*, Washington, 82 (2), 488-495.

44. Romani, S., Sacchetti, G., López, C. C., Pinnavaia, G. C., and Rosa, M. D. (2000). Screening on the occurrence of ochratoxin A in green coffee beans of different origins and types. *Journal Agriculture Food Chemistry*, 48 (8), 3616-3619.

45. Batista L. R., Chalfoun, S. M., Prado, G., Schwan, R. F., and Wheals, A. E. (2003). Toxigenic fungi associated with processed (green) coffee beans (*Coffea arabica* L.). *International Journal fo Food Microbiology*, 85(3), 293-300.

46. Taniwaki, M. H., Pitt, J. I., Teixeira, A. A., and Iamanaka, B. T. (2003). The source of ochratoxin A in Brazilian coffee and its formation in relation to processing methods. *International Journal of Food Microbiology*, 82 (2), 173-179.

47. Iamanaka, B. T., and Taniwaki, M. H. (2003). Ocorrência de ocratoxina A em café cru, café torrado e moído e café solúvel brasileiro. In *Simpósio em Ciência De Alimentos*, 2003.

48. Moraes, M. H. P., and Luchese, R. H. (2003). Ochratoxin A on green coffee: influence of harvest and drying processing procedures. *Journal of Agriculture Food Chemistry*, 51 (19), 5824-5828.

49. Prado, E., Marín, S., Ramos, A. J., and Sanchis, V. (2004). Occurrence of ochratoxigenic fungi and ochratoxina A in green coffee from different origins. *Food Science and Technology International*, 10, (1), 45-49.

50. Batista, L. R. (2005). *Incidência de fungos produtores de ocratoxina A em grãos de café (*Coffea arabica *L.) préprocessados por via seca e úmida.* Tese de doutorado, Universidade Federal de Lavras, Lavras, Brasil.

51. Gilbert, J. (1999). Ochratoxin A in coffee, *occurrence significance, sampling and analysis. Workshop*, 1999 (13).

52. ASIC (1997) Proceedings of the 17th ASIC coloquium. Helsinki.

53. Walker, R. (2002). Risk assessment of ochratoxin: corrunet views of the European scientific committee of food, the JECFA and the codex committee on foos additives and contaminants. *Advanced in Experimental Medicine and Biology*, 504, 189-255.

54. Mabbett, T. (2002). Storing up problems? *Coffee & Cocoa International*, 2002.

55. Batista, L. R. (2005). *Incidência de fungos produtores de ocratoxina A em grãos de café (*Coffea arabica *L.) préprocessados por via seca e úmida.* Tese de doutorado, Universidade Federal de Lavras, Lavras, Brasil.

56. Bucheli, P., Meyer, I., Pittet, A., Vuataz, G., and Viani, R. (2000). Development of ochratoxin A during robusta (*Coffea canephora*) coffee cherry drying. *Journal of Agricultural and Food Chemistry*, 48 (4), 1358-1362.

57. Bucheli, P. (2001). Production of ochratoxin A by Aspergillus carbonarius on coffee cherries. *International Journal of Food Microbiology*, 65 (39-44).

Bucheli, P., Meyer, I., Pittet, A., Vuataz, G., and Viani, R. (2000). Development of ochratoxin A during robusta (*Coffea canephora*) coffee cherry drying. *Journal of Agricultural and Food Chemistry*, 48 (4), 1358-1362.

58. Bucheli, P., Taniwaki, M. H. (2002). Research on the origin, and on the impact of post-harvest handling and manufacturing on the presence of ochratoxin A in coffee. *Food Additives and Contaminants*, 19 (7), 655-665.

59. Bucheli, P., Taniwaki, M. H. (2002). Research on the origin, and on the impact of post-harvest handling and manufacturing on the presence of ochratoxin A in coffee. *Food Additives and Contaminants*, 19 (7), 655-665.

60. Viani, R. (2002) Efffect of processing on ochratoxin A content of coffee. *Advanced in Experimental Medicine and Biology*, 504, 189-193.

61. Bucheli, P., Meyer, I., Pittet, A., Vuataz, G., and Viani, R. (2000). Development of ochratoxin A during robusta (*Coffea canephora*) coffee cherry drying. *Journal of Agricultural and Food Chemistry*, 48 (4), 1358-1362.

62. Pittet, A., Tornare, D., Huggett, A., and Viani, R. (1996). Liquid chromatographic determination of ochratoxin A in pure and adulterated soluble coffee using na immunoaffinity column cleanup procedure. *Journal of Agricultural and Food Chemistry*, 44 (11), 3564-3569.

63. Nasser, P. P. (2001). *Influência da separação de grãos de café (*Coffea arabica *L.) por tamanho na qualidade e ocorrência de ocratoxina A.* 2001. Dissertação de Mestrado, Universidade Federal de Lavras, Lavras, Brasil.

64. Bucheli, P., Meyer, I., Pittet, A., Vuataz, G., and Viani, R. (1998). Industrial storage of green robusta under tropical conditions and its impact on raw material quality and ochratoxin A content. *Journal of Agricultural and Food Chemistry*, 46 (11), 4507-4511.

65. Teixeira, A. A. (1999). Classificação do café. In *Encontro Sobre Produção de Café com Qualidade*, 1999 (81-95).

66. Stegen, G. Van Der, Jörissen, U., Pittet, A., Saccon, M., Steiner, W., Vicenzi, M., Winkler, M., Zapp, J., and Schlatter, C. (1997). Screening of european coffee final products for ocurrence of ochratoxin A (OTA). *Food Additives and Contaminants*, 14 (3), 211-216.

67. Buchanan, R. L., Fletcher, A. M. (1978). Methylxanthine inhibition of aflatoxin production. *Journal of Food Science*, 43 (654-655).

Chalfoun, S. M., Pereira, M. C., and Angélico, C. L. (2000). Efeito da cafeína (1,3,7 - trimethylxchanthina) sobre o crescimento micelial de fungos associados ao café. *Revista Brasileira de Armazenamento*, 1, 50-53.

Chalfoun, S. M., and Batista, L. R. (2006). Incidência de ocratoxina A em diferentes frações de grãos de café (*Coffea arabica* L.). *Coffee Science*, 1(1), 28-35.

Nartowicz, V. B., Buchanan, R. L., and Segal, S. (1979). A flatoxin production in regular and decaffeinated coffee beans. *Journal of Food Science*, 44 (2), 446-448.

68. Buchanan, R. L., Fletcher, A. M. (1978). Methylxanthine inhibition of aflatoxin production. *Journal of Food Science*, 43 (654-655).

Nartowicz, V. B., Buchanan, R. L., and Segal, S. (1979). A flatoxin production in regular and decaffeinated coffee beans. *Journal of Food Science*, 44 (2), 446-448.